Plutopia

Plutopia

NUCLEAR FAMILIES, ATOMIC CITIES, AND
THE GREAT SOVIET AND AMERICAN
PLUTONIUM DISASTERS

KATE BROWN

OXFORD

UNIVERSITY PRESS

Oxford University Press is a department of the University of Oxford.
It furthers the University's objective of excellence in research,
scholarship, and education by publishing worldwide.

Oxford   New York

Auckland   Cape Town   Dar es Salaam   Hong Kong   Karachi
Kuala Lumpur   Madrid   Melbourne   Mexico City   Nairobi
New Delhi   Shanghai   Taipei   Toronto

With offices in

Argentina   Austria   Brazil   Chile   Czech Republic   France   Greece
Guatemala   Hungary   Italy   Japan   Poland   Portugal   Singapore
South Korea   Switzerland   Thailand   Turkey   Ukraine   Vietnam

Oxford is a registered trade mark of Oxford University Press
in the UK and certain other countries.

Published in the United States of America by
Oxford University Press
198 Madison Avenue, New York, NY 10016

© Kate Brown 2013

Library of Congress Cataloging-in-Publication Data
Brown, Kate (Kathryn L.)
Plutopia : nuclear families, atomic cities, and the great Soviet
and American plutonium disasters / Kate Brown.
pages   cm
Includes bibliographical references and index.
ISBN 978-0-19-985576-6 (acid-free paper)
1. Plutonium industry—Social aspects—Russia (Federation)—Ozersk (Cheliabinskaia oblast)—History—20th century.
2. Plutonium industry—Social aspects—Washington (State)— Richland—History—20th century.
3. Working class families—Russia (Federation)—Ozersk (Cheliabinskaia oblast)—History—20th century.
4. Working class families—Washington (State)—Richland—History—20th century. 5. Plutonium industry—
Accidents—Russia (Federation)—Ozersk (Cheliabinskaia oblast)—History—20th century.
6. Plutonium industry—Accidents—Washington (State)—Richland—History—20th century. 7. Ozersk
(Cheliabinskaia oblast, Russia)—History—20th century. 8. Richland (Wash.)—History—20th century.
9. Industrial safety—Government policy—Soviet Union—Case studies. 10. Industrial safety—
Government policy—United States—Case studies. I. Title.
HD9539.P583S62   2013
363.17'99094743—dc23   2012041758

1 3 5 7 9 8 6 4 2

Printed in the United States of America
on acid-free paper

*To Nancy Bernkopf Tucker, who
could have used a little more time*

Contents

Acknowledgments

I want to thank, first of all, the people who shared their histories with me. Most stories were painful to revisit, and I appreciate the tellers' generosity very much: Sergei Aglushenkov, Aleksander Akleev, Bob Alvarez, Juanita Andrewjevsky, Dasha Arbuga, Eugene Ashley, Tom Bailie, Sandra Batie, John Blacklaw, Cindy Bricker, Ed Bricker, Rex Buck, Tom Carpenter, Bob Collie, Marge Degooyer, Annette Heriford, Roger Heusser, Crystal Hobbs, Gulnara Ismagilova, Stephanie Janicek, Joe Jordan, Rosa Kazantseva, Rashid Khakimov, Robert Knoth, Anna Kolynova, Mira Kossenko, Svetlana Kotchenko, Nadezhda Kutepova, Liubov Kuzminov, Vladyslav Larin, Natalia Manzurova, Evdokia Mel'nikova, Pat Merrill, Anna Miliutina, Natalia Mironova, C. J. Mitchell, Ralph Myrick, Vladimir Novoselov, Pavel Oleynikov, Nadezhda Petrushkina, Trisha Pritikin, Keith Smith, Karen Dorn Steele, Jim Stoffels, Richard Sutch, Louisa Suvorova, Robert Taylor, Sergei Tolmachev, Vitalii Tolstikov, Galina Ustinova, and Elena Viatkina.

For showing kindness to a stranger, I would like to thank Julia Khmelevskaia, Igor Narskii, Dianne Taylor, Don Sorenson, Michelle Gerber, Natalia Mel'nikova, Galina Kibitkina, Juli Kearns, and Sergei Zhuravlev. For help with research and archives, I am grateful to Marina Mateesky, Dorothy Kenney, Evgenii Evstigneev, James Thomas Christian Oestermann, Murray Feshback, Shiloh Krupar, Paul Josephson, Steve Wing, Robert Bauman, John Findlay, Jane Slaughter, Connie Estep, Peter Bacon Hales, Janice Parthree, and Terry Fehner. I want to extend an especial thanks to Harry Winsor, who patiently served as my guide and tutor in physics, engineering, and chemistry. Thoughtful readers and editors for this project include Sarah Lazin, Susan Ferber, Catherine Evtuhov, Rosa Magnusdottir, Maggie Paxson, Mike Faye, David Engerman, Ethan Pollock, Choi Chatterjee, Beth Holmgren, Andrew Fisher, Paulina Bren, Neringa Klumbyte, and Gulnaz Sharafutdinova. For invitations to workshops and lectures related to this project, my thanks go to Kathleen Canning at the University of Michigan; Don Raleigh and Louise McReynolds at the University of North Carolina, Chapel Hill; Mary

Neuburger at the University of Texas, Austin; Karen Dawisha and Stephen Norris at Miami University; Lewis Siegelbaum at Michigan State University; Cathleen Cahill, Melissa Bokovoy, and Samuel Truett at the University of New Mexico; Catherine Wanner and Yurij Bihun at Pennsylvania State University; Erika Milam at the University of Maryland, College Park; Peter Rutland and Victoria Smolkin-Rothrock at Wesleyan University; Frederick Corney and Hiroshi Kitamura at the College of William and Mary; Richard Wortman and Tarik Amar at Columbia University; David Cantor at the National Institutes of Health; Karen Wigen at Stanford University; Blair Ruble at the Kennan Institute; Jeff Sanders at Washington State University; Rhiannon Dowling and Victoria Frede at the University of California, Berkeley; Don Raleigh and Louise McReynolds at the University of North Carolina; Molly Nolan and Andrew Needham at New York University; Jane Costlow and Jim Richter at Bates College; Anna Tsing at the University of California, Santa Cruz; and Diana Mincyte and Christof Mauch at the Rachel Carson Center.

I am very grateful for financial support from the University of Maryland, Baltimore County; the John Simon Guggenheim Foundation; the National Endowment for the Humanities; the National Council for Eurasian and East European Research; IREX; and the Kennan Institute. I owe special thanks for guidance to Michael Benson, Rebecca Boehling, Bill Chase, Warren Cohen, Geoff Eley, John Jeffries, Kristy Lindenmeyer, Robert Self, Nancy Bernkopf Tucker, Lynne Viola, and Richard White. Finally, friends and family have nursed this project along; many thanks to Sally Brown, William Brown, Liz Marston, Aaron Brown, Julie Hofmeister, David Bamford, Kama Garrison, Lisa Hardmeyer, Leslie Rugaber, Bruce Gray, Prentis Hale, Tracy Edmunds, Leila Corcoran, Sally Hunsberger, Ali Igmen, Michelle Feige, Sasha Bamford-Brown (for the title), and Marjoleine Kars (for reading every word).

Plutopia

Introduction

This book is about the embrace of two communities, united in fear, mimicry, and the furious production of plutonium. Richland in eastern Washington State and Ozersk (the name means "Lakedale") in the southern Russian Urals were Cold War enemies, but they had a great deal in common. Nuclear weapons complexes produced far more than warheads and missiles. They generated happy childhood memories, affordable housing, and excellent schools in prize-winning model communities that became havens for the new nuclear families that inhabited them. The plutonium pioneers of Richland and Ozersk recall never having to lock their doors, children roaming safely, friendly neighbors, and the absence of unemployment, indigence, and crime. I was puzzled by these memories of safety and security in cities at ground zero of the nuclear arms race. In the plutonium cities, security agents and doctors watched residents anxiously, with networks of informants, phone taps, and mandatory medical exams. Plant engineers, meanwhile, were pushed to produce as much plutonium as possible as quickly as possible, and they polluted the surrounding landscape freely, liberally, and disastrously.

Of all the stops on the nuclear weapons assembly line, plutonium production is the dirtiest. Each kilogram of final product generates hundreds of thousands of gallons of radioactive waste. In four decades of operation, the Hanford plutonium plant near Richland and the Maiak plant next to Ozersk each issued at least 200 million curies of radioactivity—twice what Chernobyl emitted—into the surrounding environment.[1] The plants left behind hundreds of square miles of uninhabitable territory, contaminated rivers, soiled fields and forests, and thousands of people claiming to be sick from the plants' radioactive effluence.

Chernobyl is a household word. Why have so few people heard of Hanford and Maiak? How could these sites of slow-motion disaster be considered by their residents to be so lovely and desirable? Leaders of Richland and Ozersk had a habit of proudly counting the number of PhDs in their cities. Why did people so satisfied with their knowledge agree to remain in ignorance for decades about the massive environmental contamination going on around them?

Researching this book, I was surprised to find that the powerful men charged with manufacturing the world's first supplies of plutonium worried a great deal about housing, shopping, schools, and recreational programs in addition to graphite piles and chemical processing plants. Alongside reactors, they built family-centered, consumer-oriented communities where working-class people were paid and lived like the middle class. Nor was this an anomaly. In subsequent decades, the blueprints for monoclass affluence migrated to civilian nuclear projects. The city of Pripiat, next to the Chernobyl reactors, was a rare modern city of urban conveniences in an otherwise poor, rural Ukrainian landscape. After the Fukushima disaster, press reports described how Japanese power companies, though they skimped on safety, amply subsidized American-inspired "nuclear villages" and sold nuclear power with visions of middle-class prosperity.[2] I wondered about this enduring connection between nuclear power and high-risk affluence.

Ozersk and Richland were government-owned and managed by corporate bosses. Richland was unusual on the American landscape because it had no private property, no free market or local self-government. Ozersk was one of ten nuclear cities in the Soviet Union that existed secretly, off the map, behind fencing; every resident required a special pass to live there. Strangely, residents appeared to like these restrictive arrangements. In Richland in the 1950s, voters in two separate elections turned down incorporation, self-government, and free enterprise. In Ozersk in the late 1990s, 95 percent of voters polled wished to keep their city's gates, guards, and pass system. At this writing, Ozersk remains in a state of incarceration, fenced and guarded. I wondered about these choices. Why did residents of these plutonium cities choose to give up their civil and political rights? Soviet citizens had no electoral politics, no independent media, but the residents of Richland lived in a thriving democracy. Why did the famed checks and balances fail to the extent that a calamity surpassing Chernobyl occurred in America's heartland?

These are the questions that drove this book. In answering them, I found that to entice workers to agree to the risks and sacrifices involved in plutonium production, American and Soviet nuclear leaders created something new— plutopia. Plutopia's unique, limited-access, aspirational communities satisfied most desires of American and Soviet postwar societies. The orderly prosperity of plutopia led most eyewitnesses to overlook the radioactive waste mounting around them.

This is the first book to narrate in tandem the history of the plutonium disasters in the United States and the Soviet Union. After this book, I hope it will no longer make sense to tell the two histories separately. People in Ozersk used to say that if you drilled a hole straight through the earth, you would end up in Richland. That is how I imagine the two cities: orbiting each other, linked on the same axis. Richland and Ozersk were made in each other's image—deliberately, as I will show—through the careful footwork of intelligence agents and community boosters who feared the end of plutonium production nearly as much as they feared the nuclear rival.

The narrative is told in four stages. Parts One and Two telescope on eastern Washington from 1943 and the southern Urals from 1946 as migrant workers, prisoners, and soldiers put up colossal plutonium plants. Initially, American and Soviet leaders planned to produce plutonium with militarized labor in army camp settings. Horrified, however, at the boozing, brawling construction workers, American and Soviet plant managers quickly changed their minds. Operators of the world's first plutonium plants, they realized, could not be as volatile as the product they made.

The answer to disorderly, violent migrant workers unhinged from family and community was to embed plutonium operators safely within nuclear families living in well-heeled, exclusive atomic cities. Americans called Richland a "village," recalling the mythical pastoral roots of American democracy. Soviets called Ozersk a "socialist city," referring to the mythical communist future without impoverished villages. Government officials spent lavishly on plutopia, more on schools than on radioactive waste storage, far more on residents within than on citizens without. As the Cold War promises of affluence, upward mobility, and the freedom to consume materialized in plutopia, anxious residents gradually came to trust their leaders, the safety of their plants, and the rightness of their national cause. As plutopia matured, residents gave up their civil and biological rights for consumer rights.

Demographically the plutonium cities were working-class, but because of their affluence they were seen at the time and are remembered now as middle-class enclaves. In the United States and the Soviet Union, middle-class professionals created and shaped national memory by appropriating the working classes, speaking for them, and merging them in an amorphous, "classless" society.[3] As class was made to disappear, Soviet and American factory workers learned to identify with middle-class supervisors and scientists who told them their workplaces and homes were safe.

Plutopia could not exist on its own. Historians Bruce Hevly and John Findlay describe how the Hanford plutonium plant spawned a series of "staging grounds," temporary camps and garrisons for low-level workers.[4] I found the same patchwork landscape of an affluent enclave surrounded by labor camps and garrisons in the southern Urals. Alongside these plutopian cities, American and Soviet leaders founded communities of soldiers, prisoners, minorities, farmers, and migrant workers, none of whom qualified to live among the "elect" of plutopia, but who served them and paid for them. Why go to the expense of separating people into distinct communities? Why not fabricate the usual large industrial cities with posh neighborhoods upwind and upriver and working-class zones downwind and downriver? The answers to these urban history questions are bound up in the history of science, medicine, and public health as much as in the history of intelligence and nuclear security, and they tell an important story about the ways territories were zoned, dividing people by class and race, which in turn determined not just how wealthy they were but how healthy.

Although people remained within their discrete communities, plutonium and its radioactive by-products recognized no boundaries. Part Three describes the years during which plant operators produced tons of plutonium behind double-walled fences of concertina and barbed wire. The plants' hermetic security and the segregation of territory into nuclear and non-nuclear zones created what I call a zone of immunity, in which plant managers were free to run up budgets, embezzle, conceal accidents, and, most ominously, pollute. Soviet engineers in the Urals followed the American experience of dumping waste quickly and cheaply underground and into local rivers and pumping radioactive gases skyward. Over the years, plant operators struggled with many accidents; some were massive, such as the 1957 blast at the Maiak plant, but most spills were routine and intentional. As operators dumped, radioactive particles joined air currents, filtered into drinking water, and flowed down rivers.

Within the first years of research, scientists in eastern Washington and the southern Urals grasped the dangers of the fission products they were producing. They learned that radioactive isotopes saturated the food chain and entered bodies—plant, animal, and human—where they lodged in organs and damaged cells. The first factory leaders worried about "epidemics," noticeable outbreaks of stand-alone illnesses among nearby populations. As time went on, however, a clear pattern of illness among exposed workers and neighbors did not emerge. This was not entirely a surprise. Scientists experimenting on lab animals understood that various radioactive isotopes worked in a variety of ways on bodies and that when it came to radiation poisoning, no two bodies were alike.[5] Scientists also realized that at low doses it took a long time for a body to get visibly sick and die. Hopeful that latency periods would allow time for progress in science to solve the problem of leaking and spreading radioactive isotopes in the future, facility managers failed to make time-consuming and costly design changes to protect workers and neighbors.

Radioactive isotopes proved just as difficult to detect on earthly landscapes as on bodily ones. Instead of the usual map of concentric circles with contamination magnified near the source, researchers charted a mottled, dynamic map with unpredictable "hot spots" scores of miles from the plutonium plant, while areas near the reactors could be relatively clean. The cagey qualities of radioactive by-products and their health effects—both were hard to forecast, locate, and diagnose—made it easy for American and Soviet leaders to deny them. As they did, managers found it politically more popular to shift funds from safety and waste storage to consumables, services, better housing, and higher salaries in plutopia.

In the face of a secret, emergent environmental catastrophe, the partition of territory into plutopia and staging grounds came in handy. The people of plutopia—young, affluent, fully employed, and medically monitored—statistically appeared to be the very picture of health. Meanwhile, migrant workers, prisoners, and soldiers did construction work on contaminated ground. They cleaned up spills and repaired factory buildings after accidents. As temporary workers, they were not monitored. They served as what would now be called "jumpers" because they moved on after they finished a job, taking with them the radioactive isotopes they had ingested and any subsequent health problems that might leave an epidemiological trace.

Farmers and indigenous people dwelled near the plutonium plants. Unlike the residents of plutopia, they did not live solely off remote consumer markets, but in large part off the land, which downwind and downstream was increasingly littered with hot spots of radioactive isotopes. As the plutonium plants spawned regional development, more people moved into the nuclear buffer zone and into harm's way. These people, too, were scarcely monitored for exposure. Risk, in other words, was calibrated along the lines of class and affluence, which roughly matched the primary and secondary zones of the nuclear security maps.

Part Four tracks the pioneers who first discovered they lived on a radioactive frontier. After the 1986 Chernobyl disaster blew the lid off the plants' security regimes, downwind and downstream neighbors began to attribute the occurrences of chronic illnesses and high rates of birth defects, infertility, and cancers in their communities to the plutonium plants. They had trouble making their case because of the opaque qualities of the plutonium curtain that had long guarded knowledge of the plants' radioactive footprint. For decades, experts armed with classified knowledge had spoken with assurance about safety and permissible doses, while dismissing the concerns of lay people. After 1986, local farmers, journalists, and activists demanded accident records and environmental and health studies. They insisted on learning the risks that corporate and government power brokers had exposed them to. In the court battles that ensued, self-styled victims' groups organized around novel conceptions of knowledge, freedom, and citizenship.

This was a stunning new movement in which American and Soviet activists, long focused on political, civil, and consumer freedoms, demanded biological rights.[6] They railed against corporate contractors who had privatized the tremendous profits from nuclear weapons production while socializing the risks to health and environment. Activists armed themselves with rival scientific expertise and their own community-based health studies. As they did so, they invented a new brand of civic engagement, adopted also by groups in Ukraine and later in Japan.

While there are big, transnational histories of nuclear weapons and national and regional histories of atomic programs or nuclear installations, *Plutopia* unites the transnational histories of the arms race with the places and lives of the people who created the bombs.[7] From the lofty heights of the spy satellite, the narrative descends to street view to focus on the towns in the crosshairs of nuclear annihilation in order to relate what the atomic age meant for the hardworking people who built the bombs and for their farming neighbors who lived with fission products cradled in their landscapes.

During the Cold War, propagandists and pundits often compared the United States and the Soviet Union in order to absolve one side or the other of an injustice or fault. Instead, I place the plutonium communities alongside each other to show how plutonium bound lives together across the Cold War divide. I suggest that the world's first plutonium cities shared common features, which transcended political ideology and national culture and were derived from nuclear security, atomic intelligence, and radioactive hazards. The major difference between the American and Russian plutopias, the difference that so critically determined health and illness, was that people in and near Richland lived in a far wealthier country, which meant their sacrifices for nuclear security, while great, were not as encompassing as those of the people who lived in and near Ozersk.

Documents provide the framework for this book. I worked in more than a dozen archives in the United States and Russia and relied heavily on the work of historians who explored before me. The written record is astounding, describing what officials knew, what they decided to hide, how much they chose to disclose, and the reasons they did so. The words of officials who made science and policy show how intertwined the nuclear security state was with the creation of urban landscapes, public health disasters, and the widespread contamination of the environment.

The people who lived in the nuclear cities and worked in and around the plants are the protagonists of this book. Over the past five years, I have carried out dozens of interviews with an unlikely cast of characters who stepped into the footlights of this drama because of a career move or the accident of their birthplace. Many people who signed a lifetime vow of secrecy agreed to talk to me only because they were angry about a perceived injustice. The Russian Ministry of Atomic Energy did not grant me permission to enter Ozersk, so I met people in nearby towns and villages in arrangements reminiscent of a Cold War spy novel. Some spoke nervously in whispers and coded speech. Several refused to be quoted. For the few shy of public exposure, I have assigned pseudonyms.

Several people told me fanciful stories that led me to doubt their credibility. When I checked their accounts, however, many turned out to be true. I learned to look out for apparently unreliable narrators as potentially rich sources, as people who saw their surroundings with a wider gaze than the usual blinkered perspective. Because the context of an interview often determines what is said, I recount where and when I met my sources. I also describe their vulnerabilities and my cultural insensitivities to show how the interview process, like archival research, is pitted with omissions, contradictions, and intentional and accidental ignorance. Some of my interviewees met me with suspicion or distrust because in researching this book I became a kind of disaster tourist. For them, I was the unreliable narrator. Perhaps that is also true for some of my readers, and that is right and proper. I do not claim to have uncovered the truth. Rather, I hope to have illuminated a corner of it. I look forward to other accounts and interpretations.

The political hostilities that fueled the Cold War have ended, but the nuclear chapter of world history is far from over. The lethal landscapes surrounding the plutonium plants are pockmarked with landmines of percolating radioactive waste and people who are persistently sick, they believe, because of it. The homeless state of American and Japanese nuclear waste attests to the complicated problem of safely containing volatile and dynamic radioactive isotopes that self-heat to hundreds of degrees, corrode metals, and seep readily through soils to be taken up by plant life—and will do so for tens of thousands of years. The nuclear stakes are high, and the impulse to renounce and deny invisible radioactive isotopes is great. Before Chernobyl and Fukushima came Hanford and Maiak, and with them the practices of plutopia: partitioning territory into "nuclear" and "clean" zones, skimping on safety and waste management to prioritize production, repressing information about accidents, forging safety records, deploying temporary "jumpers" to do dirty work, and glossing over sick workers and radioactive territories, all while treating select citizens to generous government subsidies and soothing public relations programs. Meanwhile, whistle-blowers who tried to alert the public to accidents and public health problems at the mothballed plants have been watched, harassed, followed, and frightened in both the United States and Russia *after* the end of the Cold War. Many of these scenes were repeated in Ukraine in 1986 and again in Japan in 2011.

Plutopia is about an inheritance that many citizens of nuclear powers have yet to face, have yet to figure out how to talk about, even as leaders around the world discuss an emergent "nuclear renaissance." Nuclear disasters in sequestered militarized landscapes are easy to hide, which explains the fact that, while Chernobyl and now Fukushima are household words, few people have heard of the plutonium disasters at Hanford and Maiak. My hope is these stories, told by the people who lived in two of the world's most radiated territories, will encourage readers to take another look at the nuclear past.

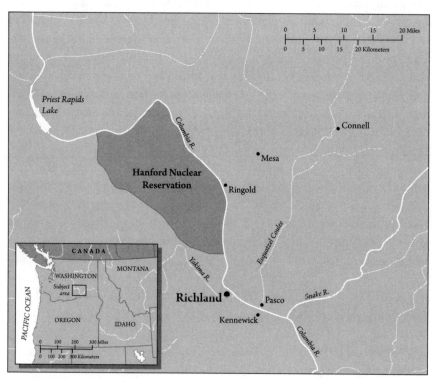

Map of Richland / Hanford area.

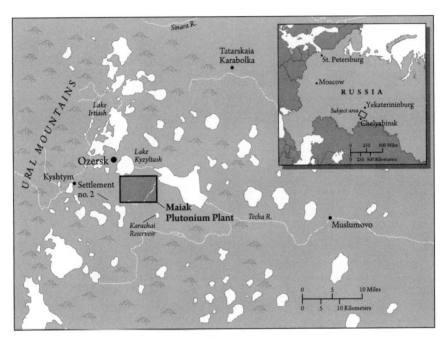

Map of Ozersk / Maiak area.

Part One

INCARCERATED SPACE ON THE WESTERN NUCLEAR FRONTIER

1

Mr. Matthias Goes to Washington

In December 1942, Frank Matthias, a lieutenant colonel in the U.S. Army Corps of Engineers, recorded his peregrinations that month across the inland American West to a number of hard-to-reach, careworn places.[1] Working with a couple of executives from the DuPont Corporation, Matthias was searching for the best site to build the world's first industrial-scale plutonium plant for the Manhattan Project. The Army Corps of Engineers was to supervise the project, while DuPont would serve as the plant's primary contractor.

The men drove in a thick fog across the Columbia Basin. They passed small family ranches that looked none too prosperous. They passed by dry-land homesteads, gray and worn, and little pump houses that fed silted-up irrigation ditches. They rode into and out of dusty little two-street "cities" of faded clapboard. The towns, lonely pinpricks of civilization on a stretched, parched country, were the remnants of the Inland Empire, a vision to transform sage and sand into prosperity. The dream held that large dams that zippered up the powerful mountain rivers flowing through the Columbia Basin would generate electricity for industrial development and water for farming. Settlers in the late nineteenth century sold themselves on the Columbia Basin as "good poor man's country," meaning that with hard work a man who had little to invest could make the land pay.[2] The powdery volcanic soil was rich in minerals but bone dry, with an annual six to eight inches of rain. Before cattle came to the country, the high hills and corrugated plains were covered with billowy bunchgrass. Cattle, sheep, and plows carried off the virgin grasses. Trains and wagons dropped seeds of "invading" plants such as Russian thistle, Russian olive, and cheatgrass, which stole moisture from other plants. After plowing or in the wake of a construction project, the soil, as light as dandelion seed, took to the air. The old-timers remember how the airborne soil got into everything—how it clogged machinery, stung eyes, and tore flesh. A poor man could afford seventy-cent acres, but farming semidesert land was capital intensive. Farmers who had no money for irrigation usually saw their dry-land farms fail to thrive.

The four visitors in the government-issue car spent the night in Mason City. The next day they drove on, over bumpy, pitted roads. Matthias climbed aboard a small plane and circled the Grand Coulee Dam and the desert beyond. The dam had been the brainchild of Rufus Woods, a local newspaperman. Though politically conservative, Woods had been inspired by the great dams going up in the Soviet Union in the early thirties. Woods won New Deal funds for his project, and in 1934 President Roosevelt made a trip to the region to trumpet the Grand Coulee Dam as a boon to what he called this "land of abandoned hopes."[3]

Boosters sold the dam as the biggest ever built—bigger, pointedly, than Soviet dams.[4] The congressman Homer Bone saw the dam as a way to make the promise of independent farmers triumph over what Bone feared was the emasculating power of big corporations and large ranching concerns. With the government building dams, independent American farmers could get the benefits of electricity and so the conveniences and productivity of the city. The dam would bring the disparate worlds of the nation together, tying washed-out farmers in the interior with prosperous middle classes on the coast. Grand Coulee appealed to American nationalism, but in so doing it acknowledged the advantages of Soviet-style, state-sponsored development, a state-led capitalism that in subsequent decades would reconfigure eastern Washington.

Detractors warned that the dam was overbuilt. Why, they asked, in the slump of the Great Depression would the government build a dam to generate electricity that no one wanted, only to reclaim two million acres that no one needed?[5] The critics had a point. The dam did not pan out as boosters imagined. Once it was built, power lines darted away from the Columbia Basin toward the west, to fuel factories and cities in the Puget Sound area. Meanwhile, the Bureau of Reclamation laid out the Columbia Basin Project, an ambitious plan for a network of pump houses and irrigation canals to turn a million acres of desert green. The plan stalled, however, because agricultural produce already glutted Depression-era markets.

But the big dam with a dearth of customers did attract Frank Matthias to its humming turbines. Matthias knew dams. In the early thirties he had worked on the Tennessee Valley Authority. In his diary Matthias did not marvel, as tourists did, at the size of the Grand Coulee Dam or the majesty of the unearthly rock formations of the surrounding coulees. Instead, he paid attention to kilowatts and the direction of power lines. And he was impressed.[6] With Matthias, the overbuilt dam finally found a customer.

Matthias settled on Hanford, Washington, for the world's first plutonium plant because it possessed the features he sought: plentiful supplies of clean water from the Columbia River, a sure source of electricity, a high percentage of government-owned land, and a certain scent of failure.[7] He rejected other sites closer to Grand Coulee because of the lush wheat fields and more prosperous-looking farms. The mild-mannered civil engineer could foresee the trouble involved in prying farmers from profitable crops on land irrigated at great effort and expense. Matthias found Hanford and White Bluffs "far more promising" in terms of the meager quality of crops and the shabby look to the ranches in the area.[8] "I was pleased," Matthias wrote to the head of the Manhattan Project, General Leslie Groves, "the total population was small and most of the farms did not appear to be of any great value."[9] Matthias found in Hanford what he was seeking: need and insolvency. Matthias' pleasure at the locals' poverty sounds pitiless, but it was a merciful gesture. Matthias had come as an "un-founding father" to the luckless communities he selected for removal.

There was a certain truth to Matthias' assertions of poverty. Half the acreage of Franklin County was foreclosed or abandoned in the twenties and thirties. By 1942, the population had dropped by 40 percent from its peak in 1910. The original ranch town of Hanford was located at a bend of the Columbia River, where it runs south and east on the Great Columbia Plain. The plain is a giant saucer set between the Cascades and the Rocky Mountains. At the lowest point of the great bowl, near Hanford, the Columbia becomes a desert river, snaking darkly through tumbleweed, sage, and scabland. D. W. Meinig, the foremost historical geographer of the region, pictured the surrounding country as a famished body, "starved as well as desiccated, with the bare ribs of the earth exposed and only thinly fleshed with soil."[10]

In 1943, the dam started to transmit electricity to the massive, secret government construction project, managed by the DuPont Company, at the time one of the nation's largest corporations. Instead of the Inland Empire of independent farmers, after 1943, local boosters at first haltingly and then passionately hitched their fortunes and prosperity to an expanding military-industrial complex backed by private corporations and sustained by federal subsidies. [11]

White Bluffs, 1938. Courtesy of Department of Energy.

This story of the American West has been penned many times before. This time, the Grand Coulee Dam and Hanford brought to eastern Washington the embrace of big government married to powerful corporations that evicted indigenous people from their land and rezoned hundreds of square acres of territory.[12] Anyone who paid attention to the patterns of dispossession in the history of the American West might have seen it coming when Matthias' government-issue vehicle arrived in late 1942. Apparently no one did. The two thousand residents of Hanford, White Bluffs, and Richland were taken by surprise when they got word in the mail in February 1943 that the federal government would be taking their land, ranches, orchards, homes, and businesses. The letters informed them that they had anywhere from a few weeks to a few months to vacate their properties. The federal appraisers offered reimbursements that in most cases did not cover the cost of the year's requisitioned crops, let alone the value of the real estate. To the locals, the whole deal felt like robbery. C. J. Barnett of Richland recalled people in shock, repeating in disbelief: "They can't do it."[13]

The expulsion of the residents of the Columbia Bend occurred nevertheless, and with great speed and efficiency. Several thousand people departed in an orderly manner, with Matthias pleased that there were "no substantial objections made by the farmers."[14] If a band of Indians had swooped in from a dime store novel and attacked these ranch towns, the homesteaders would have imagined picking up guns and fighting back. Instead, faced with invasion by their government, they had no recourse but to pack up and file a petition in court, hoping to get a better price.[15] Like Indians before them, immobilized by the transformation of sacred hunting and fishing grounds into square-mile sections, locals were caught flatfooted before the bewildering redesignation of their farms as "federal reserve." The disembodied quality of the Manhattan Project takeover meant there was no one to fight, no body to take the blows. It left local farmers paralyzed and passive as they were transformed from participants to spectators of the subsequent chapter of the Inland Empire.

2

Labor on the Lam

Federal agents and corporate managers arrived in eastern Washington in the spring of 1943 with the suddenness of a natural disaster, igniting the sleepy farm towns. Ranchers who had witnessed a steady drain of neighbors and the slow takeover of farmland by weeds saw time suddenly speed up. A classified DuPont documentary film on Hanford projects this image of effortless transformation: empty field giving way to earthen foundation, foundation covered in scaffolding, workers painting finished building, furniture rolling through new pine doors, clerks punching in at the entrance of the completed building.[1] The DuPont documentary offered to its select in-house viewers a soothing simplification of time. The film shows few bodies or faces. Combustion engines attached to claws and cranes appear to do most of the work, moving earth, piling steel, amassing concrete. The machines easily surpass the limitations of the human body. The film's producers left on the editing room floor images of the men who ran the machines, along with most every emotion—consternation, anticipation, boredom, fear— the conflicts, and the sheer sweat it took to build the massive plutonium plant.

The film, like many historical testimonies, is as much wish as reality. Machines did not build Hanford. People did, in the tens of thousands. The remote Hanford site made sense from the perspective of safety and security, but in terms of securing labor, it failed miserably. The first and most long-standing problem for Lt. Col. Frank Matthias and the DuPont managers who arrived to build Hanford lay not in commanding machines but in commanding bodies. And the foremost predicament lay in getting bodies to the remote, forbidding site and keeping them there.

As a consequence, Matthias worked largely as a procurement agent, a broker in the wholesale construction of not just the plutonium plant but several distinct communities created to build and run the plant. His first job was to staff the project, initially with engineers and civil servants, but soon he needed strong backs and skilled labor. The projections for the manpower needed to build the massive series of factories soared into the tens of thousands in the first months.[2] Matthias and DuPont managers devoted a great deal of time and resources to recruiting labor, but even so, a year and a half passed before they could get enough workers on the site.[3] During the first, critical eighteen months, the DuPont Company lacked 50–70 percent of the workers required for construction. There were not enough men to load, haul, or dig the immense foundations for the reactors or to pour the miles of concrete for the canyon-length chemical processing plant. Even after the number of employees reached its apex—forty-five thousand in June 1944—labor turnover, amounting to 20 percent of the workforce, plagued the construction site.[4]

The labor shortage persisted as the site's most dire problem because it slowed the pace of construction and left DuPont managers racing, in the end, to meet their deadlines. General Groves put a great deal of pressure on DuPont officials to deliver an industrial reactor and plutonium processing plant quickly. In 1943, Crawford Greenewalt, DuPont's project manager, overconfidently promised to deliver finished plutonium by the end of 1944.[5] Within a few months Greenewalt realized his mistake. In the summer of 1943 the construction project for the vast plant and reactors was stalled out due to a lack of carpenters. By August, DuPont workers had made no progress on the plant.[6] In November, DuPont executives were begging General Groves for more time.[7]

The biggest obstacle to attracting workers was the landscape. Nell Macgregor remembered in her memoirs traveling to Hanford, leaving behind lush coastal Oregon and crossing over to "wasteland." She cringed at the barren foothills "crouching low to earth" and offering no protection from the glaring sky or the oppression of unobstructed space. T. R. Cartnell remembered arriving by bus and hearing a woman hollering: "You take me out of here and right back to Tennessee or I'll take the children and leave you. This place isn't fit for a human being to live in."[8] Hanford suffers from a wind that blows for days and rattles screen doors, tarpaper roofs, and nerves alike, so that dogs take to barking and housewives to staring at their husbands' necks and fingering, as Raymond Chandler famously wrote, the edges of their carving knives. As bulldozers leveled ground for construction sites, great gusts seized the downy soil and gathered it in gray-green clouds. Many veterans of the Dust Bowl showed up in Hanford looking for work. When they recognized the gritty taste in the air of an approaching dust storm, saw the sky darken and visibility drop to a few feet, they turned around and boarded the next bus out of town. "They never even stayed overnight," Robley Johnson remembered, "a lot of them."[9]

In August 2006 I made a similar pilgrimage to Hanford. I had traveled the road from Seattle many times before, but I was taken again by the abrupt change in landscape to the burnished, undulating earth in eastern Washington. When I arrived in Richland in the late morning, the hot asphalt gave way underfoot and the sun had blanched the city's strip malls into two dimensions. I had an appointment to meet with Michelle Gerber, the Hanford site historian, who had agreed to show me around the nuclear reservation. After I was issued a badge, we climbed aboard a big government-issue SUV, the air-conditioning on high. We stopped to get some sandwiches because the trip would be long, Gerber warned, the expanses extraordinary.

In 1986 Gerber had been out of work and a single mother when government officials released ten thousand pages of declassified Hanford documents. These were the first cracks in the wall of secrecy around the plutonium plant. As a trained historian, Gerber immediately saw the value of the declassified files. After she saw her kids to school, she went to the Department of Energy's small new reading room in Richland to read the files. She told me that each day she expected the *New York Times* to sweep in with a team of researchers and scoop her story. They never came, and Gerber wrote the first history of Hanford informed by the new documents.[10] It was a brave act. She lived in Richland, a community proud of its contribution to national security, sensitive to criticism, and divided—into people who felt themselves part of the plant's heroic efforts and those who saw themselves as victims of the plant's criminal malfeasance. Writing a history that tried to find a middle ground between these two positions meant that she could lose friends.

Gerber steered the truck up to the guard post. A guard waved us in, and we crossed into the Hanford Nuclear Reservation. I had called for a month, traveled across the continent, and met with a team of public relations employees in order to get into the reservation. I waited for the moment when I would feel the historic significance of the place.

I waited in vain. The nuclear reservation is just not that impressive. We could have been in any industrial park, albeit a very big one. Inside the highly guarded zone was more desert land, partitioned by flawless asphalt roads and contained by miles of cyclone fencing that wrap massive buildings of raw concrete. Between the squared-off roads were zones of fenced-off beds of gravel pitted with concrete stumps, like petrified swamps in the middle of the desert. I had a sense there were workers around attending to the site's $100 billion environmental cleanup, but the distances were so vast that the human form became spectrally insignificant.

Gerber drove past the first B reactor, long mothballed, and the T plant, a vast concrete canyon where workers using robotics washed tons of irradiated uranium in chemical baths to distill grams of plutonium. We passed the site of U Pond, a dry bed of pebbles, where plant engineers had dumped radioactive waste. The spot of gravel looked no different from other sites that had been "cribs" for liquid waste, producing radioactive wetlands spooned up like alphabet soup: B Swamp, U Swamp, et cetera. Gerber drove by concrete pads, which she explained were buried "tank farms." She described how workers used to transport stainless-steel "pigs" of sample tank farm waste to the lab for testing.

Despite the terminology, there was nothing farmlike or natural-looking about Hanford. The place was alienating, sterilized, lunar, an open-air factory, curiously silent despite its great, latent volatility.

Gerber showed me with a sweep of her arm the site of the former Hanford Camp, set up in 1943 to house workers building the plutonium plant. I nodded unthinkingly and then looked again. She was pointing at emptiness: a flat plane, laser-graded, a few trees weakly prodding the sky. Staring, I began to make out the faint outline of streets converging at right angles. Western ghost towns usually have a few walls standing, foundations that outline a saloon or bank. This site had been expunged almost fully, although in its day the camp had been a city of sixty thousand people, and for a few months it had had the state's fifth-largest population.[11] Dining halls, barracks, stores, barber shops, theaters, taverns, a roller rink, a dance pavilion, a swimming pool, a bowling alley, a hospital, and the state's busiest bus depot and post office had all once stood on this site. It had been a vast camp that never slept, with round-the-clock shifts, or rather a place that always slept, as graveyard-shift workers sank their blinds and hoped for quiet amidst the constant rumble of machinery. Hanford Camp went up in a few months in 1943 and disappeared in a few months in 1945. This teeming city stood all of twenty-three months, and a half century later it had vaporized back to desert.

During the eighteen months it was populated, Hanford Camp was cut off from the surrounding area with a system of fences and gatehouses. Initially General Groves sought to invest as little as possible in the camp. He ordered only the bare essentials: Quonset huts for barracks, mess halls, and health clinics, and prefabricated olive-green buildings for offices and staging sites.[12] For entertainment, Army Corps of Engineers builders put up a barn-sized tavern. Inside the restricted federal zone, Manhattan Project officials subdivided the former range- and farmland into gated, guarded zones. Some areas were designated for living quarters and others for production, and a third category served as buffer zones.[13] Within the zones, territory was compartmentalized further. Women in the construction camp lived fenced off from men. Families lived in trailer parks separate from single workers in barracks. Management lived in quieter areas in houses in Richland, thirty miles down the road.[14] Production zones were also compartmentalized by jobsite. Workers on one site were restricted from entering other construction sites in order to limit their knowledge of the entirety of the secret project.

Security officials screened employees before they were hired. The officials looked especially for shady political backgrounds, such as membership in the Communist Party or a left-wing union. These people were turned down.[15] In the rush to find labor, however, lots of questionable types slipped in. The blue-blooded DuPont executive Crawford Greenewalt remembered that among the many fine workers, there were "utter riffraff—anybody who could hold a saw."[16] From March 1943 to August 1944, the plant police apprehended 217 employees on the lam from the law and 50 draft dodgers.[17] James Parker marveled that after he lied on his application, inflating his age to eighteen, he not only got a job but was assigned to work in the most restricted nuclear reactor site.[18]

By all accounts, Hanford Camp was an unruly frontier town. The DuPont police recorded crimes in escalating numbers: 4 suicides, 5 murders, 69 sex cases, 88 cases of bootlegging, 177 robberies, 450 grand larcenies, 1,124 burglaries, and 3,156 charges for intoxication.[19] Rape doesn't figure on the list of crimes, though apparently women were raped in considerable numbers. Robert E. Bubenzer, who supervised Hanford plant protection for DuPont, did not count rape as a crime, but rather considered it part of a free market exchange. "Most of the rapes occurred," he claimed, "because a customer didn't want to pay."[20]

With no family and no community, single workers found that drinking was one of the few distractions available. Men lined up outside the beer hall waiting to squeeze in. When the packed tavern got unruly, Bubenzer sent his men in with

Woman escorted into a paddy wagon at Hanford Camp. Courtesy of Department of Energy.

tear gas to clear it out. "Paddy Wagon" Davis would then pull up. "You threw them in 'til you couldn't close the door," Davis remembered, "and then you slammed it."[21] The plant had its own, corporate-run court and jailhouse, but DuPont officials didn't care to prosecute the drunks. They wanted the workers back on the job as soon as they sobered up. More worrisome were workers who spouted off in the mess hall about politics. Bubenzer said it was his job to quiet them down, and then these men "would gradually be eased out" of the project or "told to shut up." Bubenzer tallied up forty-four cases of this behavior, which he categorized as "un-Americanism."[22]

Bubenzer had 1,395 corporate patrolmen plus FBI agents and military intelligence working on-site, investigating, monitoring, snooping, and listening in on phone calls.[23] Vincent Whitehead, a sergeant in Hanford military intelligence, said he relied on intelligence from a network of mostly women. He called frequently upon two elderly women who belonged to just about every knitting society in the area and who soaked up information from their friends and neighbors in the guise of idle gossip.[24] Nell Macgregor called the surveillance "unceasing" and remembers fearfully watching a young nurse led away by guards for a mysterious transgression.[25] No one was above suspicion. Even the plant photographer, Robley Johnson, had to get permission to look at negatives of the pictures he had taken.[26] "Everybody," Whitehead remembered, "was spied on."[27]

DuPont officials hired Macgregor as a den mother to keep order in the women's barracks. The barracks were enclosed in an eight-foot steel and barbed-wire

Women's barracks, Hanford Camp, 1944. Courtesy of Department of Energy.

fence with one gate, patrolled by armed guards. The fence and the guards served to keep the "wolves," sex-starved men, at bay.

Hanford Camp, in short, had all the attractions of a minimum-security prison.[28] As a foray into social organization, the camp failed dismally. Workers didn't like it—the bleakness, the fences, the incessant surveillance, the lines, the brawling and stealing, the roaches and fleas (and the DDT used to kill them), the boredom, and not knowing what they were building or what cause they served. The police chief, Bubenzer, described the camp as "a depressing sort of a place. It was almost like being in prison. Wired in, barb wire . . . We had a number of nervous breakdowns of personnel. It was loneliness and depression and they hit the booze very hard."[29] As a consequence of the bleakness, workers quit and moved on. In June and July 1944, an average of 750 to 850 workers walked off the job each day.[30] The workers took with them their training and knowledge of the secret site. The camp, in other words, built to keep in secrets, leaked them daily as hundreds quit and boarded buses out of town.

3

"Labor Shortage"

Matthias and DuPont executives were seeking to build a radically new product in a plant of unprecedented size and type that required labor on a revolutionary new scale. They ran, however, from any kind of revolution in social or cultural practices that would have helped solve their immediate and chronic labor shortages. Instead, plant managers engaged in long-established, discriminatory hiring policies that were costly and inefficient.

For despite all the talk about a "labor crisis," plant managers had plenty of labor near at hand. Matthias could have tapped into deep wells of skilled and unskilled labor, much of it already mobilized in the western and southern American interior. In February 1943, for example, three hundred thousand 1-A registrants had been passed over by local draft boards. These were mostly single, healthy young men, a large labor reserve, rejected because they were African American and the segregated armed forces had little use for them.[1] Meanwhile, the Farm Security Administration (FSA) maintained migrant labor camps in the interior West loaded with mostly Mexican Americans, seasoned labor conditioned to long hours on temporary jobs in makeshift conditions—just the kind of workers Matthias most needed. The FSA even sponsored mobile camps that rolled on-site for seasonal agricultural work.[2]

Matthias dealt continually with many branches of the federal government, requisitioning supplies and cutting deals.[3] With a call to the FSA or the War Relocation Authority, Matthias could have quickly secured thousands of laborers, outfitted with their own housing, field kitchens, and guards. Despite perceptions of wartime labor shortages, there were plenty of workers available, mobilized and underemployed, had Matthias just called them up.[4]

But Matthias didn't make those calls. Most Americans in the 1940s did not think of labor as interchangeable units ranked by skill and training. Rather, racial and ethnic hierarchies overlay job sites and working communities and helped determine who could work and live where. This costly discrimination persisted even in wartime, even in the top-priority Manhattan Project.[5] Despite Matthias' desperate need for workers, despite Groves' constant inquiries about labor supply, Matthias could not hire significant numbers of nonwhite employees. Judging from his diary notes, he did not even consider it at first, and only later succumbed under pressure to hire a limited number of nonwhites.

In June 1943, the Fair Employment Practices Commission (FEPC) showed up at Matthias' office complaining that African Americans had not been hired on the federally funded project. Matthias demurred for a year, citing the prohibitive cost of adding extra quarters for black workers. Pressured, DuPont hired fifty-four hundred African Americans in the summer of 1944. To accommodate them, DuPont contractors built "colored" barracks in a separate, fenced-off section of Hanford Camp, and saved money by paying black workers less.[6]

At the start of 1944, Matthias penned in his diary, "Extreme pressure is being placed on us by the War Manpower Commission to accept Mexican laborers on the grounds that they are the only ones available."[7] Matthias again resisted, using the same economic rationale as for African Americans: "This [Mexican] labor will require a third segregation of camp facilities, inasmuch as the Mexicans will not live with the Negroes and the Whites will not live with the Mexicans."[8]

After much hand-wringing, Matthias and Groves agreed to employ clerical workers of Mexican ancestry who were "of a higher type that have proven satisfactory."[9] The last clause gives away the anxieties that Manhattan Project leaders papered over with budget concerns. To Manhattan Project officials, Mexican American and African American workers were potentially dangerous and disloyal. Groves instructed Matthias to avoid any discrimination, but in the same breath he told Matthias to carry out "careful checks as to citizenship and loyalty before these people are employed."[10] Military intelligence came up with a clearance procedure that added four days to the usual two-day protocol. Of the first group of ninety-two Mexican Americans vetted, less than half passed the security screening.[11]

Despite Matthias' worries, Mexican American labor came cheap. To avoid building a third segregated zone, Matthias had two buildings hastily refurbished in the city of Pasco. Latino workers paid to ride a bus sixty miles—a two-hour trip—to Hanford.[12] For the budget-conscious Groves and Matthias, hiring lower-paid minorities was nothing if not a cost saving. Yet they did very little of it. In total, the Hanford Works hired 100 Mexican Americans among the 125,000 employees who passed through the gates.[13]

The Manhattan Project officials introduced racial segregation to eastern Washington. The rationale for segregation was the presence of white workers from the South, who Matthias and Groves reasoned were too racist to live with nonwhites.[14] Yet only a third of the workers came from the South. The majority arrived from northern states, where Jim Crow was relatively new.[15] When James Parker landed at Hanford Camp in January 1944 from Idaho, he was shocked by the bathrooms marked "White" and "Colored" and by the separate barracks, mess halls, and theaters for African Americans.[16]

Instead of expense or workers' racism, discrimination and segregation followed established DuPont and army policy and local political pressures. DuPont, like much of corporate America at the time, did not make it a practice to hire minority labor. In the thirties, the DuPont Corporation had acquired a particularly racist reputation by backing anti–New Deal propaganda that used the threat of racial equality as a scare tactic to warn voters away from the Democratic Party.[17] When DuPont teamed up with the Army Corps of Engineers, which followed U.S. Army segregation policy, it was second nature for project recruiters to scour the country, at great expense, for white men and women to staff the construction site.[18] Racial discrimination wasn't so much a decision as a nondecision, a default practice, ingrained, trusted, unquestioned. Seeking to hire white labor for the well-paid Hanford jobs was, in the United States at the time, business as usual.[19] The only problem was that it was bad business.

To avoid hiring minority labor, for instance, Matthias set up a prison labor camp, called Columbia Camp, which proved an extremely expensive way to procure unskilled white labor. In the summer and fall of 1943, Congressman Hal Holmes appeared in Matthias' office arguing that there were two hundred tons of pears rotting on land confiscated by the Manhattan Project. Holmes said that if Matthias gave the farmers permission to pick the fruit, it would help compensate them for the low prices the government had paid for the land. Traditionally families had harvested their crops with the help of Indian and Mexican American migrant laborers. Hanford managers worried that allowing these workers inside the fenced-off construction area would threaten security.[20]

Rotting crops, however, created a problem. Locals hoped that after the war, the orchards would be returned to their former owners. Matthias was happy to keep this fiction alive, but he had no reserve labor to project a show of farming. If the orchards were not watered and pruned, they would wither and revert to "worthless desert land," causing a "public relations problem."[21]

Meanwhile, representatives from Prison Industries Inc. sought a contract with Matthias to set up a convict labor camp. Matthias realized he could use convicts to keep both the orchards and the myth of postwar restitution thriving. He consulted Groves, who had doubts about felons working near the top-secret plant. The prisoners, Matthias assured him, were trustworthy types: conscientious objectors, white pacifists—political prisoners, not criminals. Groves gave cautious assent, and by the fall of 1943, just as Matthias was refusing to hire and build separate facilities at Hanford Camp for Latinos from Texas for fiscal reasons, he was negotiating to hand over free of charge territory, barracks, electricity, and equipment to Prison Industries Inc. to set up a labor camp for white prisoners from McNeil Island.[22]

In the labor camp deal, Prison Industries managed and guarded the convicts. In return, the company retained all harvested produce. In dollars, this meant the

Columbia Camp for convict labor. Courtesy of Department of Energy.

federal government gave the private contractor an average of $313,000 a year to maintain the camp, and Prison Industries then sold the fruit for an additional $150,000 annually. The deal amounted to a generous subsidy to the private corporation to sell fruit appropriated from individual property owners a few months before.[23]

In 2008, I visited the former prison camp, Columbia Camp, with Bob Taylor, the son of the camp's commander. In 1944, Herbert Taylor arrived from McNeil Island charged with overseeing the construction of the labor camp.[24] Bob Taylor showed me the site where the camp stood from 1944 to 1947, in a gulch along the Yakima River. Pushing aside grass and thistle, we traced out the concrete foundations of the prison barracks. We walked the makeshift street where the warden and guards had lived in small prefabricated houses, and strolled down to the spot on the river where prisoners swam under guard after work. I asked Taylor how Columbia Camp, on the banks of the Yakima, got its name. Taylor said that when the Prison Industries agents first arrived they named the camp thinking they were on the Columbia River. We laughed over that. It is hard to imagine mistaking the broad, swift Columbia for the rocky little Yakima.

Prison Industries agents clearly did not possess much else in the way of local knowledge. The farmer's intimate understanding of local weather, water sources, soil, and topography escaped Taylor's father, a lifelong prison warden from coastal Washington. As warden, Herbert Taylor essentially ran a large agribusiness with stewards who were guards, not farmers, and with convicts who came from all walks of life. Judging from his correspondence, Taylor ran a humane camp. Judging from the financial records, the camp failed miserably. A 1947 Atomic Energy Commission review called the enterprise "highly uneconomical farming."[25]

The existence of the labor camp did serve one purpose, however: it helped smooth over the estrangement of local farmers from their land and crops. Convicts removed crops that would have rotted in plain sight. Prisoners carried out demolition work in 1945, taking down the construction workers' Hanford Camp, which closed a few months after the reactors started because of its proximity to the plant's radiation-emitting stacks. In 1947, bulldozers moved in and pushed down the labor camp, too. Four years after Herbert Taylor worked sixteen-hour days to build up the camp, it was gone.

The plutonium plant was billed in 1945 as the "Hanford Miracle," and in subsequent years it has been remembered as the accomplishment of a particularly American, can-do style of democratic capitalism. As Matthias wrote in 1946, the Hanford workers "are tributes to American ingenuity; they are proof that democracy can act swiftly, rapidly and with terrific directional force."[26] Yet in respect to labor, Hanford was neither speedy nor efficient. Nor was it cheap.

Bypassing nonwhite labor spelled ballooning budgets for the secret project. To get white professionals to relocate to the desert, DuPont officials argued, they needed to offer corporate, not government, wages, 30 percent above going rates. They enticed hourly workers with high wages in the form of overtime.[27] They also spent lavishly on teams of recruiters who crisscrossed the country looking for workers.[28] The cost for recruiting and incentives mounted to a whopping $7.2 million.[29] To put that figure in perspective, the entire Medical Health section of the Manhattan Project, which supported a dozen hospitals and labs, had an annual budget of less than $1 million.[30]

Federal funds also went for community-building programs to keep workers fed, entertained, and happy so they would stay on the job. Food was one way Matthias sought to win over Hanford construction workers. Groves personally wrote letters requisitioning tons of meat and poultry to the Hanford site. The camp kitchens served up massive heaps of rich, caloric food. Entertainment was another way to keep workers happy. In December 1943, while plant construction was far behind schedule, Matthias and DuPont managers devoted a great deal of time planning the camp's Yuletide Celebration. Matthias oversaw the Christmas decorations, ordered a million pounds of frozen chicken, and worried how to keep the mountain of frozen flesh cold.[31] And for two weeks in 1943 Christmas came to Hanford Camp with all the diversions and glitter of a traveling carnival. A ruby-cheeked Santa winked from behind guard posts above notices warning workers to keep mum. Loudspeakers pumped out Christmas carols. A life-sized Nativity scene radiated amidst corrugated Quonset huts.[32] White employees danced in cleared-out mess halls. A minstrel show with performers in blackface took place in the new auditorium. White workers could attend sporting events, and white children could see a show each day.[33] Plant officials were so pleased with the festival that in 1944, as operators were loading uranium slugs into the first Hanford reactor, they planned to repeat it, this time with a lesser program, for "Colored."[34]

In January 1945, navy pilots flying over the sharp bend in the Columbia River between Hanford and the cities of Richland and Pasco looked down on a territory transformed in just twenty-three months from scrub, open range, and farmland into a wholly new, carceral terrain. The fences of the Hanford Works divided the project internally, segregating each production and staging area. Those fences butted up against the barriers encircling Hanford Camp, and inside the camp ran more fences sectioning off men from women, blacks from whites, families in the trailer camp from single workers in barracks. To the south ran the barbed enclosures of Columbia Camp, confining convict labor. Flying southeast, across the Columbia River, pilots crossed the city of Pasco, where no barriers other than the powerful force of racial prejudice divided the town east from west, white from minority.

Discriminatory practices triggered labor shortages, which delayed plant construction.[35] With the news that DuPont would start the first reactor not in June 1944, as promised, but in the late fall of that year, Groves was fit to be tied. As DuPont executives tried to explain their manpower problems, Groves gave them notice that he would not accept postponed delivery of the final product.[36] He pushed "embarrassed" and increasingly compliant DuPont officials to make a series of shortcuts related to safety and waste storage. The resulting releases of radioactive isotopes would in subsequent years be imprinted on the landscape, on emergent notions of public health, and, specifically, on the bodies of the people who lived in the newly reconfigured terrain.

4

Defending the Nation

The last episode in the building of the Hanford plutonium plant concerns a group of thirty-three Wanapam Indians and their leader, Johnnie Buck, who showed up, like hundreds of other petitioners, on Matthias' doorstep in 1943. This episode offers a glimpse into how Matthias and his contemporaries felt about the territory in eastern Washington they were removing from the realm of human habitation for the foreseeable future. Sacrificing land also sacrifices people attached to that land. Matthias in 1944 sought in vain to avert that unfortunate outcome.

At the end of his life, Matthias donated his private papers to the DuPont Corporation's Hagley Museum. The papers mostly include correspondence related to the plant and drafts of speeches he made after the war when he advocated international atomic control and peaceful uses of nuclear power. One letter, hand-delivered and handwritten on paper torn from a school notebook, is addressed to Matthias from Johnnie Buck. Buck wrote in breathless English:

> Dear Sir, Why is that we can't go through to Hanford and we'd like you come Sunday because we will be all to gather for a Indians feast its that why we want you to come Sunday and then after that you not see us we might move for summer stay and we'd like to go through Hanford and Horn [Rapids] while we fish that what we want to know about and we are stay few more weeks and that why we want you to come Sunday let you see the house we stay in must close
>
> —Johnnie Buck[1]

Surprisingly, the busy Matthias accepted Buck's invitation to the Wanapam feast, which honored the five sacred foods (water, salmon, berries, roots, and game) and was part of the Wáašat (Washat) religion. At the feast, Matthias sat on the ground, listening to the Indians' songful prayer and watching them dance.[2] A photo shows Buck and Matthias shaking hands. The accommodating lieutenant

Matthias and Johnnie Buck, 1944. Courtesy of Hagley Museum and Library.

colonel, in army green, faces Buck; the chief, in buckskin and beads, is turned away from Matthias, staring beyond the camera.

In the fall of 1943 the Wanapam were barred from making their annual fishing trip to White Bluffs. For a hundred years, the nation had evaded federal Indian reservations and moved on when their territory had been overrun by ranchers and farmers, but when the huge Hanford Reservation was heaved squarely into the midst of the Wanapam's traditional fishing grounds, they found it hard to migrate around the sheer immensity of the guarded, fenced zone.[3] Buck went to see Matthias for permission to pass into the federal reservation to their autumn fishing grounds.

I wondered how Buck and Matthias conversed through Charlie Moody, Buck's interpreter.[4] The two men spoke mutually incomprehensible languages, linguistically and philosophically. On the face of it, Buck had little chance for a successful petition. Matthias had already refused deed-holding, voting, tax-paying farmers access to their confiscated land and crops inside the security zone. It seemed a matter of course that he would also deny access to the tribe, especially as the Wanapam, who had never signed a federal treaty, could make no documented claim to the land.[5] Instead Matthias responded with surprising cultural sensitivity, and, in so doing, veered far off course from his normal bureaucratic caution.

Initially Matthias told Buck to name his price for the lost salmon. He offered to either pay the tribe for their fish, ship in an equivalent stock of fish, or let them fish this year but not the following year, when the government would buy their fishing "privileges." But the chief did not want money. He told Matthias the Wanapam had fishing rights and they wanted to fish in that one spot, where they always had, at White Bluffs, right where the secret plant was going up. Matthias penned in his diary: "His [Buck's] only interest was to get the fish."[6]

Matthias dwelled in a world where almost everything was bought and sold in huge volumes for negotiated prices. He was confounded by Buck's refusal to exchange his Columbia River salmon for other salmon or for money that would buy fish and other calories. Matthias' daily life of rations, regulations, and negotiations rarely encountered the sacred and non-negotiable. And, apparently, it tickled him. Faced with Buck's diplomatic obstinacy, Matthias agreed finally to let the Wanapam onto the secret federal reservation to fish. He also requisitioned an army truck to bring the Wanapam in each day and transport out the catch each night. That way Matthias didn't have the Indians wandering around unsupervised but the Wanapam got their fish. Matthias was pleased with the solution.

Every person who entered the secret reservation was subject to a background security check; even congressmen were barred without it. Ethnic minorities underwent extra background checks. Matthias, however, granted the Wanapam access to the federal reservation with no security procedures at all. Matthias gave the chief and his two assistants carte blanche to escort in any other tribal member, also with no security clearance.[7] This was an astonishing potential breach in security. How did Buck convince Matthias to grant him this unprecedented access?

Looking for answers, I went to see Rex Buck, Johnny Buck's grandnephew and the current spiritual leader of the Wanapam.[8] Buck lives with other Wanapams in a little community on a handkerchief of land pressed between the Priest Rapids Dam and a high bluff the color of chocolate. The hamlet, consisting of a dozen double-wide mobile homes and a longhouse of corrugated steel, sits in the shadow of the dam, across from the Wanapam's submerged sacred island, flooded when the dam was completed in 1961.

When I arrived, Buck was holding up breakfast for me. He met me without a smile and introduced his wife, Angela, and twentysomething daughter, Lila. Angela looked me over broadly, skeptically. I thought I understood. Buck is the chief spokesman for the Wanapam, and a lot of people like me come through, ask questions, and leave with the story they want. I was just another in a long trail of inquisitors.

We sat down to breakfast, without formalities. The table was laid with salmon, huckleberries, and three bowls of roots, boiled and blanched white. There were boxes of breakfast cereal on the shelves, but, Buck explained, this was their special Sunday morning breakfast, "medicine food." Buck started to chant in "Indian," lifting each bowl and calling each dish. As he did, we each took a piece. After Buck had called everything but the berries, we ate. Buck talked. I asked questions.

Buck, born in the fifties, belonged to the first generation of Wanapam to go to school and learn English. Buck explained that his father knew only a little English. "He used to practice with my *Dick and Jane* books, but he didn't get very far along," Buck said.

I helped myself from the bowl of huckleberries. Angela flinched faintly and handed the bowl to her husband. Buck sang, calling the berries, and then the others took helpings. Huckleberries are gathered last in the year and so are eaten last in a meal, Buck let me know diplomatically.

Buck explained a bit about the Wanapam's Wáašat religion and its relationship to this particular stretch of the land between the Columbia and Yakima Rivers. The earth, he said, is alive; it speaks and listens. The work of gathering food is akin to prayer; the tribe's territory makes up the sacred space of that prayer. Women, for instance, gathered the roots we ate for breakfast silently, so they could hear the earth speak. If, Buck said, the tribe did not observe the ceremonies and gather their food in the proper way, the earth would not hear them and would forsake them; it would no longer give them food, and the Wanapam would perish.

Each place of the ancestors and of the past, Buck explained, had to be visited and observed each year. The tribe needed to do that. That is why his great-uncle had to get into White Bluffs to fish and tend the graves. Johnnie Buck wasn't talking about calories, market value, or an exchange of commodities; he was trying to convey the essence of the Wanapam's sacred cosmology. In other words, like Matthias, Buck was worried about his nation's survival.

On the federal reservation, Matthias turned property owners off their land and transformed independent merchants and farmers into landless wage laborers, yet he respected Johnnie Buck's requests for spiritual survival and cultural autonomy. Matthias noted that the Wanapam had camped at that bend of the Columbia River since the time of Lewis and Clark and that they had never signed any treaties or taken any government funds.[9] Matthias seemed to admire the Wanapam's self-reliance, their determination to migrate each summer to the mountains for roots and berries and to the river in the spring and fall for salmon. "In general these Indians are very independent and insist on maintaining their independence and their treaty rights to fish on the Columbia River. I do not believe that their loyalty can be questioned," Matthias assured himself, with no evidence for this assertion but the force of Buck's personality.[10] Perhaps Matthias thought the Indians too simple and local to engage in international espionage of sophisticated science and technology.

Or perhaps the Indians struck Matthias as real, primeval Americans, the kind that were disappearing quickly as the plutonium plant brought eastern Washington further into the mainstream of the American industrial economy and federal bureaucracy. To him, Buck in his ceremonial dress, speaking through an interpreter, might have personified a schoolboy image of what he had expected to find on the western plains, instead of the Quonset huts, plywood barracks, dump trucks, and construction dust he encountered daily. Perhaps Matthias exempted the Indians from the security regime out of nostalgia for the image of a pioneering West, which he was having bulldozed just outside his office window.

After the bomb made with Hanford plutonium fell on Nagasaki in August 1945, the Wanapam's pass to the reservation was revoked. In the 1960s, the tribe was granted access again. Rex Buck described how tribal members were escorted into the reservation: a jeep with soldiers and machine guns in front, a jeep with soldiers and machine guns behind, the Indians riding in a truck in between. They were on sacred land at last, but they did not carry out their ceremonies. "It was not comfortable," Buck remembered, "My uncle would say in Indian language, 'Let's look and be careful and keep going.'"

After the war, Matthias didn't forget about the Wanapam. He wrote letters inquiring about them and their claims to land and fish.[11] Nothing came of the correspondence. The white chief from Hanford was powerless to help the Indian chief and his small tribe, as much as he would have liked to. Both leaders stood impotent before the humbling new forces that had brought nuclear power to eastern Washington.

5

The City Plutonium Built

After the Army Corps of Engineers bulldozed the original ranch town of Richland, Corps officers and DuPont executives went to work repopulating Richland anew. Richland was to house plant operators, Matthias noted, "who must be kept under control for security reasons."[1] After witnessing the boozing, brawling single migrant workers in Hanford Camp, DuPont executives determined that the new operators' village would be dedicated to workers safely rooted in nuclear families in the new atomic city. DuPont and Corps employees bickered about what this new city would look like. The compromises they grudgingly made amounted to the creation of a whole new kind of community, one that banished single migrant laborers and minorities to the outskirts, displacing working classes to the cultural margins. They established a new regime that equated security with white middle-class families in a new upscale, exclusive bedroom community bankrolled by generous federal subsidies. After the new Richland took shape, it was widely promoted as a "model" community. In subsequent years, thanks to a similar alliance of federal subsidies and corporate control, many other exclusive, all-white, upzoned planned communities cropped up across the United States. The model was so successful, in fact, that Richland now appears unexceptional. Suburbs like the made-over Richland multiplied at such a rate in the postwar decades that it is now easy to overlook how novel it was in 1944.

General Groves had in mind a town akin to an army base—fenced, guarded, compact, gridded, with numbered streets and barracks-style dorms and apartments centered around a few utilitarian commissaries.[2] This was the kind of spare, fortlike town Corps engineers were already constructing at Los Alamos and Oak Ridge.[3] DuPont executives, however, rejected Groves' plan. They resisted putting up a fence around Richland because, they said, their employees would not live behind a fence. They assured Groves they had run company towns before and knew how to keep secrets and workers under control.[4] Instead of a fortress, DuPont executives dubbed Richland, in company-town fashion, a "village."[5] They hired an architect, G. Albin Pehrson, who sketched a city with gently curving streets spiraling around spacious single-family houses on large lots and a downtown business district with plenty of services and shops.[6] Groves pruned Pehrson's plan considerably. The plate-glass windows, the second grocery store, and the landscaped schools all had to go.[7] In fact, Groves didn't even want to call the hotel a "hotel," which to him implied luxury. Instead he renamed it the "transient quarters."[8]

DuPont executives did not readily submit to Groves' dictates. They wanted to build a settlement more substantial than an army base, more luxurious than a classic company town. They pointed out that in hiring for the world's first plutonium plant "they couldn't take a chance on junior men." They said they would need well-trained employees "of the highest type" to run the new plant.[9] Convincing senior DuPont employees and "good men" to live in Richland would be tough, they argued. Edward Yancey, a DuPont vice president, contended that "people out here would not be satisfied unless they had at least the bare essentials of normal, small cities."[10] "Normal" in this case meant the kind of infrastructure—housing, schools, stores—that middle-class professionals had come to expect in the East. Reasonably, DuPont managers wanted to build for themselves and their white, highly select employees a comfortable full-service city; even more reasonably, they wanted the government to pay for it.

But General Groves was a scrupulous manager who kept a close watch on the budget. Ideologically this should not have been a problem. DuPont executives shared with Groves a disdain for what they called "hegemonic" big government. They also disliked government planning, social welfare spending, and, generally, most New Deal programs. Irénée du Pont was an influential member of the governing board of the American Liberty League, which channeled corporate dollars into opposing New Deal spending to combat the Depression.[11] The Liberty League claimed that, in sheer panic, the Roosevelt administration was destroying capitalism and American democracy and that the president would soon make himself a communist dictator.[12] Instead of government interference led by the irrational passions of the electorate, DuPont leaders championed private stewardship of free markets led by clear-thinking corporate elites.[13]

Laissez-faire ideology, however, collided with DuPont's history. The company had emerged as a financial powerhouse by serving the U.S. government as a military contractor during World War I, when DuPont's annual profits escalated eightfold, earning DuPont the moniker "Merchant of Death." DuPont also stood to profit handsomely in the new war by supplying the army and navy with explosives, synthetic rubbers, insecticides, and nylon. For DuPont, war was very good business. As Lammot du Pont put it in September 1942, addressing the National Association of Manufacturers (NAM): "Do business with the government as you would with any other buyer. If it wants to buy, it has to do so at your price."[14]

The more the U.S. government spent, the more DuPont stood to gain. New Deal social welfare went against the grain of DuPont corporate ideology, but government spending that promoted business, generated profits for deserving parties, and preserved unspoken class divisions—that was the desired future, and in planning the city of Richland DuPont executives sought forcefully to push this vision along.

Initially DuPont architects submitted designs for houses exclusively with three or four bedrooms because "the employees at this station will be of a higher than normal type." Matthias objected strongly to such luxury, writing that "a temporary village under war conditions . . . opposes every principle of war economy and is deleterious to the war effort."[15] But DuPont's Yancey held his ground. He predicted that 25 percent of the plant employees would be supervisors and technical staff—"like commissioned officers," he translated for Matthias. Men with higher rank, Yancey stated, would require larger houses. The telegrams went back and forth, Matthias and then Groves demanding that DuPont submit designs for smaller houses, DuPont managers steadily refusing.[16] DuPont executives appear to have had the upper hand. As the rift widened, they had Groves and Matthias come to meet with them in Wilmington, each trip requiring for Matthias several days of travel.[17]

What were they thinking? Edward Yancey was a DuPont vice president, in charge of the vast explosives division. Groves was masterminding the entire Manhattan Project, and Matthias was charged with constructing the world's first plutonium plant. The nation was at war, and these leaders were fighting over none other than whether there should be two or three bedrooms for tract houses in Richland. Why was the question of a few extra rooms so important?

Groves was concerned about justifying to Congress after the war the expense of the Manhattan Project. At the time, three-bedroom houses were a luxury reserved for a minority of American elites. DuPont was proposing to build a town of nearly uniform largesse in the midst of wartime rationing—an appallingly extravagant notion.[18] Yet DuPont executives held fast in part because they believed that meddlesome federal officers should butt out of DuPont's contracted business, but also because they made a forceful argument that the success and security of the plant depended on housing designs and urban planning.[19] Pehrson, the project architect, argued that they needed to maintain morale among the transplanted workers. "High morale," Pearson wrote, "cannot be achieved by crowding skilled and veteran workers into inadequate dwellings."[20]

On other planning issues DuPont executives also held their ground. Despite Army Corps of Engineers orders, no fence went up around Richland. Unlike at Los Alamos and Oak Ridge, residents did not wear security badges or pass through a guardhouse to get home. Groves wanted houses more cheaply nestled together, within walking distance of the town's amenities. Pehrson spaced the houses far apart, which increased the cost of sewer and electrical lines, as well as making the city residents more dependent on cars and bus service.[21]

Groves was shocked at DuPont's plan to locate houses of a certain class together by type, so that residents of Richland would be clustered in neighborhoods by rank on the corporate flow chart. In a society where popular rhetoric held that citizens were equal, this spatial dramatization of class was too much.[22] Over Groves' objections, however, DuPont laid out the best houses on the most desirable lots along the river for the top brass.

Yancey made only one major compromise. He agreed that one-third of the houses would be either low-cost prefabricated houses or duplexes, but he still insisted that most of these houses would have two or three bedrooms. The prefabs were small, cramped, drafty affairs, with plywood furniture, pipes that froze, and roofs that had to be tethered because they took off in the fierce desert winds.[23] For the same price, apartments or row houses could have been built that were more cost-effective, spacious, and durable.

In fact, in this yearlong argument, Groves was right: there was no point constructing large, expensive housing in a sprawling layout when the "village" was supposed to be temporary (a fiction used to cover up what was projected to be the long-term project of building up the U.S. nuclear arsenal) and building supplies and labor were in short supply. Yet Groves, reputed to be a willful, arrogant "sonavobitch," largely gave in.[24] DuPont executives held their ground and built a community unique at the time on the American landscape—a wartime company town, paid for by the federal government, that resembled a private, upscale, postwar suburban development.

Clearly, for DuPont executives, freestanding houses bore a cultural meaning that overran practicality, even during a war, even on the Manhattan Project. DuPont managers' compromises in themselves point to this fact. The cheap prefabs for blue-collar employees were shoddy, but they sat on their own lots and did not look like working-class accommodations. The freestanding, suburban-style prefabs spelled middle-class respectability and tranquility, even if no middle-class people would live there.[25] DuPont managers glossed over the fact that 75 percent of plant employees were to be blue-collar workers.[26] Yet if most workers were blue-collar, why did DuPont managers argue so stubbornly for middle-class housing?

DuPont managers promoted Richland's master plan while engaged in a larger ideological battle on the national level for what they described as the survival of the "American way." Working through the DuPont-supported NAM, propagandists argued that, in contrast to New Deal social programs, American business would deliver a uniquely American "abundance," which would serve up a uniquely American freedom—the freedom to consume. NAM advertisers promised that in a laissez-faire economy, abundance would flow to all Americans, uniting the common worker with the middle-class professional in a shared, classless surfeit of consumer goods.[27]

In Richland, the concrete-and-drywall solution to this vision of a classless society was cheap, mass-produced working-class housing that *looked* middle-class. In insisting on middle-class housing, DuPont executives argued that only a community united in middle-class abundance would deliver plutonium safely and securely. Yet to run the vast plant they had to stock Richland with working people. So they simply called the proletariat "middle-class" and in that way co-opted it.[28] The scheme worked. Although Richland was a city with a working-class majority until the 1970s, it was seen and is remembered as a middle-class town of scientists and engineers, a homogeneous, monoclass society.[29] Disappearing the working class and recharacterizing Richland as "classless" helped muzzle the voices of labor and suppress unions, while coaching workers to identify with their managers in the interests of both national security and their own financial security.[30]

Once DuPont and the Army Corps of Engineers had settled on Richland's design, the city went up quickly, in less than eighteen months. DuPont managed to build swiftly by mastering assembly-line building techniques, in which workers were assigned simple, specific tasks and moved from site to site constructing a series of uniform houses. Prefab houses went up even faster. They came assembled in sections. Cranes lifted the walls and roofs off truck beds onto foundations, and crews bolted the walls together and attached the roofs.[31] Transforming a leveled terrain into a residential area in a matter of months was a

New construction in Richland. Courtesy of Department of Energy.

revolutionary new development, one that after the war shaped the emergent suburban landscape. Bill Levitt, the founder of Levittown, learned how to mass-produce communities as a wartime army builder on projects similar to Richland.[32] In this too—assembly-line residential developments—Richland was a trendsetter.

As DuPont executives increased the size of houses, they correspondingly raised the bar on who could live there. As the price of larger individual houses escalated, Groves, sweating over the rising tab, reiterated the need to reduce costs and so provide housing "only for those people who are required to live there for security reasons." To keep costs down, Groves decided that low-level workers would be barred from living in Richland.[33]

But where would the low-level workers live? Because of the massive influx of construction workers, housing throughout the region was impossibly scarce and expensive. The Corps and DuPont executives decided that unskilled plant workers who did not qualify to live in Richland would commute from neighboring farm towns, where they would live in existing housing or in new federally funded (FHA) housing, which, though rudimentary, Yancey pointed out, would suit these "service and low level employees" because they "will be people whose housing standards are none too high."[34] Groves and Yancey specified which lower-ranking plant operators—"laborers, janitors, and other manual workers"—would be excluded from Richland.[35]

This reiteration of Richland's exclusivity occurred shortly after news came in from nearby Pasco that the overtaxed little city, which had tripled in size with wartime construction workers, was a threat to public safety. In December 1943, Matthias penned in his diary: "The situation at Pasco with respect to crowding and general lack of control of workers is one which shows potential danger." Pasco had a "ghetto," one of the few places in the region where nonwhites could rent a shack, park a trailer, or pitch a tent. Pasco also had a strip of cheap eateries, bars, and bordellos. The "danger," Matthias reported, was that "irresponsible workers" were "flagrantly disregarding the local law." Matthias planned to get additional state troopers assigned to Pasco, and he worried: "If this condition is serious now, it will undoubtedly be more serious in the near future when this project begins to terminate employees who are undesirable." Something would have to be done, Matthias continued, "to see that these people actually leave Pasco and this area to avoid a concentration of undesirables and an unbearable load on the facilities, both social and law enforcing, of the Pasco area."[36]

Pasco's working-class volatility so near the emergent plutonium plant presented a major national security threat. Washington's governor, Arthur B. Langlie, went to see Matthias, worried about the problem. He and Matthias came to an agreement to eject laborers who were no longer needed, "particularly the negroes."[37] As work slowed down in 1944, supervisors first laid off African

Americans from the construction site.[38] Matthias ordered more state troopers to Pasco in 1944 to help disperse "vagrants" and unemployed drifters.

Pasco served as an example, one laced with a threat, of what Richland should not become. Part of the task of "securing" Richland involved quarantining it against the bellowing, brawling, shack-dwelling working classes and minority laborers of Hanford Camp and Pasco. Building respectable single-family housing with multiple bedrooms ensured that upright white family men, rather than explosive working-class bachelors, would work at the plant. DuPont officials won the debate over housing basically by making it a security issue. They successfully argued that the operators of the world's first plutonium plant had to be securely embedded in nuclear families in an exclusive atomic city.

After the war, journalists piled into Richland. They had limited access to the plant behind the gates but could range freely in Richland, and they loved it. The *San Francisco Chronicle* described the "self-contained, shiny new village" as "Paradise."[39] *Business Week* called it "utopia." The *Christian Science Monitor* hailed it as "a model city . . . to be carefully studied by urban planners for years to come."[40] Yet Richland was a puzzling creation in American society—a collection of what appeared to be private homes, private businesses, and grassroots organizations that were centrally planned, managed by a corporation, ethnically segregated, federally subsidized, and closely watched and controlled.[41] This model echoed deeply in postwar America as all-white, highly subsidized suburbs sprang up wherever prosperity allowed.[42] DuPont executives' success derived from the fact that they focused not on building for a community but on building for individuals as loyal and valuable employees to the corporation, as consumers, and also as objects of security, safety, and surveillance.

By charting onto the landscape (invisible) zones of class and race, by offering financial security alongside military security, DuPont executives managed to hit these multiple targets without needing guard posts, identity cards, and fences, as at other Manhattan Project installations—without creating the appearance that Richland was a closed nuclear reservation for white male workers of a higher type. Oak Ridge and Los Alamos, fenced and patrolled, leaked nuclear secrets to Soviet agents. So far, no evidence of an espionage breach from Richland has surfaced in Soviet archives. Richland had no incarcerated people, just incarcerated space. It was quite an accomplishment.

6

Work and the Women Left
Holding Plutonium

The plutonium plant also differed from Los Alamos in that it was not a lab but a bomb factory, a very large one. Few, however, of the common laborers from the Hanford construction site were hired as permanent employees. Instead, DuPont recruiters set out hiring anew two classes of workers—blue-collar operators and the white-collar supervisors and managers who would direct them. Access to knowledge about radioactive hazards was portioned out on a sliding scale. Those who worked most closely with radioactive solutions were often the most scantily trained and least informed.[1] Ignorance and anxiety rode shotgun up through the hierarchy, dividing workers by rank and gender. The higher up on the corporate hierarchy an employee was, the less that employee had to fear.

In hiring operators, DuPont had an attachment to values derived from the du Pont family's old-line Protestantism.[2] There was no talk of hiring black and Mexican American workers, whom the company had been forced to hire for construction. Some divisions of the corporation discouraged hiring non-Christians. With this selection process, the term that officials of DuPont and the Army Corps of Engineers used—"higher type"—takes on an Aryan weightiness. The first (classified) census of the new Richland revealed that all residents were white. The vast majority were Protestant. Fifteen percent were Catholic. Ten employees were Jewish.[3]

DuPont recruiters set up two categories of employees—exempt and nonexempt. Exempt workers were paid a salary and tended to be transfers from other DuPont plants. They had a higher education, worked in supervisory and technical positions, and were for the most part already "DuPont men."[4] The second category, the majority, was nonexempt workers, who were paid weekly or hourly wages for shift work. These workers tended to have no more than a high school degree. DuPont managers sought to hire these workers locally.

In Richland, I went to see some of the people they call "old-timers," hired at the plant in 1944. I met Joe Jordan in his comfortable ranch house, the furniture circa 1960, neat and mod. DuPont hired Jordan in 1941 after he graduated from Georgia Tech with a degree in chemistry. In 1943, Jordan got transferred to Chicago. There he reported to his new supervisor, who flipped on his desk a uranium fuel slug, used to power nuclear reactors, and laid out the whole Manhattan Project mission. Jordan's new job would be to take fuel slugs after they were irradiated in a reactor and dip them in a series of chemical baths to strip them down to grams of plutonium. The plutonium extract would be used to make a very powerful bomb.

For several months Jordan trained at the Met Lab at the University of Chicago. In October 1944, Jordan arrived in Hanford and toured the automated, remote-control plant under construction. As a chemist at Hanford's T plant, Jordan's job was to analyze samples of irradiated solutions along the plant's assembly line. Jordan oversaw a group of lab technicians who did the actual work of gathering the radioactive solutions, handling and measuring them, and moving the solutions through the production process.

When I met Jordan in 2008, he was ninety years old, one of those individuals whose longevity defied the talk about Hanford's radioactive legacy. Jordan was a little bent, but his step was quick. He had a full head of glossy white hair and a ready laugh. Jordan made old age look easy.[5]

As a college-trained, salaried employee, Jordan was in the minority. Most T plant workers clocked in for shift work in blue-collar jobs. DuPont sought people

T Plant, the ship-sized chemical processing plant at Hanford. Courtesy of Department of Energy.

who could be trusted to operate machinery and follow instructions precisely. In labs with special hazards, they needed people with more than "the usual attention to work."[6] As the recruitment drive started, there emerged a gendered division in hiring. DuPont recruiters hired men to staff the plant's three reactors, considered the most important and dangerous workplaces. Initially DuPont officials did not imagine hiring female plant operators at all because they feared genetic damage to women of childbearing age. Manhattan Project officials insisted, however, that because of the presumed labor shortage, "women should be substituted wherever practicable."[7] Work in the chemical processing plant, where workers would distill irradiated uranium down to drops of plutonium, was considered to be safer and less complicated than work in reactors.[8] That guess proved wrong. The chemical processing plants turned out to be as hazardous for workers as the reactors were.[9]

DuPont records offer no further explanation as to why chemical processing jobs were gendered female. Cost might have been a factor. It was cheaper to hire women because women were paid less and did not qualify for subsidized housing in Richland.[10] Jordan, who supervised many female lab assistants, said DuPont hired women because they were good workers. They did just as they were told and followed directions precisely. The best lab technician he knew was a woman who had been a short-order cook. She was good at following the same recipe, exactly the same way, over and over.

DuPont recruiters were looking for high-school-educated white women between the ages of twenty-one and forty, of "good health, pleasing personality, alert and intelligent."[11] In 1944, female applicants asked recruiters a lot of anxious questions—especially about the hazards of working in the mysterious plant. Locals guessed that DuPont was making chemical weapons. Rumors went around that people were being killed inside the plant and their bodies were being brought out under the ruse of removing Indian graves.[12] DuPont executives felt they had a "moral" obligation to disclose to their workers the hazardous nature of the plant's product. They maintained that even low-skilled workers could guess anyway, and full disclosure made for safer, more intelligent operations.[13] But Groves strongly objected to informing workers of the hazards.[14]

After accepting employment, women had to pass a health exam and a background security check. Unlike male operators, women were not sent for training in Chicago or Oak Ridge, but underwent a rushed six-week apprenticeship, consisting of only the essential skills and procedures, with no science or theory.[15]

Marge Nordman DeGooyer was one of the new DuPont recruits. DeGooyer grew up on a struggling farm in South Dakota, the kind of place where the farm can't provide an adequate living, so family members find work wherever they can. DeGooyer learned how to fly planes and worked as a crop duster, then a cab driver. In 1944, she followed her father, who was pursuing rumors of jobs, to

Woman on the job, Hanford, 1953. Courtesy of Department of Energy.

Richland. DuPont hired DeGooyer as a secretary, but a recruiter noticed her aptitude for math and told her that if she worked in the technical area she could get an education the likes of which no university in the world could provide. DeGooyer said she took that challenge.[16]

After a long bus ride to the plant, thirty miles past the entrance gates, DeGooyer arrived at the chemical processing plant, a massive "canyon" with no windows in its seamless concrete exterior. On her first day, the shift manager asked DeGooyer if she preferred to cook or sew. DeGooyer, confused by the question, replied that she didn't like to do either, but if pressed, she would cook.[17] So she was sent to the analytical chemistry lab to work with liquid chemicals, greenish "hot" solutions that the female lab assistants pipetted into beakers in exact, minuscule quantities.

DeGooyer was told how to do things, but not why. Her supervisor explained that the chemicals she worked with were dangerous, but he did not mention radioactivity. He also did not want the women to wear gloves because they hindered working quickly and precisely.[18] DeGooyer, however, was clued in to the dangers of her work by the behavior of her supervisors. She described how the chemists, "with their college degrees," would come to the door to give them new formulas. "They wouldn't come into our lab," DeGooyer remembered. "They'd stand on the threshold and hand the paper through the door, and then they'd run off."[19]

You can't blame the college-trained chemists for taking care based on their knowledge of the hazards, which security regulations prevented them from sharing with lab technicians. Chemists such as Joe Jordan who did analyses of the radioactive solutions knew a lot more than DeGooyer about the dangers involved. They also knew that because of the many problems DuPont had hiring workers to build the plant, the production of plutonium had fallen behind schedule. In order to catch up and have a bomb before war's end, in the summer of 1945 Groves ordered DuPont managers to shorten the cooling time for irradiated fuel slugs, thereby speeding up production. That meant workers pulled the highly radioactive slugs from underground cooling ponds after only a few weeks, rather than the two to three months necessary for the radioactive components to decay to safer levels. This "green" fuel sent up radioactive isotopes in great, toxic belches, the likes of which the planet had never experienced.[20] The decision to speed production to make up for lost time meant that the young lab techs were exposed to higher concentrations of radioactivity.

DeGooyer and the other lab techs measured and poured these highly radioactive solutions using bare hands. Spills were not uncommon. Each night as DeGooyer left work she placed her hands and feet in a counter. If her hands were not clean, she went back to the lab and rinsed them off, again and again. Radioactive solutions have a persistent quality that stands up to soap and scrubbing. DeGooyer got the nickname "Hotfoot Marge" because once the radiation monitors noticed that her clothes locker set the dosimeter ticking furiously. When they found that DeGooyer's work shoes were highly radioactive, they confiscated them and buried them in a radioactive-waste dump.

As I talked to DeGooyer, it was clear she was in pain. Her hand kept worrying a spot on the right side of her neck. She had a Band-Aid on her nose. "I've had cancer everywhere," she said as her hand flew around her body, "on my legs, hands, face, and then I had a mastectomy." DeGooyer's husband had also worked at the plant as a blue-collar operator on the F reactor—the reactor that over the years experienced the most leaks and other "incidents." While still young, DeGooyer's husband developed a problem with his heart valves. He had surgery and a long, incomplete recovery. Then he fell from a ladder and broke a leg, which mysteriously never healed. He retired early, and DeGooyer became the family breadwinner. She had a head for numbers and acquired a reputation at the lab for solving problems. Scientists sought her out to ask advice. Supervisors wanted her in their labs. DeGooyer worked her way up and eventually came to run the mass spectrometer at the plant. She was proud of that accomplishment.[21]

Before we parted, DeGooyer told me one last story. After news broke in 1945 about Hiroshima, a team of photographers came to tour the plant. They wanted to have a look at plutonium. DeGooyer's boss asked her if she would serve as a model. DeGooyer was flattered. She went to the bathroom, where she took off her coveralls and freshened her makeup. The photographers set her up at a glove box, into which she slipped her hands to hold a vial of plutonium solution. Then, to her horror, her boss told the journalists to leave the room, just to be on the safe side. He said they were not sure if the cameras' flash would make the solution go critical, sending out a lethal blue shower of neutrons. The photographers set their timers and hurried out, leaving DeGooyer alone to wait for the flash, holding the test tube, heart pounding. Years later, DeGooyer was most upset that her brave act was not recorded in the newspaper article. The photographers cropped her body from the photograph, which showed only her gloved hand holding the plutonium—a fitting parable of how many histories of the Manhattan Project have trimmed from memory the stories of the working people who took the most immediate risks.

7

Hazards

Defenders of the Manhattan Project medical record argue that in the 1940s researchers knew little about radiation's effects on the human body. Managers, they argue, placed workers such as Marge DeGooyer in harm's way unwittingly, and they were as careful as possible, given the wartime emergency.[1] With these arguments in mind, I set out to discover what Manhattan Project medical researchers knew about radiation and when they learned it. The answers show how managers and researchers discovered within the first years of research most of the critical dangers of the fission products they were creating. This realization, however, scarcely altered plant design, plant operation, or, most critically, the dumping of radioactive waste.

In the Atlanta branch of the National Archives, I came across a puzzling medical file for Don Johnson, a young DuPont chemical engineer. The file illustrates the vanishing qualities of the record of radioactive contamination, qualities that have since caused so many polarized views on the safety of the nuclear industry. In the fall of 1944, Johnson began to feel ill. He had nausea and severe gastric pain. His gums bled. His legs ached. He was fatigued and had night sweats, a mild fever, and, his Richland doctor reported, a pallor. The following week, doctors at the Richland medical center diagnosed acute leukemia. Within a few months, Johnson, age thirty-seven, who had been given a clean bill of health a year before, when he started work on the Manhattan Project, was dead.

DuPont officials acknowledged that Johnson had been exposed to radioactive sources at the Met Lab in Chicago and in Oak Ridge before coming to Richland, but at levels, they noted, below the then established tolerance. Researchers had set a "tolerance dose" of 0.1 roentgen a day in the thirties. They knew at the time that ionizing radiation from both gamma rays (electromagnetic waves of very short wavelengths), exposure to which comes from external sources, and beta and alpha particles (released from an atom's nucleus), exposure to which could come from ingested or inhaled substances, could damage cells, causing cancers and genetic problems.[2] Johnson's case caused a lot of anxiety in DuPont circles. His wife learned through a third party about Johnson's exposure to mysterious toxins and sued for compensation. DuPont lawyers were not about to admit liability, but they did recommend to General Groves that the federal government quietly pay her a settlement.[3] DuPont executives Roger Williams and Crawford Greenewalt, in charge of building the massive plutonium plant, had already been nervous about worker safety. Johnson's death elevated their anxieties.

DuPont managers were no strangers to workplace hazards or sick workers. In the early thirties, a DuPont chemical dye plant had an outbreak of bladder cancer among its workers. DuPont officials hired Wilhelm Hueper, a German scientist specializing in toxins, to figure out what was giving the workers cancer. Hueper isolated a new chemical agent, beta-naphthylamine, used in dye production, which, he said, caused bladder cancer in rats. Rather than pull the chemical from the line, DuPont officials took Hueper off the research project, and when he refused to drop the issue, they fired him. Fearful that Hueper would broadcast his findings, they assigned another scientist, Robert Kehoe, at the company's Kettering Lab, to carry out research that would discredit Hueper's findings. For the next twenty years, DuPont workers continued to use beta-naphthylamine, which caused bladder cancer in nine out of ten employees exposed to it.[4] For the subsequent two decades, DuPont officials harassed and censored Hueper in his work as director of the environmental cancer program of the National Cancer Institute.[5] Because of this experience, DuPont officials were more keenly attuned than Manhattan Project directors to the long-term consequences of workplace toxins and the threat of liability.[6]

In 1943, Williams and Greenewalt asked Army Corps of Engineers officers a lot of questions about the possible hazards of the reactors and processing plants they were designing.[7] The queries reveal their anxiety about sending forth into the earth's biosphere the world's first industrial-sized quantities of man-made radioactive isotopes. The executives asked: "What advantage would there be in hiring women beyond the age of menopause or older men? Would 0.1 rad [the daily tolerance dose at the time] be safe from causing genetic changes in offspring of workers? What is the natural mutation rate in humans—number of monsters, percent of spontaneously defective children; percent miscarriages?"[8]

In 1942, Groves had set up a Medical Section within the Manhattan Project with an eye to health and visibility. Groves and his chief medical officer, Stafford Warren, worried that workers would get so much contamination as to "produce physiological damage," which might undermine secrecy and production.[9] Ensuring production was both the main purpose of the new Medical Section and its essential shortcoming. As Hymer Friedell, a chief medical officer, put it, "the services of the medical organization are an accessory function. The primary interest is to maintain the health of the operators at a level which will in no way interfere with operations."[10] In other words, the medical division was there to keep workers healthy enough to produce, but not to solve the mammoth questions concerning the impact on human health of radioactive isotopes. In the steadily bloating Manhattan Engineering District bureaucracy, the medical research division was a needy stepsister, employing, at its peak, all of seventy-two medical officers to conduct research and to monitor and care for tens of thousands of employees, as well as to look after the environmental health of the surrounding air, rivers, lakes, and agricultural livestock and produce.[11] With few resources to spare, Stafford Warren instructed scientists to engage only in studies that would produce quick results and protect the agency from liability.[12] But Warren rarely had quick answers. His replies to DuPont executives' anxious queries about safe doses and genetic consequences were usually the same: *Researchers are studying these questions. We'll get back to you.*[13]

DuPont's top brass were not content with ignorance. By the 1940s, scientists had known for decades that radioactivity caused infertility, tumors, cataracts, cancer, genetic mutations, and general symptoms of premature aging and early death. Researchers in the 1910s and 1920s showed that X-rays produced cancers in animals.[14] In the twenties, American newspapers headlined the story of several hundred young women in New Jersey employed to coat watch faces with luminous paint that contained radium. The women had strange symptoms, as if they had sped into old age in a half dozen years. Their hair thinned and grayed, they became stooped and had to rely on canes, and their bones cracked with sudden movements. Their gums swelled and bled, and they lost teeth. They took to their beds too fatigued to walk in the park, go out on dates, or do the things other young women did.[15]

DuPont executives worried about the radium example, more so after September 1943, when Dr. Robley Evans published photographs of a radium worker with the lower half of her face consumed by a softball-sized tumor.[16] Evans reported that some of the autopsied radium workers had as little as 1.5 micrograms (0.0000015 gram) of radium in their bodies, minuscule amounts when compared to the tons of radioactive waste the plutonium plant would soon produce. A month later, DuPont executives sent the radium handbook to Groves' office and asked, again, for answers about the effects of uranium and its radioactive by-products.[17]

Natural uranium radiates only weakly, and a body has to be near it for long periods to incur damage. But when uranium is bombarded in a reactor, the result is a tremendous discharge of energy plus neutrons and new radioactive elements. This energy can affect the structure of any atoms it encounters. After the war, Atomic Energy Commission (AEC) scientists emphasized the "natural" radiation in the environment—from sources such as the sun's rays and minerals in the earth.[18] But there was nothing natural about the new radioactive isotopes produced in the Manhattan Project's reactors and cyclotrons, radioactive isotopes such as iodine-131, strontium-89, cesium-137, and plutonium-239. The new plutonium plant promised to generate these and many other man-made, hazardous isotopes in excess. In 1943, scientists could only worry what would happen when these new fission products entered living tissue, jolting with great energy the molecules, cells, and genes that support life.[19]

Not content with Corps promises, DuPont's Crawford Greenewalt started his own research programs, unique in the Manhattan Project, into the particular environmental panorama of the Columbia Basin. Greenewalt asked for a fish specialist to look at hydrology and habitats in the Columbia River before designing effluent pipes that would send radioactive waste into the river.[20] He called in a meteorologist to study the powerful winds that swept past the plant smokestacks.[21] DuPont executives requested their own medical health staff and asked for more and better doctors and researchers.[22] Corps officers found these safety precautions "excessively expensive and elaborate," but they paid for them.[23] At the same time, Stafford Warren contracted with researchers at several universities to carry out studies of the short-term effects of various radioactive isotopes on animals and humans.

At the Crocker Lab at the University of California, Dr. Joseph Hamilton was offered the job of researching how the fission products produced at site W (Hanford) would be metabolized in animal and human bodies as well as in plants, and what would happen when they entered soils. Hamilton eagerly accepted the assignment, apparently thrilled to be on the cutting edge of research on the biological effects of radiation.[24] Hamilton had long been one of many enthusiastic boosters of radioactive isotopes as a new diagnostic tool and cure-all for human ailments. In the 1930s, he stood in front of audiences and swallowed radioactive iodine to demonstrate how a few minutes later his thyroid set the Geiger counter ticking furiously.[25] In 1936, he and his colleague Robert Stone tested radioactive sodium on willing leukemia patients. In 1939, Stone treated wealthy cancer patients, who arrived drinking champagne in limousines, with neutron baths in the cyclotron; nearly half of these patients died within six months, suffering horribly from the side effects of radiation. In 1941, Hamilton injected six volunteer bone cancer patients with radioactive strontium, also with disappointing results.[26] With reputations as the leading researchers and promoters of radiobiology, in 1942 Stone and Hamilton were invited to work on the high-priority Manhattan Project in the medical division.

Hamilton set to work on the metabolism of radioactive isotopes, but his research agenda mutated strangely when an army general called to ask if it would be possible to poison an enemy population with radioactive by-products. Although Hamilton's lab was very short of money, staff, and especially time, Hamilton took a puzzling and costly detour in 1943 into this question of the "tactical" uses of radioactivity. On the general's suggestion, he investigated how Hanford radioactive waste could be used for "offensive purposes." Hamilton injected radioactive solutions into mice and turned solutions into smoke and food pellets for mice to inhale and ingest, trying to figure out the surest and swiftest ways to induce the mice to die.[27]

Manhattan Project security compartmentalized knowledge on a "need-to-know" basis, and Hamilton's reports went up his chain of command in the Medical Section, bypassing DuPont.[28] This security wall created a bizarre parallel correspondence within the Manhattan Project. In the summer of 1943, for instance, DuPont executives exchanged anxious letters with Groves about the health effects of workers' daily exposure to Hanford's radioactive isotopes, while Hamilton corresponded with Stone in Chicago about the best ways to use Hanford waste to "make everybody [in an enemy population] nauseated, vomiting and incapacitated within 24 hours."[29] While DuPont executives were worrying about air currents swirling in the topographic bowl around the Hanford plant, which created an inversion, trapping radioactive dust over local towns, Hamilton worked with a meteorologist to determine how they could best use the same inversion effect to confine radioactive dust in air currents over an enemy city. While DuPont officials grew anxious about the highly radioactive nature of Hanford waste, Hamilton was estimating the number of curies in a hundred pounds of the same waste, which, he imagined, could be spread on the ground, allowed to seep into well water, or turned into a gas for "offensive purposes."[30] The waste was so potent, Hamilton's assistants gushed, that "[radioactive] strontium smoke would be over a million times more lethal than the most deadly war gases."[31]

Hamilton, like many of his compatriots, was caught up in winning the war, but his research program inadvertently upended DuPont's concerns for public health. Instead of looking at ways to increase safety, Hamilton studied how to manufacture greater radioactive hazards. Instead of determining how to preserve life, Hamilton researched how best to bring about death. Hamilton's subordinates suggested building a plant specifically to process radioactive waste for weapons—a proposal that, considering the Mt. Vesuvius of radioactive effluent Hanford would soon produce, now reads as cruelly sardonic.[32]

At the time, however, Hamilton's results were encouraging from a military standpoint. A conventional bomb, Hamilton pointed out, is dropped, does its damage, and ceases to be destructive. Radioactive bombs, on the other hand, ensure destruction long after they are detonated. Hamilton reported, "A person who has become internally infected [with radiation] will be subjected to internal irradiation for many months after exposure," and "that a very large proportion of the long life fission products are retained for protracted periods of time in the lungs." He found that many of the radioactive by-products emitted at Hanford—strontium, barium, and radioactive iodine—were readily absorbed by the digestive tract and moved into the bone marrow.[33] In other words, radiation, once ingested by the enemy, was like a ticking time bomb buried deep inside the body. Hamilton reported optimistically that it was quite easy with relatively small amounts of radioactive substances inserted into the proper environmental conditions to incapacitate or even kill whole communities.[34]

Radioactive dust or smoke trapped by temperature inversions or dispersed by swirling currents, fission products unleashed in rivers and groundwater, radioactive particles dusted over crops—these were the scenarios that haunted the nightmares of safety-conscious DuPont executives who by 1944, as start-up approached, began to worry more audibly about the "super-poisonous" nature of the product they would soon produce.[35] At DuPont's pilot reactor at Oak Ridge, scientists were amazed at how "a minute quantity of hot material" could cause "widespread contamination."[36] In late 1943 and early 1944, DuPont executives joined others at Manhattan Project sites asking ever more urgent questions about safety and health.[37]

Mercifully, DuPont engineers had no access to Hamilton's hair-raising monthly reports on the offensive uses of radioactive waste. But they also had no real answers from the medical division about how to safely launch the world's first plutonium plant and dispose of its millions of gallons of radioactive gas and liquid. Despite the general sense of urgency, two years into the medical research program neither Hamilton nor his colleagues in labs in Rochester, Oak Ridge, and Chicago had useful answers.[38] Little wonder answers were missing: Manhattan Project researchers could not publish their work, discuss it at conferences, or even solicit the help of fellow scientists working on different areas in the Manhattan Project.[39] Meanwhile, Hamilton's program, most directly concerned with the problem of Hanford waste, had squandered a year studying the military applications of radioactive by-products.

In December 1943, Stone gently steered Hamilton back toward the Hippocratic Oath: "We have no authorization for investigating offensive radioactive warfare, but we have a responsibility to know as much as possible of the action of the dusts that might be around a plant resulting either from normal operations or accident."[40] Hamilton, with his characteristic perceptiveness about how best to achieve professional success, quickly recalibrated. Just three weeks later, he sent Stone a new proposal to study radioactive smoke and dust, much as his research group had before, but now in the context of an "accident or normal operations" at a project plant.[41]

Hamilton's yearlong research calculating how to induce a slow radioactive death reveals that researchers had a good idea about the killing qualities of the products and by-products Manhattan Project plants would produce, even before they produced them in industrial quantities. Hamilton's correspondence also shows that there was no real ideological division between military medical officers (such as Stafford Warren and his loyal deputy Hymer Friedell) and civilian research doctors (such as Stone and Hamilton). All were eager to serve the war cause in the most direct way.

Hamilton's foray into radioactive weaponry reveals, too, something about the nature of the Manhattan Project's medical program in the midst of a genocidal war: its cool appraisal of death and destruction, its surfeit of imagination about masses of enemies "nauseated, vomiting, incapacitated within 24 hours," and its deficit of imagination to envision the same scenario among Americans near Manhattan Project plants. Perhaps this initial martial gleam helps explain what followed in the history of medicine on the Manhattan Project.

8

The Food Chain

In 1943, Manhattan Project medical radiologists predicted that plutonium would not be a very dangerous material because plutonium differed from radium in that it emitted few gamma rays, the kind of radioactive energy that travels great distances and can penetrate through walls, clothing, and skin into the body. Instead, plutonium was an alpha emitter. Alpha particles do not travel more than the width of a hair and can be stopped by a sheet of paper. As a consequence, researchers estimated that plutonium would be fifty times less dangerous than radium.[1]

In February 1944, Hamilton received one of the first allotments of liquid plutonium, eleven milligrams, enough to begin lab experiments on the effect of this new isotope on the body. Hamilton's group started experimenting on mice, then moved on to rats, rabbits, dogs, and monkeys. The researchers smeared plutonium on skin and injected plutonium-laced solutions into blood and muscle tissue. As the first results were tallied, the picture of plutonium grew increasingly dismal. Hamilton discovered that once inside a body, plutonium lodged in the skeleton and bored into the vulnerable blood-cell-generating bone marrow. Hamilton had hoped to find ways to flush plutonium from a body, but he had no luck.[2] Plutonium, the researchers found, had an uncanny knack for bioaccumulation, concentrating in organs and insinuating itself into the biochemical processes the body uses to thrive. Thyroids, for example, greedily drank up radioactive iodine. Plutonium and strontium-89 imitated calcium and quickly migrated to the skeleton. Strontium-89 also traveled with speed and ease from placenta to fetus, from mother's milk to newborn.[3] John Wirth, the Oak Ridge medical director, was fascinated with how radioactive isotopes inserted themselves into biological processes. He marveled at the "ease with which it [radioactivity] seems to get about as though it were a living creature, trying to spread itself anywhere."[4]

Hamilton's exposed lab animals grew listless, their hair grayed, and their livers deteriorated. They developed lymphomas, bone sarcomas, and precancerous cells.[5] At Columbia University, researchers exposed mice to fast neutrons.[6] The mice lost weight, hair, and white blood cells. They became anemic, grew sterile, and developed cataracts. Their lungs became inflamed and clouded with bacteria. Strangely, the mice suffered these symptoms in different ways, no two alike. After thirty-four weeks most of the mice had died. On autopsy, the doctors could not determine a specific cause of death—not a tumor, a cancer, or organ failure—and attributed death to "a general malfunctioning."[7] Researchers found the random, vague qualities of these symptoms troubling. They had hoped to determine telltale signs that a body was approaching an overdose of radiation, but they discovered that different kinds of radioactive isotopes behaved in particular bodies in their own exceptional ways, and produced symptoms that were difficult to differentiate from symptoms in a body suffering from a more conventional illness such as pneumonia, anemia, or tuberculosis. It would be relatively easy, in other words, to mistake a death from radiation for a conventional one, or radiation illness for a general malaise and vague complaints of infirmity.

The researchers' experimental doses were high, such as an employee might experience during an accident or explosion. On a daily basis, most employees and bystanders would be exposed to far lower doses, but this exposure might continue for months, possibly years. Long-term, low-dose experiments took time and required the ability to measure minute levels of radioactive isotopes in the body, a skill Manhattan Project researchers had not yet mastered in 1944–45.[8] There were only a few studies that looked at the long-term effects of the new radioactive isotopes. Researchers at the University of Rochester conducted a two-year study of the effects of chronic radiation on mice, monkeys, rats, and dogs. The animals were given X-ray doses equivalent to the accepted tolerance dose for workers in the Manhattan Project. Much of the experiment failed, however, because epidemics of typhoid and tuberculosis overtook the mice and monkeys, killing them and skewing the results.[9] The researchers were looking for tumors, cancers, or disintegrating bones—symptoms suffered by workers exposed to radium and X-rays in the twenties and thirties. They were not looking for immune disorders, which might trigger a susceptibility to common illnesses. If they had, then the epidemics among the mice and monkeys might have been taken as results rather than as a sign of failed experiments.[10]

Meanwhile, a team of geneticists irradiated 73,901 fruit flies (genus *Drosophila*) starting at 25 rads (the annual tolerance dose for workers at the time) and finishing at 4,000 rads. Since the twenties, geneticists had been aware of radiation's effect on genetic mutations. In 1925, the geneticist H. J. Muller won a Nobel Prize for studies showing that X-rays caused damage to fruit fly chromosomes. Subsequent investigations determined that in all species radiation triggered mutations.[11] In Manhattan Project studies, researchers found that even the lowest doses directly affected the rate of mutations in offspring. The researchers moved on to mice and found that the higher the dose a mouse received, the better the chance for mutation.[12] The geneticists concluded their study by questioning the daily tolerance limit for Manhattan Project workers: "We are forced to wonder whether a human exposure of 0.1 rad/day is acceptable." The researchers doubted that any radiation dose was safe because of the random quality of chromosomal damage, which triggered in offspring changes ranging from superficial differences in eye color to a worrisome and vague "reduction of general vigor or of life span."[13]

Most Manhattan Project researchers were focused on the immediate goals of winning the war and minimizing the loss of American lives. They lined up potential casualties in nuclear weapons plants alongside the greater risks for battlefield soldiers and judged nuclear risks to be comparatively negligible. But the small group of geneticists working on the margins of the Medical Section took a different perspective. They reflected on the impact of long-term, large-scale deployment of atomic energy "in terms of society and the human race."[14] Their report, tossed in a large file labeled "Medical Summaries," displays an uneasiness about the way radioactive isotopes, so quick to lodge in the body and linger there, affecting biological systems, would, once distributed on an industrial scale, no longer remain an external feature of human existence, but would become a lasting detour (or cul de sac) on the path of human evolution.[15]

The consequences of these grim medical findings could be averted, of course, if humans minimized contact with radioactive isotopes. With that goal in mind, Manhattan Project researchers sought out the paths by which plutonium and other fission products might enter the body. The scientists found that these radioactive particles migrated outdoors, to the grasslands, into the rivers, and into air currents. The idea in locating the Hanford plant in the wide-open, sparsely populated Columbia Basin was to use the local territory as a vast sink into which engineers could dispose of hundreds of thousands and eventually billions of gallons of radioactive and toxic waste. With a vast reach of territory, the scientists figured, radioactive isotopes would scatter into the air, soil, and water to the point where they would be so diluted as to be harmless everywhere to everybody. The strong winds would carry away radioactive gases from high smokestacks. The swift, high-volume Columbia River would speed off liquid waste to the Pacific Ocean. The earth in the miles-wide buffer zone around the plant and the sandy sediment under the plant would easily absorb radioactive waste and make it vanish. The sink was an application of nineteenth-century notions of industrial waste disposal to twentieth-century garbage—one of those ideas that sounded good at the time, because radioactive garbage is undetectable by the senses. Passing one's eye across the rambling Columbia Basin made visible sense of the notion of the sink.

DuPont engineers did not approach the despoilment of the Columbia Basin cavalierly. Greenewalt realized quickly that Hanford was at the very nadir of the basin. Consulting meteorologists, Greenewalt learned that local air currents unfortunately would not evenly disperse Hanford's effluent. Warm air flowing over the top of the basin often formed a ceiling trapping cold air below, which then circulated and flowed liquidlike near the ground, heading south over the Columbia River toward Pasco and Richland, where a bottleneck of "high concentrations of radioactivity" could occur.[16] Greenewalt learned that when conditions were favorable, stable currents indeed held stack gases high, emitting them over many miles, but these emissions most often traveled southeast to the region's major population points—Richland, Pasco, and Kennewick—and on to Walla Walla, sixty miles away. At other times, downdrafts deposited emissions, hardly diluted, within a few feet of the stacks.[17] Frank Matthias learned of these disquieting scenarios in 1944, but by then he could do nothing about the plant's design or location. Instead, Matthias noted in his diary the desperate, optimistic belief that once the plant was up and running, engineers would hold up production while awaiting favorable weather conditions. Matthias and Greenewalt had a high tower built to forecast good weather for production.[18] Meanwhile, the plant ran around the clock in good and bad weather.

Hamilton had on staff two soil experts, R. Overstreet and L. Jacobson.[19] They tested the soils under the Hanford reservation and found that the soils in the Hanford area showed an amazingly high capacity for holding on to fission products. Overstreet and Jacobson packed soil into vertical glass columns and poured in radioactive waste from Hanford. They noted that 80–90 percent of the waste did not percolate down, but settled in the first few inches of topsoil.[20] These results were disconcerting because, like Greenewalt's meteorology studies, they directly contradicted the notion of the sink. If radioactive isotopes combined readily with Hanford soils, if most of the radioactivity settled in the topsoil, and if wind currents cycled inside the Columbia Basin toward population points, then the result would be not diffusion but concentration of radioactive isotopes in just the places where humans, flora, and fauna were most likely to come in contact with them.

Reflecting on this problem, Hamilton, writing from the floral splendor of Berkeley, California, penned to a colleague in dusty, dry Pasco: "There is one question which I think is very important that was probably not emphasized too strongly in the report, and that is the unhappy state of affairs that will take place should fission products in any large amounts ever come in contact with the top soil. Under such circumstances, unless the contaminated dirt is properly buried or otherwise disposed of, such material could be transported considerable distances by action of the wind."[21]

Placing fission products in contact with the topsoil is just what DuPont engineers were up to when Hamilton wrote his letter. DuPont engineers designed a waste disposal system in which they piped the most dangerous waste into underground storage tanks, while they mixed low-level waste with well water and poured it into depressions in the ground, creating open swamps and ditches of radioactive mud, liable to evaporate in the dry air and send particles airborne in one of Hanford's frequent dust storms.[22] The DuPont medical team took readings of the swamps and found the radioactivity to be high (6.5 millirems an hour). Stone, in Richland for a visit in February 1945, tried to put a stop to the practice, but it was too late. Workers continued to dump low-level waste into open trenches for decades.[23]

DuPont engineers also dug "reverse wells," deep holes into porous underground strata, to dump medium-level waste. Overstreet and Jacobson were concerned about this plan, too, and arranged to consult with DuPont engineers on the wells. DuPont engineers were happy to have the help and eagerly provided information and more soil samples. Overstreet and Jacobson saw problems in pumping waste into the ground near underground aquifers and foresaw that the soil would draw in and hold radioactive isotopes for as long as it took them to decay.[24] The two scientists experimented by growing peas and barley in contaminated topsoil. They found that plants eagerly drank up radioactive isotopes. Overstreet and Jacobson found to their surprise that there were higher concentrations of fission products in the plant roots than in the surrounding soils and that even relatively small concentrations damaged plants. "Contamination of the soil," the scientists warned, "may result even at very low levels in dangerous amounts of radioactivity in edible crops."[25]

All of this news ominously contracted the diffusion theory upon which Hanford waste management was premised, but plant managers made no changes in design or practice. While Jacobson and Overstreet in Berkeley studied the problem of reverse wells, Stone visited Hanford and learned that they had already been installed. He wrote Hamilton: "They have no present intention of changing this [reverse well design] in any way unless tests of waters from various wells indicate that contamination is occurring."[26] Just a few months later, the first incident of radioactive contamination of drinking water occurred, as Overstreet and Jacobson had predicted. Even then, however, engineers made no changes to the reverse wells. The head of Hanford's Health Instrument Division, Herbert Parker, pledged only to monitor the wells more closely. In subsequent years, plant operators continued to dump radioactive waste into deep holes, and the soil studies were forgotten. A decade later, Parker characterized the Hanford site as one "admirably suited to the disposal to ground of large volumes of liquid wastes," as if Jacobson and Overstreet had never walked the earth.[27]

Symbols sometimes play larger in human imagination than complex realities do. When people from afar thought of the Columbia River Basin, they thought not about factories burping high-tech contaminants. Rather, they thought of salmon—the majestic, determined fish that made their way against the crashing waterfalls of the mighty Columbia to their spawning places deep in the abdomen of the arid interior West.[28] If something were to happen to the salmon, then the game would be up for the Hanford Engineering Works, for DuPont, and for the Army Corps of Engineers.[29] Plant designs called for large pumps to channel river water through the reactors to cool them. This volume of water was colossal: thirty thousand gallons a minute flowed through a reactor core. After the water became effluent, warm in temperature and hot in radioactivity, it was allowed to cool for a few hours and then pumped back into the Columbia River. Aware that the Hanford plant would be the only upstream polluter on the Columbia, Greenewalt requested in 1943 that an ichthyologist come to Hanford to study the effect of radioactive effluent on salmon that spawned near the plant.[30] A few months later, Lauren Donaldson started a program in his lab at the University of Washington where he radiated salmon, as eggs, as spawn, and as mature fish.

In a random file in the National Archives I came across a series of small photos, glossy three-by-five snapshots, of Columbia River salmon exposed to X-rays. At 100 rads, the fish in the first-month alevin stage, in which salmon fingerlings live off their yolk sacs, appear normal.[31] At 250 rads, there is something funny about the fingerlings. Scientists reported "evidence of disorganization." The photos show yolk sacs bulging, the fish thinning.[32] At 1,000 rads the bodies of the fish have shrunken radically, given over to a tumor-like growth in the abdomen. At 10,000 rads the fingerling's eye is blotted out, blanched from cataracts. The twiglike body holds up a swollen yolk sac. Inside bobs a shiny black growth. The fingerling swims mouth agape, gasping. At all levels of exposure above 500 rads, the fish soon died.[33]

But 500 rads is a high exposure, far higher than salmon would get swimming directly downstream from the plant's effluent pipes.[34] The first results, though sad to look at, were good news for fish researchers, showing that it took a high dose of gamma rays to harm the valuable salmon.[35] As Stafford Warren admired the complicated halters Donaldson had devised in his Seattle lab to pinpoint gamma rays on fish, researchers in Chicago tried something experimentally less elegant but more to the point: they dumped goldfish in various diluted solutions of Hanford waste and watched the fish as they sucked the effluent through their gills and fed on microscopic algae and plankton in the water. The Chicago researchers found that fish concentrated radioactive elements in their bodies at levels a shocking ten to forty times higher than the amounts in the water in which they swam. This was troubling news, since once inside the body, radioactive particles could do much more damage to vulnerable organs and cells than from outside the body.[36]

Donaldson replicated the experiment with Columbia River trout and salmon. On the high bank over the Columbia, each of the three reactors had large basins, where reactor effluent cooled before descending to the river. Donaldson set up

Fish laboratory at Hanford. Courtesy of Department of Energy.

fish troughs outside the basins and pumped in effluent mixed with clean river water in various dilutions.[37] Dumped directly into reactor effluent, the fish died. But in diluted water, the fish initially thrived, multiplying rapidly and outgrowing the tanks. Richard Foster, Donaldson's assistant, sacrificed some of the salmon. His autopsies showed that the fish behaved just like peas, barley, algae, bone marrow, and thyroid glands: they sucked in the radioactive isotopes hungrily, so eventually the concentration of radioactivity in the bodies of the fish exceeded by up to sixty times that of the water in which they swam.[38]

In the summer of 1945, Foster reported that the fish in the test troughs had external parasites and bacterial infections. Then Foster reported that on two days—July 27 and again on August 31—the fish died in mass "kills" in effluent diluted with three parts river water.[39] At the time, racing to produce plutonium before the war ended, the reactors were issuing as much as 900 curies a day into the river. Foster had no idea about the production speed-up and consequent spike in radioactivity in the river, as this was classified information. He puzzled over the substance in the effluent that was lethal to fish. Apparently he never figured it out. Perhaps if Foster had talked to the mice and monkey researchers, who by then had communicated with the doctors treating the TB outbreak at Oak Ridge, the scientists together might have discerned a pattern of immuno-logical weakness. As it was, however, only a handful of top Manhattan Project

officials were getting the whole arsenal of medical reports from various locations, and these men were inclined to judge the results in the most optimistic
light.

So Foster's fish studies were filed away, as were Overstreet and Jacobson's soil
reports, buried near the troubling studies on fruit fly genetics, meteorological
surveys, and the metabolism of plutonium in mice and dogs. All these reports
landed in the vast textual reverse wells of the Manhattan Project, into which information went and never came out.[40] If anyone had had the time and stamina in
that harried wartime era to read all the reports, he or she might have noticed that
the studies showed researchers across the medical research division coming independently to similar conclusions: that radioactive isotopes sought to attach
themselves to living organisms, making their way up the food chain. This was
bad news for those creatures at the top of the chain.

9

Of Flies, Mice, and Men

By early 1945, Manhattan Project leaders knew quite a bit after these first years of research. They knew the parameters of damage from exposure to various radioactive isotopes and the pathways those isotopes took into the body. They learned that the most troubling radioactive isotopes were those with long half-lives, and that once they were in the soil and in living organisms, they were difficult to detect and dislodge. Once those isotopes had made their way into bodies, the researchers learned, their radioactivity destroyed cells; caused cancers; resulted in problems in the immune system, the digestive system, and the circulatory system; and accelerated aging and death—all in random, unpredictable ways. They grasped that this research must be closely held, for once contractors and employees, already worried about "unknown amounts of product in the body," found out about the studies, they might panic.[1]

In the summer of 1944, DuPont's Roger Williams wrote to General Leslie Groves in just such a panic. Williams noted that in the previous months they had come to an astonishing realization—that the "most extreme health hazard is the product itself." "It is now estimated," Williams wrote, "that five micrograms (0.000005 grams) of the product [plutonium] entering the body through the mouth or nose or by skin absorption, will constitute a lethal dose. The poisonous effect of the product is cumulative, i.e., product entering the body is permanently absorbed and effective, like radium."[2] In the margin next to this passage, a Manhattan Project medical officer, probably Hymer Friedell, wrote, "Wrong."

That was the contested issue—how large a dose was "lethal." Williams was writing based on preliminary research results from Manhattan Project labs, and at least from rumors of research like Hamilton's on mice and dogs and Donaldson's in-house work on fish.[3] Williams concluded that if plutonium and other radioisotopes accumulate in sensitive areas of the body—bone marrow, thyroid, liver, kidney, lungs, and spleen—then even the smallest dose stood a chance of damaging cells and triggering the growth of cancerous tissue or genetic mutations. Manhattan Project medical officers, however, looked at the same results another way. Researchers blasted the bodies of fish, mice, and dogs

with increasing doses of radioactive isotopes and noted that the lab animals succumbed only after very high doses. At medium doses, the scientists detected
cellular changes and a certain weakness in the "material specimens." At low
doses, they could detect no changes other than the presence of radioactivity in
tissues, organs, and bones.[4] Extrapolating from these experiments, they reasoned that there was a "tolerance dose," below which it was reasonably safe for
humans and animals to dwell. This was the only conclusion Army Corps of Engineers officers could draw in order to continue their unswerving course toward
winning the war with nuclear weapons. If they were to conclude, as geneticists
insisted, that no dose was safe, then the whole nuclear enterprise was folly.[5]

Although the lab results were grim, the experience in the new nuclear zones
was less so. Workers were not falling ill, at least not in epidemic numbers. There
was no great spike, though they watched for it, of stillborn babies or babies with
deformities.[6] Animals and birds did not disappear from the sites.[7] Fish continued to swim up and down the Columbia River, even through exceptionally
warm water that was "milky in appearance."[8] It's true that there were fewer fish
from one inspection to the next, but the cause was "impossible to account for."[9]
There was an outbreak of tuberculosis at Oak Ridge, along with cases of workers
with skin ailments that refused to heal, which Warren learned were "definitely
related to work" at the Oak Ridge plant.[10] Several uranium miners mysteriously
died.[11] Two soldiers were rushed from Hanford to a Walla Walla hospital for
kidney pathologies in the summer of 1945.[12] And there was a rash of employees
whose lab reports were couriered about Medical Section offices. Medical
personnel discussed whether to reassign these workers to safer work or terminate them "to protect the interest of the Government and Contractor against
possible claim for compensation."[13] But these cases are just footnotes in the
highly cleansed, declassified files, relatively minor incidents in the larger scheme
of the vast project.

When, just to make sure, Manhattan Project doctors injected plutonium into
eighteen unwitting human subjects and polonium into five more unsuspecting
patients, first at Oak Ridge and Rochester, then in San Francisco under Hamilton's direction, the previously healthy, "normal" subjects did not die.[14] Their
white and red blood cell counts dropped dramatically, and tests showed that
their bodies accumulated plutonium with greater efficiency than the bodies of
rats and mice, but the human subjects lived on, and that, too, was promising
news.[15] The researchers measured radioactivity in urine and feces, but they did
not record how the subjects felt with 50 micrograms of plutonium-239 in their
bloodstream or 18.5 microcuries' worth of polonium slipped into their food and
descending through their digestive tract. Symptoms and treatments were not the
point. The researchers hoped to learn how to measure ingested doses by studying
the subjects' urine and feces. This was a "medico-legal" research question related

to guessing exposures and thus liability for workplace damages.[16] Family members later told of the intense pain, weakness, depression, vague complaints, and malaise that clouded the subjects' lives afterward, but the human subjects carried on ("very nearly a normal individual," as Hamilton boasted about his "experimental material," the house painter Albert Stevens), and that was a medical triumph as well.[17]

What did Manhattan Project leaders do in the workplace with the research results? They simply carried on, almost as if there had been no research. Robert Stone recommended to DuPont doctors in the fall of 1944 that they completely exclude premenopausal women from work in the plant.[18] Army Corps officers, however, stipulated that DuPont hire women because they feared hiring racial and ethnic minorities.[19] And so in 1944 DuPont recruiters placed young women in the most hazardous jobs in the chemical processing plant.

Once the reactors started up in the fall of 1944, DuPont executives worried about an explosion spreading radiation to the populous Hanford Camp.[20] They sought permission from General Groves to tell workers that they were exposed to radiation on the job and to hold practice evacuation drills.[21] Groves was more worried, however, about security and retaining workers. Groves asserted that if hourly workers learned of the potential dangers, they might quit. To make his argument Groves pulled out a trusted rhetorical device: shifting the scale from DuPont's local and individual concerns to the affairs of the nation ("the best interests of the United States"). The deployment of scale was a common tool in the Manhattan Project—the hazards at nuclear plants were described as no greater than those of the chemical industry and the risks "compatible with the overall urgency of the Manhattan District."[22] Most famously, Groves used scale after the war, arguing that the death of more than two hundred thousand Japanese civilians at Hiroshima and Nagasaki "saved [American] lives."

Manhattan Project workers were subject to regular "medical surveillance." Doctors had permission to inform employees of medical abnormalities only if the maladies were not related to radiation. To contain this knowledge within the trusted group of plant doctors, DuPont managers built up a full-service, low-cost medical clinic in Richland.[23] This kind of New Deal–style medical program directly contradicted DuPont's conservative philosophy, but in this case DuPont managers argued that a subsidized medical program for Hanford employees and their families would be advisable both to maintain control over the plant medical staff and to cover up the distinction between occupational and regular illnesses. The service plan would pay for both and thus "avoid embarrassing situations" and patients' "undue alarm." Writing with that knowing wink of Manhattan Project officialdom, a DuPont manager concluded, "The important value of this feature can be readily understood."[24]

Left in ignorance, however, workers worried "continually" about the reason for the urinanalyses, the blood tests, and the teams of safety monitors passing through the sterile-looking cement halls with ticking equipment.[25] Workers guessed that something was awry, all the more so because of the secrecy, mystery, compulsive cleaning regimen, fences, gates, alarms, and guards. So once Groves decreed that workers would be left in the dark, those workers had to be convinced that they were safe, and in this way public relations gradually overtook public health.

Instead of education about jobsite hazards, in 1944 Matthias started an annual extravaganza, called the Safety Exposition, that repackaged the dangerous plant as a beacon of safety. The Safety Exposition combined entertainment with exhibits promoting jobsite safety. To get workers to attend, the exposition featured concerts, dance troupes, door prizes, and a beauty contest electing a Safety

Hanford Safety Exposition, 1952. Courtesy of Department of Energy.

Queen. The chief purpose of the event, Groves emphasized, was to "build up morale" so that workers would stay on the job.[26]

In the fall of 1944, when DuPont engineers started up the first B reactor, six months behind schedule, Groves was in a great rush. By that time, it was clear that Germany would be defeated. Manhattan Project intelligence also reported that German physicists would not produce an atomic bomb.[27] Nonetheless, Groves raced to produce a bomb before the war ended so that he would not be left with a $2 billion tab and nothing to show for it. In his haste, Groves demanded that DuPont engineers begin transforming spent uranium fuel into plutonium before the processing plant was finished. DuPont engineers had designed the processing plant for safety, with robotic devices, underground chambers, and massive cement walls to shield workers from dangerously hot radioactive solutions. Beginning the processing early would mean working in "make-shift laboratories," DuPont's Roger Williams pointed out, exposing workers to radioactive hazards in "chance-taking" operations.[28]

In resisting Groves' urgent demands, DuPont had a problem. In 1943, Crawford Greenewalt had promised Groves that he would have the entire plant finished by the end of 1944. As I have shown, however, discriminatory hiring practices had slowed construction progress, and by the summer of 1943 Greenewalt knew he would miss this deadline, as he had failed to meet the earlier construction targets for the first reactor. DuPont managers repeatedly tried to delay the completion dates, with Groves resisting: "I am still unwilling to accept such a setback." For DuPont executives, the construction delays were nothing but "embarrassing," which gave them little room to negotiate when Groves asked for shortcuts that sacrificed safety.[29] Consequently, in the fall of 1944 DuPont executives agreed to manufacture plutonium before the processing factory was finished, employing young women in makeshift labs, with all the extra hazards that would entail.[30]

Groves was still unhappy. By February 1945, the processing plant, finally up and running, was only producing 250 grams of plutonium a day. Desperate for a bomb, he ordered DuPont managers to speed production by pulling irradiated uranium fuel rods out of the cooling ponds after only five weeks, rather than the three months required to allow short-lived radioactive isotopes to decay. That decision meant that the plant issued four times the usual amount of deadly radioactive isotopes, all of which spilled onto the ground, into the Columbia River, and into air masses that floated south and east over the Columbia River, across farmland, and then on to Walla Walla and Spokane. The most troublesome short-lived radioactive isotope was iodine-131, a problem because it selectively deposits in the thyroid.[31] Because of the shorter cooling times, plant releases of I-131 from the stacks soared in the first half of 1945, from a few hundred curies a month in January to 75,000 curies a month by June.[32]

Herbert Parker, in charge of radiation monitoring at Hanford, noted that iodine vapors built up on surfaces downwind from the stack, but Parker was not one to get easily alarmed. He added that the extraordinarily high releases were not a "critical hazard." "In the interests of morale," he concluded, "it may prove more desirable to restrict the evolution of fumes under certain atmospheric conditions."[33] Not surprisingly, Parker's mild recommendation went unheeded. Emissions of I-131 climbed the rest of the summer and soared even higher after the war ended—inexplicably higher, since all managers had to do was hold irradiated fuel cells in the cooling bins an extra month, presumably not a problem because Japan had already surrendered.

Once released, the plumes of radioactive iodine traveled great distances largely undiluted. In December, Parker's monitors recorded levels of radioactive iodine on shrubs and trees in Richland and neighboring Kennewick that were six times higher than the already liberal tolerance dose.[34] In Walla Walla, they found that ground contamination from radioactive iodine equaled that of the soil right next to the processing plant.[35]

Repeatedly there was an institutional impenetrability, as if research on the biological effects of radiation was conducted in isolation from plant operations. Why did Manhattan Project managers bother with research if they were likely to ignore the results? There are some clues as to what they were thinking. In 1960, Matthias wrote Groves about the genesis of the fish program. Matthias erroneously attributed Greenewalt's fish program to Groves, calling it "a brilliant tactical move." "I am convinced," Matthias continued, "that we would have had a very bad time with the fish people after August 1945 if we had not been able to demonstrate so conclusively that we had considered the salmon problem a serious one and had produced much evidence to show the effects were not serious."[36]

Likewise, in the summer of 1945, Herbert Parker was reluctant to begin regular urinanalyses of Hanford workers because at other Manhattan Project sites, such tests had "led to considerable alarm" among those who received positive results. Nonetheless, Hanford workers worried about plutonium in their bodies. It would be good for "plant morale," Parker reasoned, to begin a testing program, but if the results were positive, then he would have an even greater morale problem. Parker worked his way out of this snare by devising a plan to test workers after long weekends.[37] Testing after several days' leave gave workers time to urinate the most radioactive samples safely into their toilets at home—another brilliant tactical move.[38] With urinanalysis, workers felt safer, especially when the results were negative, and that was good for morale. Like the early environmental studies, medical research had a validating public relations function, useful when dealing with nervous workers toiling on the hazardous frontier of the atomic age.

At other times, Manhattan Project officials deployed medical research because of concerns with liability. Take the case of Donald Johnson, the DuPont engineer who came down with acute leukemia after just eighteen months of working with radioactive substances. During his employment, Johnson's urine and blood had been monitored. Medical Section doctors could confidently show that Johnson had received doses no higher than the tolerance dose at the time. Johnson's first autopsy showed significant radioactive contamination, but the second set of tests came back negative. Stafford Warren, chief of the medical section, was happy with the second report, and, in a rare recorded moment he revealed the urge not to see by ordering that the report of the first, troubling autopsy be purged from the files. Without that report, there was no evidence that Johnson's leukemia had anything to do with radiation.[39]

When Johnson's wife later attempted to sue for compensation, she did not know that her husband's case would never—could never—come to trial. In June 1943, DuPont and the Army Corps had made a secret deal with officials of the Washington State Department of Labor, who pledged to redact from workers' files information that would threaten the plant's secrecy. They also agreed that workers' lawsuits would not go to civil court but would be heard before a special tribunal consisting of representatives of the federal government and the contractor.[40] In the tribunal, the federal and corporate counsel, thanks to the medical research division of the Manhattan Project, would have submitted a wealth of carefully selected and edited reports by reputable doctors at prestigious universities, making for a bulletproof defense.[41]

By the spring of 1945, Matthias, Greenewalt, and other officers and corporate managers assigned to produce plutonium had accomplished a great deal. In the course of two years, they had built a series of factories and the world's first industrial reactors for plutonium production. They had demolished three towns and built in their place two new cities and a labor camp from the ground up, the larger city, Hanford Camp, already bulldozed as well by mid-1945. They had created Richland, a new kind of community of white nuclear families, subsidized by federal coffers, managed by corporate lawyers with a planned economy and carefully controlled access. They had also created a medical and environmental monitoring program that produced worrisome but classified studies. In the public realm, on the other hand, public health and public relations programs successfully placated anxious workers.

In just two and a half years, Manhattan Project leaders furiously invented new technologies, new communities, and novel ways of living that would radically alter postwar American society. One fact, however, General Groves and his staff did not learn. They did not know in 1945 that many of the secrets they had worked to contain had already left the country. The Soviet allies, who were a major target of Manhattan Project counterespionage, already knew a great deal about the American bomb and the cities created to build it.

Part Two

THE SOVIET WORKING-CLASS ATOM
AND THE AMERICAN RESPONSE

10

The Arrest of a Journal

The Moscow offices of the scientific journal *Questions of Natural and Technical Sciences* did not see much in the way of excitement until one day in 1992, when an elderly man walked into the chief editor's office with a thick packet of documents marked "top secret." The man introduced himself as a former spy and a decorated veteran of the Committee for State Security (KGB). He even gave his real name—Anatolii Iatskov. He said that during World War II, he was the assistant "resident" of the Soviet espionage ring in the United States and that he served as a handler of atomic secrets, passing them to the Soviet Union.[1] Iatskov's visit sparked a debate in Russian society in the 1990s over the origins of the Soviet atom bomb. The idea that plutonium, a new element on the periodic chart, migrated via coded messages from the United States to the USSR dropped like a bombshell on Russian society.

The chapters in this section describe how the Soviet plutopia was conceived and made in the American image. Like their American counterparts, Soviet leaders also created a community of select plutonium workers secured both physically and financially, which was orbited by lesser communities of workers, prisoners, and soldiers, servicing both plutopia and the spreading radioactive contamination flowing from the plant. The Stalinist regime may seem like it was ready-made for the kind of surveillance, submission, and obedience demanded by the nuclear security state. But that was not the case. Due to sheer poverty and disorganization, it took more than a decade to build the first Soviet plutopia, and it cost the nation dearly.

Iatskov's appearance in the office of *Questions* was one of those formerly unimaginable moments of the immediate post-Soviet period, when a KGB agent, after decades of silence, gave up his quiver of aliases and came in from the cold. Iatskov asked the editor, Boris Kozlov, if he would like to publish some documents on atomic espionage. Kozlov jumped at the scoop. Little did he know he was blindly entering a contested terrain between spies and scientists over who played the decisive role in producing the Soviet Union's first atomic bomb.

With the revelations in the late eighties and early nineties of the crimes of the Soviet state and Communist Party, the image of the KGB had taken a significant beating. The post-Soviet press portrayed the KGB, and the National Commissariat for Internal Affairs (NKVD) before it, as carrying out the dirty work of a corrupt, power-hungry communist regime. Iatskov sought to fight back. The documents he handed over described the pivotal role the NKVD had played in the arms race by gathering crucial atomic intelligence to defend the nation. The history of atomic espionage was well known in the West but had been a banned topic in the USSR. Even top Soviet scientists who worked on the atomic bomb had not known their bosses were consulting technical data purloined from the West.[2] So in Russia the story of Soviet atomic espionage promised to be big. If, as the common Soviet wisdom went, the Americans would have dropped a bomb on the Soviet Union had the Soviets not rapidly armed themselves with a "nuclear shield," then Soviet spies had saved the nation from certain nuclear annihilation.

The journal's staff rushed to prepare the documents for print. They called the major Russian newspapers to alert them to this media bombshell. Then the phone started to ring. The renowned physicist Yuri Khariton asked why the journal was publishing materials that could jeopardize state secrets. A person from the "First Department" (the security wing) of the Russian Academy of Sciences chewed out Kozlov for attempting to publish sensitive information that violated the anti-nuclear-proliferation treaty.

With threats coming in, Kozlov sought out Iatskov again to make sure the documents really were cleared for publication. The aging KGB agent, from a hospital bed, confirmed that archivists had legally declassified the documents.[3] Despite these assurances, the Russian Academy of Sciences, not the KGB, ordered Kozlov to withdraw the special issue from distribution. No one had seen anything like it in the post-Soviet period of unbridled freedom of the press. Two years passed while copies of the special issue languished under arrest in a Moscow warehouse. During that time, retired physicists battled in the media with retired KGB agents over who could take credit for the first Soviet atomic bomb.

There is no doubt that espionage played a major role in Soviet industrial development in the thirties and forties. In 1933, after diplomatic recognition between the United States and the Soviet Union, Soviet diplomats in the United States created "residencies" to maintain a stable of intelligence agents. Soviet spies usually posed as diplomats, journalists, or trade representatives.[4] They sought informers in Congress and the State Department to learn of international developments, and they targeted corporations such as DuPont, General Electric, and Westinghouse to purloin technology.[5]

Soviet leaders were hungry for information because they were acutely aware that the Soviet Union lagged behind. Haunted by fear of another war, Stalin launched in 1928 an industrialization drive to provide the socialist economy with a footing for national defense. His timing, at the start of the Great Depression, was unfortunate. Soviet businessmen had great difficulty financing machinery and raw materials. European and American political and business leaders, suspicious of communists, put up trade barriers against Soviet imports, offered loans only at high rates for short terms, and, in 1933, began to withdraw credit and trade access altogether as Depression-era national economies sank into protectionism and autarky.[6] With this externally imposed isolation, Soviet leaders sought to absorb as much information and technology from the West as possible, using any means available. They also learned a lot about American technologies of surveillance.

Soviet agents were quick studies. One agent, "Magnate," described in a classified telegram his work for the U.S. Army on miniature bugging equipment that was so small it could be concealed in a briefcase. "The results are very good," the station report noted in January 1937. "We hope to use this apparatus for our work here. We'll send you by the next post a complete set of blueprints and details. If we are interested in this type of briefcase, we can organize its production in our own country."[7] The Soviets did manufacture their own miniature bugging devices.[8] Ironically, the device that became an emblem of the Soviet Big Brother—a state always watching and listening—was initially an American export.

Soviet agents were also fascinated by the intelligence practices of J. Edgar Hoover's FBI. In 1937, agents from New York admiringly described Hoover's techniques, which combined intelligence with political control: "Hoover is keeping files on almost all major political figures: Congressmen, Senators and businessmen. He gathers compromising material on everybody and uses it for blackmail. In the course of the latest hearings on financing the FBI, Hoover blackmailed those Congressmen who tried to stand against [full funding of the FBI].... He used even cases of casual sex."[9] In that same year, 1937, agents of the NKVD carried out mass investigations into statesmen, leaders of industry and culture, musicians, scholars, and writers in a purge of the Soviet political and cultural elite. In these purges, NKVD agents were instructed to work in ways similar to Hoover's, by keeping card catalogs of compromising materials to call on when needed. By the forties, in short, Soviet leaders had a history of looking to the United States for ideas not just for technology but for policing and political control.

In 1941, the new Soviet-American alliance brought thousands of Soviet citizens into the United States to work on wartime trade agreements called Lend-Lease. At the start of the war with Nazi Germany, the Soviet Union's image improved greatly, and many Americans were drawn into left-wing politics closely connected with the American Communist Party. At the same time, the "Center," intelligence headquarters in Moscow, began demanding information about military hardware more urgently from agents in New York and Washington. Short on staff because of the purges, Soviet agents had to draw on the services of members of the American Communist Party as couriers and handlers.[10]

That winter, when the British were just starting to discuss the feasibility of a nuclear bomb, the USSR's young director of intelligence, Pavel Fitin, wrote his London and New York agents, asking them to stay abreast of developments in nuclear research, which he noted was well under way.[11] In September, "Vadim" (A. V. Gorskii), the station chief in London, reported on an important April meeting in Great Britain where British scientists agreed a nuclear bomb was feasible. "Vadim" later forwarded a thick packet of corroborating reports from Donald McLean and Klaus Fuchs.[12]

Fitin sought to verify his intelligence before he sent it to his boss, Lavrentii Beria, head of the NKVD. The NKVD maintained a massive network of gulags, or labor camps. Some were special camps, called *sharashki*, which held convicted engineers and scientists, who were put to work in prison labs and workshops. Fitin passed the materials on to Valentin Kravchenko, who forwarded them to several prisoner-physicists.[13] The incarcerated scientists reported that the materials were genuine. Kravchenko put his stamp of approval on the intelligence and sent the materials to Beria, recommending that the Ministry of Defense set up a special commission of leading scientists to investigate the possibility of building a Soviet bomb.[14]

Beria and his superior, Joseph Stalin, were famously suspicious men. When Beria received Kravchenko's report of the British nuclear bomb, Beria growled at Kravchenko, "If this is disinformation, I will throw you in the cellar!"[15] Beria suspected that British agents had planted the intelligence to trick Soviet leaders into squandering millions of rubles on a fanciful "superbomb." Beria, still leery, sent the information on to the NKVD's science and technology specialist, a young engineer named Leonid Kvasnikov, who drafted a letter in March 1942 from Beria to Stalin also recommending a committee to coordinate uranium research and to acquaint Soviet scientists with the intelligence materials.[16]

Beria, however, took a pass on this intelligence windfall. He did not sign Kvasnikov's letter. He did not tell Stalin about the nuclear bomb programs abroad. He did not form a committee to look into it. Perhaps it was personal. Some say that Beria did not trust Kvasnikov after an episode between the two of them in Poland in 1940.[17] What is clear is that Beria misstepped. He asked only for more information, cross-checked and systematized, from his agents in New York and London.[18]

Meanwhile, Soviet military intelligence (GRU) learned of the bomb project independently.[19] GRU agents informed the Ministry of Defense, which sent the nuclear intelligence documents to the chair of the uranium project, Vitalii

Khlopin, for verification. Khlopin replied that he did not think a nuclear bomb was feasible before the end of the war, so military leaders put the question aside.[20] As a consequence, for almost a full year Soviet government and military leaders sat on the greatest intelligence coup of the century.[21]

To be fair, in the fall of 1941 Soviet leaders had their hands full. By October, the Red Army and portions of Soviet society were in full flight, pursued by sun-tanned German soldiers marching through Ukraine looking for a fight. Soviet political and business leaders were frantically directing a massive evacuation of factories, businesses, educational institutions, scientific labs, and labor camps from European parts of the Soviet Union eastward. The Russian scientists who might have formed a Soviet bomb committee were busy, assigned to conventional weapons research. In short, Beria's yearlong hesitation shows that it takes more than knowledge to build an atomic bomb. The development of a nuclear weapon relies on having some room to maneuver, plus a mammoth reserve of labor and materials, as well as a certain level of security—all of which were in dire supply in the Soviet Union during the "Great Fatherland War."

While Beria and Soviet military leaders underestimated the importance of the new weapon, Pavel Fitin acted independently. He requested in August 1942 that the Ministry of Defense form a committee to apply NKVD intelligence materials to the creation of a Soviet bomb.[22] Only after forming the committee did officials at the Ministry of Defense inform Stalin of the year-old intelligence. Immediately Stalin summoned leading physicists to his office and questioned them about the possibility of Germany building this terrible bomb.[23] On hearing their opinion, Stalin decreed that the Soviets, too, must have this bomb. Stalin handed the project over not to Beria, who had failed to inform him of Western intelligence, but to Molotov, his foreign minister, who had little experience in science and production.[24] Stalin selected a young physicist, Igor Kurchatov, to serve as scientific director.

In November, Molotov gave Kurchatov a room, a desk, an armed guard, and hundreds of pages of stolen documents.[25] The Ministry of Defense granted Kurchatov a small lab in a former dog kennel, a few scientists, four grams of radium, and a miserable 30,000 rubles to initiate the Soviet version of the Manhattan Project. Kurchatov could bring as many scientists as he wanted to Moscow to work in his lab, but he was limited by a housing shortage so acute he had no place, not even barracks, for his colleagues to live.[26]

In July 1943, Kurchatov was given a new packet of information from Klaus Fuchs in London. This intelligence included an astounding new discovery that could save a great deal of time and effort. The documents described how American scientists had come upon the idea of bombarding uranium in a reactor to produce a whole new element, one, Kurchatov wrote, "not found on earth," element number 94 on the periodic chart. Kurchatov wrote, "The prospects of such a direction are unusually fascinating.... [A]s one can see, with such a solution to the whole problem, the necessity of separating uranium isotopes is eliminated."[27] Building a bomb with plutonium, Kurchatov noted, would require less time and, just as critically, less uranium. Kurchatov continually worried about uranium supplies. At the time, the Soviet Union had one source, in Central Asia. The mine, however, had long been abandoned, and orders to start it back up had gotten nowhere.[28]

Kurchatov, like any good scientist, used the stolen evidence of a rapidly advancing nuclear program in the United States to pry more resources from the state. With the German invasion, he wrote, nuclear physics in the Soviet Union had ground to a halt, while "the opposite occurred abroad." "There," Kurchatov continued, "instead of stopping, they assigned a huge number of scientists to the uranium problem."[29] As a result, "Soviet science has considerably fallen behind scientists of England and America."[30] He counted just how behind they were: in workers ("they have 700 scientists, we have 30"), in equipment ("they have ten powerful cyclotrons and we have one working cyclotron in [besieged, inaccessible] Leningrad"), in funding ("the Americans have designated $400,000 to the project"; Kurchatov had 30,000 rubles), and in uranium ("America has access to thousands of tons of uranium," while the Ministry of Defense had pledged to deliver to Kurchatov all of twelve tons of uranium by 1944). A graphite reactor to produce plutonium, he pointed out, required one hundred tons of uranium. At this rate, Kurchatov calculated, his lab would have a bomb in ten to fifteen years.[31]

Giving the Americans an opening to share in the development of the nuclear weapon, Soviet leaders went begging. In February 1943, the Soviet Purchasing Commission requested 100 tons of uranium through the Lend-Lease program. The Americans sent 700 pounds. The Soviet Academy of Science requested an exchange of scientific information, but the Americans politely declined.[32] Soviet leaders started to discern a pointed division of labor in the alliance. The Soviets, engaged against three-quarters of the Axis armed forces, took the brunt of firepower that destroyed Soviet cities and factories and decimated the Red Army. Meanwhile, the Americans and British delayed opening a second front. During the delay, Americans, largely safe from war at home, built up stocks of weapons, arms manufactories, and scientific expertise, all of which would contribute to a stronger postwar nation.

In the midst of this stonewalling, Stalin sent Winston Churchill an angry letter in June 1943 complaining that the British leader had repeatedly broken promises to open a second front. The delay, Stalin wrote, threatened Soviet "confidence in its allies, a confidence which is being subjected to severe stress."[33] A year later, a Soviet intelligence agent sent a political report to Moscow musing on the shape of the postwar global order. Before the war, the agent noted, Germany,

France, and England had worked to isolate the USSR and control Europe. With Germany out of the picture and France and England weakened, the agent wrote, the USSR would emerge as the major European power. Because England and the United States would not stand for a strong Soviet Union, the agent continued, the United States would "seek to establish American domination in the world, and their policy will be directed against the USSR most of all."[34] The agent penned these lines at the height of the Soviet-American alliance, when Soviet and American businessmen and military leaders collaborated on many levels and when political leaders were publicly talking of postwar cooperation and peace. Inside intelligence circles, however, distrust pushed against the political current.

Soviet agents became even more wary of their American allies when they learned in 1945 that a swashbuckling Russian American, Boris Pash, a former White Army soldier in the Russian civil war that followed the 1917 revolution, was racing around the European theater. Pash, morbidly anticommunist, directed General Groves' special Alsos Mission, set up to gather intelligence on the German nuclear program. As Pash traveled around conquered Germany, he came to admire many of the Nazi scientists he detained. He founded a program, Operation Paperclip, to secretly red-carpet into the United States some twelve hundred German scientists, several score of whom were accused of crimes against humanity. Soviet intelligence tracked Pash as he detained German scientists and appropriated stocks of uranium in Soviet zones.[35] Soviet agents took evidence of the Alsos Mission and Anglo-American atomic secrets as further proof that the Americans were plotting against them.

Intelligence on the American bomb hurtled Soviet and American leaders toward postwar rivalries on the cusp of their joint victory. In July 1945, the new president, Harry Truman, revealed to Stalin news of the atomic bomb. Truman was hoping to see fear or new respect on the Soviet leader's face. Instead Stalin nodded, smiled, and wished the president luck. Truman was vexed by Stalin's unruffled reaction. He thought Stalin did not understand the import of atomic energy. Truman had no idea that, thanks to espionage, Stalin had known of the American bomb project since 1942—two and a half years before Truman. While American, British, and Soviet diplomats in Potsdam lifted their champagne glasses to eternal friendship, Stalin eyed his rivals knowing he had a team of geologists hiking through the mosquito-rich forests of the southern Urals in search of a site for the first Soviet plutonium plant.

No amount of intelligence, however, could have helped Kurchatov build a bomb faster in the wartime years. With only trace amounts of uranium, a handful of scientists, and limited funds, a Soviet bomb was a futile dream. Only a nation with a global commercial grasp, with a surplus of labor and materials, and free from a homeland war, such as the United States, could conceive of building nuclear weapons. The Soviets, incarcerated within an autarkic economy, short of hard currency, and penned in by Axis forces, had no recourse.

In this context, Soviet leaders' desire after the war to become an atomic power takes on depth. For Soviet leaders, isolation meant having to beg; it meant allies keeping secrets from them. Isolation also spelled backwardness, and backwardness, as Stalin pointed out, meant losing, even in victory. Espionage, then, was a vital conduit out of Stalinist society's seclusion.

11

The Gulag and the Bomb

A week after Nagasaki, Stalin summoned from the city of Cheliabinsk, just east of the Ural Mountains, his commissioner of ammunition, General Boris Vannikov. Stalin and Vannikov had a complicated history. In June 1941, NKVD officers had issued an arrest warrant for Vannikov, who was then the minister of defense, and brought him to the feared Lubianka Prison in central Moscow. There, guards stripped Vannikov of his general's bars, took his belt, snipped the buttons from his fly, and silently ushered him, shuffling and holding up his pants, along underground passages to a basement cell.

The crime alleged to have been committed by the general, a longtime communist born to a family of Jewish workers, was the unlikely charge of spying for the Nazi government. Refusing to confess, Vannikov endured three weeks of nighttime interrogations.

It was, ironically, the Germans who saved Vannikov from sharing the fate of thousands of other Red Army officers arrested in the purges of 1939–41. When the Axis forces attacked the Soviet Union on June 22, 1941, the Red Army was caught with its pants down, because hundreds of generals who had been arrested and executed as spies, saboteurs, and traitors had been replaced by inexperienced young men who were afraid to make decisions. These events worked in Vannikov's favor. With the war suddenly on, Vannikov found himself cleaned up, with his gold bars restored, and being driven out of the prison gates and down the street to the Kremlin, where he was rushed to Stalin's office. Stalin greeted Vannikov as if he didn't notice the stink and bruises of incarceration and unceremoniously entrusted him with munitions for the country's war effort.[1]

Crippled and sickly from his stay in Lubianka, Vannikov wasted no time. He commanded factory directors to pack up their machinery, while bombs dropped, and stuff it into waiting trains, which would take it east. Vannikov hastily decamped to Cheliabinsk and set up his ministry on Revolution Square. There he built an industrial arsenal shored up by military investment and forced labor.

In August 1945, Stalin told Vannikov that Lavrentii Beria wanted to locate the Soviet atomic program within "his own institution."[2] Stalin noted that the idea had some merit. Beria's NKVD had the country's largest construction and design concerns. NKVD generals were among the Soviet Union's most powerful captains of industry. They commanded factories, power stations, trucking companies, lumberyards, mines, agribusinesses, and railroad lines throughout the country, fueled by several million unpaid manual laborers and educated specialists stacked up in labor camps and ready for any job. This vertical integration meant that NKVD commanders could get supplies and labor to the right place, where they could design, build, and produce quickly and secretly, or so Beria argued.[3]

As Vannikov recounts in his memoirs, Stalin, leery of handing Beria too much power, asked Vannikov what he thought of the idea. Vannikov, treading lightly, expressed his reservations to the dictator. Work on an atomic bomb, Vannikov surmised, would be so difficult and complicated and would require such a massive scale of operations that it would likely be unmanageable within any one organization, even one as colossal as the NKVD. The atomic bomb needed a national grasp, Vannikov told Stalin, and he doubted that free scientists and engineers could work well with prisoners. Stalin asked, "What do you recommend?" Vannikov told Stalin that it would be useful to find out what the Americans had done. Then Vannikov proposed a special civilian committee to oversee the program.[4]

Stalin called for Beria, who arrived in a few minutes with a deputy, NKVD general Avramii Zaveniagin. According to Vannikov, Stalin appropriated his idea, telling Beria that they needed to form a special civilian committee outside the NKVD. He placed Beria in charge of the new committee. "You think up the name," Stalin reportedly told Beria. "The committee needs to be under state control and it has to be top secret." Stalin asked Beria whom he would name as vice chair of this special committee (which Beria later called the Special Committee). Beria suggested his two deputies, Zaveniagin and Vasilii Chernyshev.

"That won't work," Vannikov remembers Stalin saying. "Those two have a lot of work already in the NKVD." Stalin turned to Vannikov and appointed him vice chair of the Special Committee.[5] A few days later, Beria's office issued a draft plan.[6] The Special Committee was to oversee a new First Main Department that would supervise the daily development of uranium mines, nuclear plants, and research institutions, much as the Manhattan Project had in the United States. The NKVD construction division would oversee the building of the nuclear installations, in a fashion similar to the Army Corps of Engineers' supervision of construction for the Manhattan Project. The First Main Department employed local NKVD construction enterprises to serve as the prime contractors.

In drawing up the plans, Soviet leaders had clearly studied the Smyth Report, the official U.S. government history of the Manhattan Project, obtained covertly by Soviet agents in the spring of 1945.[7] Other features of the Soviet program mirrored the Americans' organization. Soviet leaders created the Scientific-Technical Council with Vannikov in charge, similar to the American Office of Scientific and Research Development (OSRD), run by Vannevar Bush. Beria, like General Leslie Groves, was the shadow force behind the project, ultimately responsible for coordinating all branches and serving up a bomb. With the structure for the Soviet Manhattan Project in place, Stalin gave Beria a deadline of two years—a year less than the Americans.

Despite Stalin's attempts to dilute Beria's control of the bomb project, the Soviet bomb was largely the offspring of the NKVD. At the monthly meetings of the Special Committee, the men sat around a table: Vannikov, still sickly from Lubianka; Zaveniagin, terminally chilled from commanding Gulag camps above the Arctic Circle; and Beria, the feared and hated chief of the NKVD, who would die before a firing squad, accused of betraying the Soviet Union.[8] With them sat colleagues who had come up through the ranks of the security services, who had built the penal and exile system and were inured to Gulag conditions. At the end of the war, most of these men had been awarded medals for cleansing liberated territories of anti-Soviet resistance in mass campaigns of arrest, execution, and terror.[9] They all answered to Stalin, who was himself a convicted terrorist and ex-con who had spent time in tsarist prisons. These men—of the Gulag, from the Gulag—were used to reaching the goal at any cost. They possessed an unblinking stoicism before the suffering of others. Surefooted, they could scramble just as easily over scaffolding as over the rolling terrain of a mass grave. These were the men, aided by a team of scientists, who forged the Soviet nuclear shield.

It was a strange choice to make, placing the Gulag at the nucleus of the atomic project. Wisdom has it that slavery is incompatible with technological progress, that forced labor retards mechanization, technology, and innovation.[10] Vannikov hinted at this problem when Stalin sought his advice, saying scientists and engineers would find it difficult working with prisoners—and, he might have added, with their scarcely educated wardens. Moreover, Gulag prisoners were to serve as the labor force, yet many had been arrested for treachery and crimes against the state, making them just the kind of people one would least want to trust with atomic secrets. The NKVD appeared to be a good choice only because of the strafed economic landscape of late 1945.

In truth, Soviet leaders had no business building a nuclear weapons complex amidst the postwar rubble. In terms of devastation, the Axis powers had thoroughly defeated the Soviet economy. Twenty-seven million Soviet citizens were killed in the war. Twenty-five million more were left homeless, while Axis bombs destroyed thirty thousand enterprises and twenty-five thousand villages. Unlike defeated Germany, Soviets were left to reconstruct their country alone, without foreign aid. Before the conflict ended, Truman terminated the Lend-Lease program that had shipped direly needed machinery and food stocks to the Soviet

Union. The requests Soviet diplomats sent in 1946 to the U.S. State Department for loans for postwar reconstruction went unanswered. Meanwhile, in 1945, the NKVD was the country's largest corporation. According to the Gulag's own statistics, the Gulag produced more cheaply and efficiently than free Soviet enterprises.[11] Most important, the Gulag had labor. In 1945, industrial managers coveted nothing more than workers, and the Gulag commanded 23 percent of the country's non-agricultural labor force.[12] The Gulag possessed not only millions of manual laborers but also a national repository of intellectual capital—holding in special white-collar prisons physicists, chemists, engineers, and designers. For Stalin and Beria, turning to the Gulag to build an arsenal of nuclear reactors, chemical processing factories, and labs made sense.

But that was in 1945. A year later, indicators showed a sharp decline in Gulag productivity, brought on by a collapse in organization and an alarming spike in prison uprisings. By 1947, it became clear that NKVD wardens were losing control of the Gulag.[13] Hitching the nuclear wagon to the Gulag left troubling, costly, and lasting consequences.

12

The Bronze Age Atom

After deciding who would build the Soviet bomb, Vannikov and Beria had to determine where to build it. As in the United States, siting the plutonium plant in a remote territory led to problems procuring, sustaining, and retaining construction workers. The resulting delays led to rushed and shoddy construction and dictated that construction bosses had to economize on everything—housing, wages, safety, even security and secrecy.

Due to a lack of roads, the first plant scouting party traveled on foot. They scouted in the foothills of the southern Ural Mountains in a closed Gulag zone encircling a German POW camp between the industrial cities of Cheliabinsk and Sverdlovsk. Much like the way Matthias was drawn to the American West, the scouts were attracted to the Urals for the sparse population, free-flowing rivers, and substantial government presence. The Urals also had trees for cover and lay deep in the continental interior, safe from the reach of enemy planes.

It was a gorgeous area, the scouting party agreed. The lakes sparkled under racing cumulus clouds. Fields of yellow mustard and green clover were spliced against forests of pine and birch. The low mountains rose like a purple bruise in the distance. The men came across a collective farm on Lake Kyzyltash next to the ruins of a nineteenth-century mill. The twentieth-century villagers, living more primitively than their ancestors, fished from dugout canoes with grass nets.[1] This was the spot, the scouts reported back to Moscow.[2]

Such were the Bronze Age beginnings of the Soviet atom.

Vannikov placed NKVD general Yakov Rapoport in charge of building the plutonium plant. In the 1930s, Rapoport had cut his teeth working on one of Stalin's pet projects, the Baltic–White Sea Canal.[3] The canal had cost dearly in human lives and was a colossal failure, but it was a media success, an icon of how the Gulag penal system could reconfigure geography, human beings, and society itself in service to socialism.[4] In Cheliabinsk, Rapoport had at his command the large Cheliabinsk Metallurgical Construction Trust with forty thousand workers, most of them prisoners. Half of his workers were exiled ethnic Germans, detained during the war in mobile labor squads called the Labor Army.[5]

Rapoport also had the use of thousands of other exiles. Since the thirties, the Soviet government had shipped hundreds of thousands of inmates and deportees to the Urals to work in heavy industry and resource extraction.[6]

In November 1945, Rapoport issued an order, vague and urgent, to create a new construction division, No. 11, "at junction T, a remote crossing of two footpaths in the woods."[7] In 1945 there were no paved roads in the greater southern Urals. In the area chosen for the plant, there were no roads at all. The paucity of roads explained the sparse population. To live in an area cut off from supply lines in Stalinist Russia meant to toil ceaselessly, occasionally go without food, and perpetually live in fear of frost and hunger. On old Soviet road maps, the wide-open territories innocent of roads and settlements often indicated landscapes of incarceration. In Russian history and literature, the frontier is frequently depicted not as a place of independence and freedom but as carceral territory for people in chains or ensnarled by barbed wire and legal restrictions, a place for convicts, ex-cons, deportees, and runaways.[8]

In January, General Rapoport sent the first one hundred worker-prisoners to site T.[9] The construction crew disembarked from a train in Kyshtym in deep snow long after dark and made their way through the forest to the southern bank of Lake Kyzyltash. They stopped at the remote Handshake collective farm, four small huts surrounded by a commotion of sheds on the marshy edge of the lake. A few elderly villagers and children emerged sleepily. They were the only ones left in the village: during the war, working-age adults were conscripted from farms to work in factories or serve in the army.[10]

The fact that the scouting party chose a remote, roadless region determined two decisive factors in the creation of the Maiak plutonium plant. First, before Rapoport could build the plant, he had to build with primitive tools an entire infrastructure to service the construction site. Second, because of the scarce population, Rapoport had to rely on conscripted labor, meaning he built with largely unskilled, famished workers living in brutal conditions. In short, Rapoport was doomed to fabricate the Soviet Union's first industrial reactor with the blunt instruments and simmering insolence of the Gulag.

Vladimir Beliavskii remembers an autumn day in 1945 when he first heard about a new construction project going up nearby. Beliavskii was an engineer in Rapoport's large Gulag-run construction trust, but even with a managerial job, his wages were low. Obtaining food and goods was difficult. His infant son slept in a cardboard box. Beliavskii applied for a job at the secret, high-priority project, hoping for better rations. He got the job and was immediately detailed to headquarters in the beaten-down little city of Kyshtym.[11]

In the postwar Soviet Union, construction jobs paid especially poorly and usually meant working outdoors. Because these jobs were so undesirable, most migrant construction workers were either prisoners or exiles.[12] For Beliavskii, working with prisoners was nothing new. He had supervised the whole run of Gulag inmates, including exiles, deportees, interned national minorities, teenagers in labor training camps, and POWs—all part of the labor caste system of late Stalinist Russia. Freely hired, trained workers like Beliavskii were the Brahmins.

The first job was to build a road into the site. The chief of the road-building crew was Otto Gorst, an interned ethnic German. Gorst supervised a crew of Red Army soldiers who had been repatriated from Nazi POW camps.[13] Stalin considered any soldier who had allowed himself to be taken prisoner (including his own son) to likely be a traitor. NKVD security agents filtered returning POWs, questioning them. Most repatriated soldiers failed the loyalty test and were sentenced to ten years' hard labor in the Gulag.[14] Gorst recalled: "These so-called 'repatriated' were mostly adult men, 45 years old, some older. I can still remember their clothes; lightweight pea coats, shot-up, charred overcoats, worn-out boots with dirty foot wraps. Everything was tattered to the last thread."[15]

To haul in building materials over the rough terrain, the construction brigade was issued three heavy decommissioned tanks. But the tanks were no good. They bogged down in the swamps hidden under the snow. The soldiers had trouble mastering the vehicles, and the tanks sometimes slid sideways into ditches. For damaging an expensive machine, soldiers were thrown in the brig.[16] Dumping the tanks, the brigade ended up using surefooted horses and carts to lay down a primitive log road on the way to Europe's first plutonium plant.[17]

Getting supplies to haul on the new road was also a major undertaking. Rapoport had to supply the site with all necessary materials, which the construction firm largely produced on its own.[18] The trust owned steel works, mines, quarries, and logging concerns to supply raw materials, which then had to be fashioned into wheelbarrows, hammers, and sledgehammers at a machine works established on the site. To feed workers, the enterprise took over two local collective farms and staffed them with exiles. Cement and brick were scarce commodities, needed for postwar reconstruction across the country. As a consequence, Rapoport started a lumber mill, where inmates made housing (barracks and prefab cottages), scaffolding, litters, barrels, drainpipes, poles, sidewalks, furniture, and lab equipment. In the summer of 1946, there was no electrification. Wood and charcoal were used for heat, candles and pine splinters for light.

Making almost everything on-site demanded a great deal of labor. The Gulag could serve up thousands of bodies, but Rapoport had few places to house them. Working hastily in 1946, prisoners and soldiers built five military garrisons and eleven corrective labor camps for a growing crush of incarcerated labor: 10,000 conscripts, 16,000 ethnic German internees, and 8,900 Gulag prisoners.[19] However, Rapoport's superior, General Sergei Kruglov, was furious when he learned about the new camps. Why spend so much on temporary housing, Kruglov demanded to know, when it would be needed for only two years, no more? Kruglov ordered that residential construction focus on building two small settlements for future plant operators, who he imagined would be soldiers. Otherwise, crews and supplies should be dedicated to industrial construction, he commanded, not housing.

Kruglov's parsimony placed Rapoport in a vicious circle of failure. With poor housing and slim food rations, Rapoport had to house soldiers and prisoners in tents and mud dugouts and so had tremendous trouble keeping them healthy and meeting production targets. Rapoport, in charge of the nation's highest-priority site, could not get barracks and overcoats to keep workers warm. There was no getting around the nation's endemic poverty.

The soldiers' garrisons and prisoners' camps were often interchangeable. Rapoport shifted prisoners to soldiers' quarters and soldiers to prisoners' barracks, as if there was no distinction.[20] Conditions in both were miserable. Soldiers and prisoners had no hot food. They were given rations of rotten potatoes out of buckets. They lived in barracks, sod dugouts, or yurts with little or no bedding.[21] In poorly heated barracks, the walls were coated with ice in winter and with mold the rest of the year. From Rapoport's perspective, prisoners and soldiers were one and the same: "mobilized labor."[22]

Rapoport needed healthy bodies working at the jobsite, and healthy bodies were a statistical minority in his construction trust. Less than half of those arriving were fit to work.[23] In July 1946, crews of several thousand gaunt workers embarked on the job of digging a foundation for the first Soviet industrial reactor. Soviet engineers had decided to bury it deep underground so that it would look like any other official Stalin-era building, be it a Soviet palace of culture or a Communist Party headquarters.[24] This foundation became the stage of a gargantuan struggle between man and the environment. Originally estimated to take six months, the project dragged on for eighteen months as prisoners and soldiers burrowed down 175 feet. No one had ever dug a foundation so large or deep.

August 1946 was an especially rainy month. The soldiers and prisoners dug in a persistent cold rain. Groundwater ran through the bottom of the pit. The prisoners and soldiers worked with little mechanical help. In the first two years, there were no bulldozers, diggers, or earthmovers, just a half dozen American Studebaker trucks, some decommissioned tanks, and muscles, aching and cold. The men hacked through slabs of bedrock. After the aging conveyor belt broke, crews ran the sodden earth in wheelbarrows up ramps. Emaciated crews fulfilled 14 percent to 37 percent of their daily targets.[25] The strongest and healthiest men lasted a few months in these conditions before succumbing to boils, persistent coughs, and tuberculosis.[26] "The workers," a Gulag official reported, "are broken."[27]

Behind schedule, supervisors raced back and forth, fuming. The deadline, originally January 1947, was repeatedly pushed back. In March 1947, Rapoport decreed that specialized workers had to fulfill their daily production quotas, no matter their state of health. If a worker didn't finish, then he or she didn't get even the meager camp ration; for people working shifts of ten to twelve hours, with no days off, that was fatal. Prisoners and exiles tried to escape after that rule was passed.[28] They were usually recaptured or found dead and frozen, and then brought back and displayed to the horror of the rest.

In October 1946, worrying about rumors of famine for the coming winter, Rapoport came up with a novel and cruel idea. He relaxed the restrictions on prisoners receiving care packages from home. Rapoport's assistants then ordered Gulag bosses to start up literacy classes to teach prisoners how to pen a letter home asking for food and clothing. Rapoport also requested an express bus service and a warm waiting room for family members to deliver goods in person to their imprisoned kin.[29] It was a heartless scheme. On average, a Soviet family purchased one leather shoe, and, for each person, one pair of socks and a quarter of a pair of underwear each year. In the famine of 1946–47 about a million and a half Soviet citizens perished from hunger-related diseases.[30]

Among the regulations about packages, there were no directions to censor letters or search for contraband. The appeals inviting family members to the secret camp weren't good for security, but at the time no one at the construction site was thinking much about security. In the United States in the late forties, pundits were writing about Stalinist Russia as a highly secretive "totalitarian order" where, out of fear, citizens kept secrets and obeyed unquestioningly. What American commentators failed to grasp is that order and security are luxury items, ones that came naturally at Hanford but took years to serve up in the Stalinist Urals.

13

Keeping Secrets

Ideally, Gulag camps were surrounded by a double row of cyclone fencing topped with barbed wire and punctuated with guard towers and floodlights. According to regulation, no one could enter or leave a camp without a pass. Russian historians have argued that Soviet atomic installations naturally appropriated these defining features of the Gulag.[1] They depict the closed nuclear city as the Gulag's crowning achievement, the very fortress of the disciplined, punitive, coercive Stalinist police state. Who else but Gulag bosses would build special cities, surround them with fencing, and lock in free citizens for years?

But there are two problems with the equation of the mature, closed nuclear city with the Gulag. First, Soviet leaders and construction managers like General Rapoport were so taken up in the first two years with organizing a colossal nuclear infrastructure amidst the postwar ruin that they largely forgot about security and secrecy. Second, despite the popular image of a Soviet labor camp as a place of totalitarian order and control, where prisoners meekly submitted to the power of guards and wardens, that reputation is grandly mythical. No one acquainted with Gulag installations in 1947 would have taken them as a model for safety, security, order, or efficiency.

Rapoport had orders to build the plutonium plant under strict specifications to guarantee security. Beria commanded that Rapoport employ only the most trusted prisoners—no ethnic Germans, POWs, recidivists, hardened criminals, or political prisoners, and no individuals who had lived on territory occupied by Germany during the war.[2] Soviet leaders were especially worried about "traitorous nations," ethnic groups suspected of betrayal—particularly Ukrainians and Balts, who had fought against the installation of Soviet power in bloody, ongoing civil wars.[3]

Rapoport, however, found it impossible to carry out these security orders. Half of his construction workers were ethnic Germans, and they were often the most skilled.[4] As a result, Rapoport sidestepped Beria's edict by promoting ethnic Germans to the rank of "special settlers," a group that served at the top of the Gulag's penal hierarchy.[5] Rapoport also wanted only healthy, "labor-ready" prisoners, but the healthiest prisoners were hardened criminals who bullied rations out of weaker prisoners and forced them to do their work. As a result of the security regulations, half of the first arriving prisoners were invalids, useless as a labor force.[6] So, in violation of the security orders, Rapoport accepted hardened criminals.[7] According to Gulag regulations, dangerous criminals had to be guarded at all times, but Rapoport didn't have enough guards, and so the majority of maximum-security prisoners were sitting at the camp all day and not working, because of a lack of escorts.[8] Rapoport solved that problem, too. He issued orders to reclassify the maximum-security prisoners as minimum-security so that he could send them to work without a guard.[9]

The labor crises Rapoport faced grew out of the Gulag's chronic problems, which in turn made Rapoport's mission breathtakingly supersized. He was tasked with fashioning a behemoth industrial landscape with Gulag labor just at the moment when the Gulag was collapsing from mismanagement. In the postwar years, the Gulag swelled with more prisoners and exiles than ever, reaching a total of 5.25 million. Because of a chronic shortage of prison staff, prisoner warlords, running their own gangs, took over the day-to-day administration of the camps. The warlords were powerful and dangerous men. The rivalries between gangs often exploded in violent brawls. When wardens tried to reassert power over unruly "hooligan" elements, prisoners turned on their guards, rioting and fighting in skirmishes that could take months to subdue.

Ivan Butrimovich was a civilian foreman working on the reactor foundation. He remembered a prison lord, called in Gulag slang the *pakhan*, who ruled the foundation pit. Instead of the usual prison-issued padded jacket and pants, the *pakhan* wore chrome-colored boots and a business suit with little studs. In the morning, he spread out a felt rug at the bottom of the massive crater, where he would sit all day while his crew of inmates worked around him. The *pakhan* had one assistant to play cards with him and another to take orders. The *pakhan* himself did no work, but if he heard that a brigade was slacking off, he took measures right away. He would call in a brigade member and warm him up along the spine with a stick. Sometimes merely speaking a few words served to speed up the brigade.[10]

Butrimovich recalled that the *pakhan* never spoke to the civilian supervisors. But if supervisors kept the *pakhan* happy and supplied with food, clothing, and vodka, which he could then distribute among his gang, then a supervisor had no discipline problems, and the prisoners kept up a good tempo.

Butrimovich often stayed down in the foundation pit for two to three days at a time. He would spread his winter coat in a shed and sleep. One evening, while taking a stroll, a prisoner grabbed him by the arm and led him to a ledge, where he complained about Butrimovich's record keeping. Rapoport, trying to speed construction, had issued an order granting shorter sentences for completed daily work quotas.[11] The reward system relied on supervisors to keep track of work, a complicated and arbitrary process. Prisoners often contested the record with their foreman.

The prisoner, "a big, healthy fellow," Butrimovich recalled, started to question him as he pushed him toward the edge of an overhang. "'Do you know how many accidents have happened here? No? Well, I'll tell you—plenty. Give me your word that you will correct the records, and that kind of accident won't happen to you.'"

Thinking fast, Butrimovich grabbed the prisoner's padded jacket and held tight, promising to take the prisoner along on his fall. At that, the inmate let him go, saying, "Nice job. You're no coward," and walked back to the campfire.[12] Standing there on the edge of the precipice, Butrimovich understood he was lucky. He knew prisoners wasted no time dispensing with supervisors they didn't like. He had heard of a missing foreman who was found after a year, walled up in a cement foundation.[13]

Alexander Solzhenitsyn popularized the image of the hungry and scurrying inmate Ivan Denisovich, happy, at the end of a day of hard labor, with an extra ration of bread.[14] Butrimovich's story illustrates how Gulag camps and prisoners were, on the contrary, increasingly disorderly and uncontrollable. A March 1946 inspection of the plutonium sites' labor camps turned up missing and broken fencing, no floodlights, and no electricity to power them. Morning and evening roll calls were only formally carried out, meaning they weren't. The daily inmate count changed frequently, and record keepers lost track of the number of prisoners and exiles. Prisoners disappeared and went into town, showing up again only after several days. In March 1947 two prisoners robbed a kindergarten pantry, and when caught they put up a good fight. Inside the camps, prisoners fought, stole, drank, and sold contraband. A search of a camp of 650 prisoners turned up 101 weapons. Inspectors reported there was no supervisory staff at the camp on a day-to-day basis. They all slept in town. Prisoners apparently ran the camp on this high-priority, top-secret site largely on their own, largely unsupervised.[15]

In early 1947, a women's camp opened at the construction site to accommodate the flood of women arrested for petty theft during the famine.[16] Gulag rules forbade fraternizing between male and female prisoners, but the rules were ignored. Men and women had consensual sex, and men raped women. The problem got so bad that Rapoport set up a guard to stand watch over barracks for civilian women.[17] In 1947 alone, thirteen hundred prison babies were born. The two doctors on staff had to quickly fix up a makeshift maternity ward and nursery.[18]

Prisoners were unruly and violent, but so too were soldiers in construction brigades. The soldiers dressed like prisoners, lived like prisoners, and ate like prisoners, slops served up in buckets out in the open. Cold and hungry, the soldiers took to foraging. "They feed us 'really well,'" one soldier wrote home, "soup so 'thick' that at night I walk six kilometers and sometimes more for potatoes. Otherwise I'd go hungry. I go not to buy (for I have no money), but to take what I can without paying. Just so no one sees it."[19] Inspectors reported massive desertions of soldiers and officers leaving the garrisons to pillage neighboring villages. During the harvest season, officers directed their soldiers to steal crops and fodder. Farmers who caught them fought back, or tried to, until some of the officers murdered a few farmers.[20]

Soldiers stole from each other, too, and got into fights in the barracks. "The commanders are powerless in these fights," one soldier wrote home. "They don't even try to do anything."[21] Like prisoners, soldiers drank, raped, brawled, and left their garrisons to carouse. Rather than a well-ordered police state, the first plutonium settlements reflected the Gulag all too well—chaotic, dangerous, barely supervised, miserably inefficient, and spilling outside the boundaries of the Gulag zones.

Security arrangements for civilian workers were little better. As the project grew in the spring of 1946, Beria sent Rapoport orders to secure the site by setting up a "special regime zone" that included a roadblock with a guard, searchlights, and passes.[22] This order, like the ones banning dangerous prisoners, went unheeded. In the first year and a half of construction, Vladimir Beliavskii remembered, there were no formal security restrictions at the site at all.

"Anyone," he recalled, "who wanted to, could go there without trouble."[23] A train from Cheliabinsk ran daily to a stop a few miles from the construction site. Most able-bodied people who showed up were hired on the spot with no background check. People in the nearby town of Kyshtym talked openly of the "secret atom factory" going up in the woods. One worker arriving in Kyshtym remembered asking an old woman how to find the personnel office. She replied, "If you need to go to Lake Irtiash where they are making the atomic ship, then go to that hill in front of you and you will see the trucks that bring the workers to the underground factory."[24]

Housing arrangements for hired workers were equally haphazard and unplanned. Engineers and supervisors traveled daily six miles from Kyshtym, where they rented rooms, attics, and basements.[25] Hired workers mostly made do with what they could fix up for themselves from objects borrowed and found. Exiles wrote asking their family members to join them and scavenged scraps of lumber and sheet metal to build huts and dugouts.[26] If people had money, they bought cows, chickens, and goats and started gardens. Civilians and officers arranged, illegally, to have prisoners work as household servants to tend their gardens and livestock, cook, and clean for them.[27]

This first Soviet plutonium settlement was no Richland. There was no master plan, no scheme to segregate unreliable prisoners and soldiers from the upstanding, trained employees who were entrusted with state secrets.[28] As the conscripted workers caroused and lived off the surrounding countryside, they spread word about the expanding "secret" installation going up in the forest. The plutonium site was, according to contemporary accounts, a security disaster.[29] In short, the Gulag branded the Soviet nuclear project with its shabby, infectious disorder, with insubordination, violence, theft, and inefficiency. In its particular way, the Gulag imprinted the Soviet plutonium project, dooming it to a future of calamity. Doomed as well was Yakov Rapoport, the bespectacled, acerbic Gulag general in charge of the construction site. He was a tragic figure, sitting at his desk day after day, penning more and more orders, fielding ominous calls from Moscow, trying, I imagine, to forestall his downfall as he violated security regulations and missed deadline after deadline.

Deploying the Gulag to build the bomb meant for Soviet leaders the smallest investment in return for a nuclear arsenal; it meant, as with other Soviet industrialization programs, a prioritization of industry over consumption, factories over cities, bombs over butter. In other words, the atomic Gulag was business as usual for the Stalinist state. The quasi-civilian/quasi-incarcerated settlements, born of the fracturing Gulag, contained few traces of the future plutopia. Instead, the genesis of the ordered, highly controlled, closely watched, quarantined nuclear city came from another, unlikely quarter—from the leader of the free world.

14

Beria's Visit

In the fifties, physicist Igor Kurchatov, recalling the origins of the Soviet bomb, opined: "We were all alone. The Americans and English, who were ahead in science, did nothing to help us."[1] As scientific director of the Soviet nuclear program, Kurchatov read nearly ten thousand pages of documents purloined from the United States and Great Britain.[2] His scientific committee met regularly with agents engaged in atomic espionage in the West. Kurchatov had a great deal of help building the atomic bomb, but the enduring sentiment that the Allies had forsaken the Soviets persisted. The sense of betrayal contributed greatly to souring diplomatic relations.[3]

In February 1946, after Winston Churchill delivered his "Iron Curtain" speech, shoppers in Russia made a panicked run on food stores.[4] In March 1947, President Harry Truman traced the outlines of the emerging Cold War yet more clearly, stating that the United States would defend "free peoples" from subjugation by "totalitarian regimes" around the globe. The Truman Doctrine seemed to Soviet leaders like a glove thrown at their feet, especially as FBI agents charged about in a manhunt for Soviet spies in American labs and universities.

Despite these tensions, Stalin and other top Soviet leaders held out hope of American aid as a goodwill gesture for Soviet sacrifices on the eastern front. Andrei Zhdanov, a rising Politburo member, dreamed of the USSR becoming a market for American postwar surpluses. He anticipated this trade would bolster Soviet-American ties.[5] His hopes were dashed in June 1947 when General George Marshall announced a plan for massive aid to rebuild what became Western Europe in order to make it economically secure from the contagion of communism. The fine print of the Marshall Plan made it clear that the Soviet Union would not be included in the generous American aid program. In the summer of 1947, Soviet leaders came to understand the world would be divided into two camps, and one, they figured, would have to perish.[6]

After the unveiling of the Marshall Plan, all caution slipped from official Soviet propaganda. Stalin ordered up an "anti-cosmopolitan" campaign leveled at the United States and Great Britain. In Soviet newspapers, pundits berated

Soviet intellectuals who "bow and scrape with a slave's obsequiousness before Western science and culture."[7] The xenophobic crusade gained momentum in July after Stalin heard that two Soviet researchers had passed on to American scientists a miracle cure for cancer. Stalin ordered the scientists arrested, and before a packed auditorium in Moscow, prosecutors used the high-profile trial to instruct Soviet intellectuals on the toxicity of contact with the West.

In the meantime, Lavrentii Beria, in charge of the secret atomic First Main Department, frequently briefed Stalin about the American nuclear program. He reported that Soviet agents had acquired copies of U.S. Air Force plans for a nuclear attack on fifty Soviet cities.[8] From London, Klaus Fuchs estimated the Soviets had until 1949 before the Americans could generate enough bombs to wipe out the USSR. Soviet leaders continually worried about the need for a weapon to defend against what seemed to be an inevitable American strike.[9] Stalin frequently goaded Beria to speed up Program No. 1.

Yet the Soviet A-bomb's progress was braked by the plutonium plant in the Urals, where every construction deadline was months overdue.[10] Fretting, Beria sent to the site in January 1947 a top deputy, Sergei Kruglov, who learned that Rapoport's camp commanders were meeting less than half of their monthly production targets.[11] Rapoport ran not just the plutonium site but all other Gulag construction projects in the province, and the pressure was getting to him.[12] At a roll call, Kruglov, tormenting his subordinate, asked Rapoport if the prisoners had clean underwear. Rapoport rushed up to the line and started anxiously unbuttoning prisoners' jackets. Later, the prisoners laughed at how the boss's hands had quaked.[13]

Rapoport had many reasons to be terrified. He was supposed to deliver a functioning plutonium plant by November 7, just eleven months away, yet no part of the five major facilities was finished. Some hadn't even been designed. I don't have a photo of Rapoport's office, but I picture a large, utilitarian clock hanging on the wall above his desk. The clock ruled the office and the vast construction site more than anything else. The clock dictated Rapoport's work, sleep, and his growing anxiety as the hours passed indifferently, moving Rapoport clockwise toward his demise. There was no way Rapoport could beat the clock, for he had to build an industrial empire before he could construct the plutonium plant, and he had to do it with forced labor in a remote forest.

As his impatience mounted in early July, Beria boarded a specially armored train and headed for the secret site, No. 859. The petite, balding minister disembarked in the muddy, log-cabin city of Kyshtym. Soldiers unloaded his cement-lined Cadillac, a war trophy, to drive the last miles to the construction site. The heavy Cadillac slipped and lurched on the mossy log road and a few miles later sank into the mud. Furious, Beria switched to a lighter, Soviet-made car. Along the way, Beria saw other travelers on the road, spattered and weary, pushing their cars through the mire.[14]

Because of the road, workers had difficulty getting supplies and heavy machinery to the site. Beria saw soldiers and prisoners working with hand tools, rickshaws, pickaxes, and an army of horses. He grew more agitated when he realized the nation's first industrial reactor was still just a seeping pit in the forest, while other projected buildings lingered on the drawing board. Labs and workshops were renovated rough-hewn barracks and barns. Trained chemists and technicians, brought in to work at the new factories, were detailed to menial tasks while they waited for their factories to be completed. In short, two years into the project, there was little to show for it.[15] With Stalin breathing heavily into the phone, Beria had no time to build the first Soviet plutonium plant with the tools and technologies of an Egyptian pharaoh.

Beria also inspected the workers' residential quarters, called Camp Construction. Beria couldn't have missed the aerial photos, projected in American magazines, of the orderly Hanford Camp—miles of geometry and symmetry traced out in army-green particleboard.[16] Camp Construction, in contrast, looked like the aftermath of a natural disaster. Settlements were scattered about as if just spat from a cyclone's vortex. Garbage and excrement followed the slippery plank paths between the barracks, tents, outhouses, and mess halls. Cows moaned on short tethers in stinking yards, flies swarming. Goats and chickens pecked underfoot, while gnats hovered in dim, hungry veils. Prisoners, waxen and sharp-boned, sprawled around campfires. The prisoners stank. The gruel they ate smelled bitter; bitter, too, were the looks they gave the minister in the fine suit as he passed, followed by a murder of anxious generals.

As Beria toured the site, his pique exploded into a mute, vein-bulging fury that had even the hands of the famous physicist Igor Kurchatov shaking.[17] Beria was a very suspicious man. He had seen how scientists walked off Manhattan Project sites with fistfuls of secrets. If American scientists were so treacherous, how could he trust his own? Beria had intelligence on how Americans had been arming and training former Soviet citizens to return to the Soviet Union to spy and sabotage.[18] What if some of them got into the secret plutonium site?

Beria was horrified to find that workers traveled freely in and out of the ungated compound and that all kinds of people without clearance wandered around.[19] The local district attorney had taken Rapoport to court over the "willful appropriation" of farmers' land.[20] Townspeople in nearby Kyshtym, well aware of the "secret" atomic plant, came and went, trading and delivering goods. Worst of all, Beria discovered, the majority of the workers were dangerous prisoners and ethnic deportees—proven enemies of the Soviet state, many sentenced for collaboration, treason, sabotage, and espionage. Like vipers, these traitors crawled everywhere, Beria fumed, freely fraternizing with employees and townspeople. It all had to end.

What was Beria expecting? What kind of security regime had he had in mind before arriving in the Urals? Surely he was not imaging the Gulag to be a model for order and security. As the former chief of the NKVD, Beria understood what it meant to work with prisoners. He knew how volatile and dangerous they could be; how prisoners were often armed, threatening, defiant, and desperate.[21] The anarchic, postwar Gulag was not, never could have been, the model for the first plutonium citadel.

Beria had at his disposal far more successful models of security and labor management for the massive nuclear installation. As chief of the NKVD's wartime foreign intelligence branch, Beria had been one of a handful of Soviet leaders who had access to intelligence reports on the Manhattan Project, in which he was intensely interested.[22] Soviet leaders closely imitated the Manhattan Project wherever they could. Beria and Vannikov, for example, ordered Soviet scientists to drop their own (sometimes better) designs and duplicate the American bomb and reactor blueprints precisely.[23]

During the war, Beria had inquired repeatedly about the security arrangements in the gated and locked nuclear cities of Los Alamos and Oak Ridge, where the NKVD had informants.[24] He sought information in order to try to breach the American security regime, but he was also looking for organizational ideas. Ted Hall, David Greenglass, and Klaus Fuchs wrote back, describing how Los Alamos was cut off from the world. In 1944, Hall wrote: "Center Y [Los Alamos] is cordoned off from the outside world by wires and guards and outposts. The workers live within the confines of the fence. Mail is strictly censored. Only recently have they permitted trips further than 75 miles from the center, but for this one needs special permission from the military authorities."[25] From Harry Gold, Soviet agents reported about employee background security checks and how in Santa Fe immigration agents patrolled a buffer zone around the lab, checking the identities of passengers at the bus depot.[26]

On returning in 1947 from the Urals, Beria demanded an American-style security regime at the Soviet plutonium plant. Like Los Alamos, he wanted a separate town to segregate loyal and chosen plant operators from unruly, untrustworthy construction workers. And he wanted the whole territory, the production area as well as the exclusive residential settlement, surrounded with a double row of fencing, punctuated with watchtowers and guardhouses. Around the nuclear installation, he ordered an even larger, secure buffer zone, which the Americans called a "control zone" and Beria called a "regime zone." Like the Americans, he wanted only select workers employed at the plutonium plant, people whose backgrounds had been checked. He required them to have identity cards and demanded that their movements in and out of the enclosed zone be controlled.[27] Other, less critical working-class construction workers Beria sought to have zoned off in their own separate, compartmentalized areas. It appears, in other words, that Beria wanted the American way.[28]

Elevated guard post, Richland, 1944. Courtesy of Department of Energy.

It is counterintuitive that Beria, head of the notoriously secretive NKVD, would turn to the open society of the United States for security models. Yet despite the invective against Soviet intellectuals aping the West, Soviet leaders had long made a study of American industrial development, management, and urban architecture. In fact, imitation and intellectual pirating served as the crucible of Soviet industrialization. At the start of the industrialization drive, for example, Soviet engineers used the company town Gary, Indiana, as a model for the Urals steel town Magnitogorsk, and Soviet leaders relied heavily on American manufacturers such as Ford, General Electric, and DuPont for factory plans, machinery, and foreign managers to run them.[29]

Because security regimes require a considerable degree of abundance and affluence, the United States glided effortlessly ahead of the strapped postwar Soviet Union in this realm, too. The security technologies Rapoport could not afford—electric searchlights, alarm systems, miles of fencing, planted informers, censored mail, background security checks, and the personnel to make it all happen—came naturally to the U.S. Army Corps of Engineers and its corporate contractors in the mid-forties. Prosperity graciously bore aloft the American nuclear security regime.

Beria had personally made the long trip to the Urals because he worried that Stalin might upbraid him for the plutonium plant's construction delays. He was anxious to find someone else to blame for the delay—that was how a good Soviet boss stayed on top, by holding others accountable.[30] Beria fingered his old acquaintance General Rapoport, which spelled the general's swift termination in the summer of 1947.

After Rapoport left, the changes came quickly. Within weeks of Beria's return from the Urals, the First Main Department issued a series of commands to cut off the plutonium plant from the rest of the world. Beria set up a security division and put in charge another lifelong NKVD general, Ivan Tkachenko, to report directly to him.[31] For a general, Tkachenko was young (not quite forty), dark and handsome, vigilant and forbidding. During the war, Tkachenko had taken part in deporting Chechens and Ingush from the Caucasus to Kazakhstan. After the war, he served in the security organs cleaning up postwar Latvia, a merciless affair of mass arrests and mass executions. The NKVD leadership noticed Tkachenko's zeal in Latvia, and he was called to Moscow to work in the First Main Department.[32]

Upon arriving in the Urals, Tkachenko wasted little time. He established an enlarged, twenty-two-mile regime zone encompassing ninety-nine villages and hamlets. He decreed that no one could enter the regime zone without a pass and passport, and he cleared out villagers living within three miles of the site. He closed the roads between towns to regular traffic. Air traffic was banned overhead. Tkachenko instituted new rules subjecting all employees to background security clearances and issued passes that they had to keep with them at all times.

To make the regime zone stick, Tkachenko commanded that a double-row fence be put up around the area. Then, one day in October 1947, with no advance warning, Tkachenko closed the zone. Workers went to work in the morning, expecting to return to their families at the end of their shift, but they did not come back, or so memoirists remember it—not that night or the next, or the following nights for the next five to ten years.[33] After the barrier went up, employees and their families could come or go only with permission.

Tkachenko started an office to censor mail. He recruited informers and instituted a system of spot security checks.[34] Tkachenko then segregated residents of the new security zone. He ordered the creation of a series of walled ghettos for prisoners and soldiers. He banned civilians from talking to prisoners and soldiers without official permission.[35] Tkachenko created a new landscape, zoned along a sliding scale of freedom. Civilians were to live in two settlements, surrounded by a single row of fencing; soldiers were to live separately in their own walled-off garrisons, prisoners in their barbed-wire camps.

From Moscow, Beria ordered that political undesirables be cleansed from the cadres of working prisoners. Beria especially wanted repatriated soldiers and ethnic Germans removed from the secret site.[36] But Tkachenko couldn't do that. The construction site was heavily dependent on the labor of exiles, repatriated soldiers, and prisoners. Instead, he issued decrees assigning prisoners to nonsensitive jobs in residential construction and machinery shops. Tkachenko also had industrial construction areas fenced off, compartmentalizing workers' knowledge of the secret project. Tkachenko largely implanted General Leslie Groves' system whereby the security zone was conceived as a series of walled-off camps and compounds, dividing workers by class and legal category, while work too was segregated by function and security rating.

When one reads the new security decrees, they ring with a certain irrevocability. The zones were to be impermeable. The threats of arrest for violations assume absolute obedience. Tkachenko's new security regime appears to be hermetic, impervious, and total—nothing if not Stalinist. Did it work? Were Soviet leaders able to come up with enough fencing and guards, sufficient paper for passes and electricity for searchlights? The Communist Party archives in Cheliabinsk contain transcripts of an early meeting in a log house of sixty-two communists of Base 10, the newly cordoned-off settlement for plant operators. The record offers a rare glimpse into life soon after the fence went up, when the plant, still under construction, started processing the first grams of plutonium. On a frigid January evening in 1948, General Tkachenko gave a lecture on site security and vigilance. The transcript records the subsequent discussion. Here are some excerpts:

> Comrade Chaplygina said that the procedure for meeting newly-arriving employees is poorly organized. The new staff end up waiting a long time in either Kyshtym or at the checkpoint, which makes people disgruntled and inspires superfluous conversations. She asked the checkpoint be inspected because there have been occasions when people have shown up and had to search to find a guard.
>
> Comrade Bochkov said that the question of "vigilance" was raised in good time, but what must be added is the fact that our workshops are wholly disorganized and that makes it impossible to work normally. For example, Central Accounting is located above [the top-secret plutonium processing lab] and a whole stream of people file up there to get their pay and they stand in line and disturb work.
>
> Comrade Kondrat'eva says we need to be careful how we keep documents, accounts, and inventories of train cars. When trains are unloaded we have to always remember that the stevedores are unsupervised prisoners, and sometimes even Germans, and they might pick up extra words about the cargo.
>
> Comrade Pozhidaeva: I personally think the question of state secrets has been raised just in time, all the more so because there are a lot of

communists who arrived at the factory a long time ago and they make mistakes. It is true that some of these blunders are not their fault. For example, the desks in our division do not close. We have no safes. The office doors have no locks. Besides that we need to keep to a minimum the arrival of relatives for short visits. It is difficult to keep secrets in these work conditions.[37]

It is easy to read decrees issued by powerful leaders such as Tkachenko and Beria and confuse the orders with reality. This glimpse of the party meeting reveals how the vision of a sealed-off regime zone, segregation of prisoners and exiles in nonsensitive work, locked offices and labs, and the swift and efficient processing of workers was fantasy in early 1948. Family members came and went. Employees lined up outside secret labs to get paid, while workers spilled secrets in stolen phone calls to friends and family members. Meanwhile, the original villagers continued to live in the regime zone despite repeated orders for their evacuation. Instead of the number of suspect prisoners being reduced, sixteen thousand more convicts arrived in the summer of 1947 to help speed work along. Prisoners in the banned categories continued to work, as before, on sensitive industrial sites.[38]

It is natural to assume a Stalinist political culture of whisperers submissively conditioned to keep secrets, but in 1948 party members complained that fear and vigilance still needed to be internalized among residents of the special regime zone. For bosses, expediency and the need to make production schedules far outweighed the security regulations, which often slowed and complicated a job. The topsy-turvy imprint of the Gulag was difficult to shake. It took several years, many threats, arrests, repeated cleansings, long meetings, and many, many more lectures on vigilance to get the message across. Yet threats only got so far. In the end it took a great investment in homes, consumer goods and community programs to win the loyalty of plutonium workers, and that investment was a decade away. For Tkachenko, it was a long, slow struggle against gravity.

15

Reporting for Duty

The regime zone Tkachenko created in 1947 took the existing territory of indigenous villages, forest, and farms and installed a new spatial regime that dictated where people lived, what they did, and what opportunities their futures held. More than a singular leader or general, it was the dictates of incarcerated space that commanded people's lives after 1947 in the Southern Urals.

Natalya Manzurova told me her life story quickly, in great gulps, about how she came to grow up in the closed city of Ozersk. She narrated how her parents' chance arrival in the regime zone in 1947 determined her fate, and explained the tragedies and disappointments that circled her adult life. Manzurova's mother, just after the war, worked as a technician on an electrical substation near Sverdlovsk (now Yekaterinburg). One day in 1947, she was given a ticket for a train to depart the next day. She was not told where she was going or for what reason. Obediently, she boarded a train heading south.

Manzurova's father was a driver, also living near Sverdlovsk. One day the road was slow and he was delayed. When he returned to the depot, his boss was storming: "Where have you been? They've been waiting for you!" Manzurova's father was told to gather his belongings. He had an hour to report for a new job. The recruiter gave him a ticket for a train to a coded destination. Manzurov ran off to pack.

There wasn't much question of refusing to go to what the recruiters called a "post office box," meaning a secret site of military importance. With people who hesitated, recruiters threatened to take their passport (without which a Soviet citizen became a vagrant, vulnerable to exile or arrest), their ration cards, or their labor book (which they needed to get paid).[1]

Like characters in a late forties romantic comedy, Manzurova's parents, the young truck driver and the female technician, met on the train heading to Kyshtym. They were dispatched to a lakeside sanatorium, where others were waiting. Food was scarce in those days, and at the sanatorium officials gave the new recruits a generous advance and very good, free meals. No one asked questions, not with those perks.

Manzurova's parents were part of a major recruiting drive to staff the plutonium plant. Recruiters made several trips in early 1947 to surrounding cities
looking for trained workers with at least five years' work experience, including
drivers, machine operators, electricians, plumbers, sanitation staff, janitors, carpenters, lab workers, technicians—anyone, really, with a few years' education
and a clean record.[2]

In a long questionnaire, the new employees were asked to describe themselves
and close relatives in great detail. They had to disclose whether they or family
members had been sentenced or indicted, and whether they had ever departed
from the general line of the Communist Party or been in a Trotskyite organization. Alexander Saranskii, a security officer, was in charge of the personnel division. "I personally accounted for the purity of each arriving recruit's biography,"
Saranskii recalled. "We issued security passes only after a protracted background
check. We reviewed a person's file and sent out special agents to investigate. We
made sure no one had a criminal record, or had been in occupied territories. Even
if a third cousin had been in Germany, we rejected them." Saranskii remembered
his job fondly: "My work was really interesting. I read everyone's personal dossier.
I knew things about people that they didn't even know about themselves."[3]

For memoirists, the day they arrived in the regime zone was a pivotal point in
their lives. One plant veteran remembered arriving by train and seeing the fence
"that stretched out in barbed wire for a long, long time." A fellow passenger
leaned over and said conspiratorially, "That is where they are making atomic
bombs."[4] Others remember being met in the train depot and escorted to the back
of a truck with windows covered to disorient passengers. When they reached
their destination and saw the fence and guard towers, they were sure they had
landed in the Gulag under arrest.[5] Angelina Gus'kova, a doctor, described how
her mother, hearing no word from her daughter, became convinced that she had
been arrested and started to write letters to the local district attorney petitioning
for her release.[6] Nikolai Rabotnov related how his parents, afraid of a search on
entering the closed city, burned all their private letters and diaries. Later, his
mother bitterly regretted those lost letters.[7]

Most people emphasized their immobility on entering the zone. The new
technical workers were banned from leaving the regime zone without the permission of General Tkachenko. For many employees that meant they stayed put,
through family weddings and funerals held elsewhere, through vacations and
sick leave. Even as corpses, workers remained, buried in the new settlement's
cemetery. Manzurova described how after her parents arrived, they married, had
a child, and divorced, all without leaving the zone for a decade. For five years
they could not correspond with their relatives outside. All they could tell them
shortly after their arrival was that they were fine and on a job and would be in
touch later—much later, it turned out.[8]

In telling the arrival story over the years, its finality has been magnified, as if in memory the clank of the gate to the walled city still echoes. Yet there is plenty of evidence to indicate that employees left the city for personal and professional reasons or illicitly to trade or visit relatives. Even so, every memoirist remembers that he or she remained for nearly a decade, as if a prisoner with an indeterminate sentence.[9] It was this feeling of entrapment, as much as the actual fact of it, that left an imprint on memories.

In addition to the fences, much about the secret settlement in the first years was reminiscent of a prison. Living conditions were prisonlike. When recruited, employees were promised good housing and pay, but these promises initially were not kept.[10] Arriving workers were assigned to barracks, where they slept on wooden planks, up to eighty people to a room. Bosses begged for tents to shelter their employees.[11] Some bosses were especially neglectful, failing to secure firewood in time for the first October snow, or building barracks with walls so thin the wind blew through them.[12] The barracks and tents lacked the basics— bathhouses, kitchens, and laundries. The cafeterias served bad food for high prices, so young people cooked in the barracks over wood stoves. They washed in their rooms and dried their laundry over their beds, in corridors, and in entryways.[13] The conditions for free workers, in other words, were little better than those for the prisoners and soldiers in the nearby camps and garrisons.

With the pressure on, bosses focused their attention and resources on meeting production targets. Supervisors were under intense pressure because the new security regime further held up construction schedules. In the fall of 1947, Beria, ever more impatient, sent in more generals, who were followed by even higher-ranking ministers and top scientists, all heading for the forest camps. In the autumn of 1947, Beria sent the chief executor of the atomic program, Boris Vannikov. Vannikov was recovering from a recent heart attack, but Beria made him go anyway, with orders to stay on the site until the plant was finished. Igor Kurchatov, also sickly from radiation exposure, joined him.[14] Vannikov and Kurchatov moved into frigid train cars near the reactor foundation. Beria's right-hand man, Zaveniagin, made frequent inspections.

Beria also appointed a new general, Mikhail Tsarevskii, to direct construction. Tsarevskii was not as well educated as his predecessor, Rapoport. He had no training as an engineer and did not care to sit in his office reading reports. Tsarevskii preferred to be on the jobsite. His first act was to move the construction headquarters from Kyshtym to a birch grove right next to the reactor foundation, where he had soldiers put up a number of green prefab buildings for offices and apartments.[15] There, watching the reactor site, Tsarevskii got down to work.

Tsarevskii was dismayed to see the mammoth foundation looking like a big-budget film set, with a cast of thousands of workers in padded jackets crawling, shouting, and straining in the deep pit. He was appalled that the engineers hadn't bothered to install conveyor belts or any other labor-saving devices. The Gulag bosses shrugged. "Why bother with machines?" they asked. "If we need more prisoners, we just order them up." One of Tsarevskii's first acts was to request a bonfire to burn the wooden litters the prisoners were using to haul wet cement. Without them, the engineers had to fix up conveyor belts, which took a couple of days, but freed up workers for other jobs and increased productivity.[16]

While such changes helped, the tempo did not step up dramatically. There was only so much one general could do. The construction work continued with no bulldozers or excavators.[17] The cement factory, relying on prison labor, could not keep up with demand. Prisoners and soldiers walked to jobsites, and if the sites were far, they often arrived on the job late and tired.[18] The new gates and guards held up the movement of goods and labor between zones.

In November, as the snow and temperatures descended, Beria returned to Kyshtym. He was there to inspect and pressure, to sniff out a conspiracy or sabotage, to get answers about the continuing setbacks.[19] His appearance was torture for the minions who had to service his daily needs. He refused to stay with Kurchatov and Vannikov in their train cars, so he had to be taken back and forth over the icy logs to a hotel in Kyshtym, specially outfitted and supplied for him. Beria studied the situation coolly and decided that the factory boss, Slavskii, on the job a few months, was to be replaced with Boris Muzrukov, the director of a large Urals machine factory that had produced the swift and powerful T-34 tanks during the war. Beria chose Muzrukov because he had experience running a successful factory, but also because Muzrukov knew Stalin personally, which afforded the nervous Beria some protection.

Muzrukov, in his forties, was in poor health. After the war, Muzrukov had fallen ill with tuberculosis, and in 1947 he was still weak and had only one lung. Beria ordered that either Muzrukov or Tsarevskii be on the job night and day, overseeing construction of the reactor, Site A, and the plutonium processing plant, Site B.[20] He made the two men personally responsible for the new deadlines, and in so doing put in motion an administrative engine that finally managed to transform the muddy anthill of human exertion and misery into Europe's first plutonium plant. Beria did so by cementing the fate of the plant to the personal destiny of his leadership. He made it clear that if the plant failed, so would they. The big bosses, fearing arrest, then rushed to attach the same consequences to their harried subordinates. They placed foremen in charge of distinct projects—lumber mill, tool factory, water treatment plant, reactor, processing plants—and made it clear that failures to meet deadlines, as well as mistakes or accidents, would be criminally prosecuted.

This system of personal responsibility was institutionalized in language, as each project was named after the boss in charge. "Dem'ianovich's Establishment" was the radiochemical plant. "Alexeev's Establishment" was the metallurgical shop. Beliavskii's was the cement factory. That is how these businesses were listed in the phone book when one was later published. Each boss was given a pool of labor, including trained civilians, security personnel, and a camp or two of exiles and prisoners. The boss could do with his workers and prisoners as he saw fit. He could reward them, and he could punish them.[21] The head of an establishment received allotments for housing, food, clothing, supplies, and a production schedule with daily and monthly targets. How the boss distributed the goods and made quotas was up to him, but production deadlines had to be met.[22]

The bosses handed down personal responsibility to overseers of shops and labs, who in turn made shift supervisors accountable, each smaller boss holding his subordinates to production targets in a downward-sliding, anxious chain of command. If a shift didn't finish the day's work, then the foreman made the staff stay, toiling into the night to finish. As the bosses cracked their whips, employees drove themselves to exhaustion. Their bosses told them the long hours were right and proper because they were on the front lines defending the nation from bloodthirsty capitalists bent on destroying the world. There was no room in this war for personal lives or physical limitations.[23]

Over this regime of personal accountability ruled Tkachenko and Vannikov, both men greatly feared. Vannikov, who had narrowly escaped the Gulag in 1941, had a special knack for personalizing terror. Inspecting the labs, he would politely ask a worker if he or she had children. When the worker replied in the affirmative, Vannikov would nod, smile, and say, "If you don't finish your assignment, you won't be seeing those children anymore."[24] Once Vannikov turned to an engineer named Abramson (pronounced in Russian "Abram-zone"), who had made a major mistake in a design. Handing over his arrest order, Vannikov reportedly joked, "You are no longer Abramson, but Abram in the zone."[25] I imagine no one but Vannikov thought it very funny, watching Abramson, flanked by Vannikov's two personal marshals, head for the paddy wagon. The story got around and reminded people of the consequences of errors.

Meanwhile, Tkachenko labored like Sisyphus, obsessed with vigilance, spies, and purity. His mission was to root out spies and saboteurs, a job that depended on keeping afloat rumors of spies and saboteurs. Taking security to new heights, he insisted, for example, that workers refrain from holding traditional holiday parades. If they were all to appear at a parade, he reasoned, then a potential spy could count the number of workers on the project, a closely guarded secret. No one argued with him. No one wanted to come into conflict with Beria's personal representative.[26]

In 1948, Tkachenko received orders to cleanse from the security zone "cosmopolitans," people too enamored with the West. These orders went out to security bosses throughout the Soviet Union, but the decree was particularly difficult to carry out at the plutonium plant. Soviet scientists had to balance their orders to copy the American atomic bomb exactly—the chief physicist, Igor Kurchatov, was closely following blueprints stolen from the Manhattan Project, and he regularly requested permission to see the pinched materials—with not betraying a slavish devotion to the West.[27] The answer to this challenge was to target ethnic minorities, especially Jews.

Moishe Pud, for example, the hardworking director of the lumberyard, unexpectedly inherited several thousand dollars from an uncle in New York. Pud quickly renounced the money, donating it to the Soviet government, but that was not enough for Tkachenko. He fired Pud and had him removed from the site.[28] In the subsequent years, Jews gradually disappeared from the security zone for having relatives abroad, or just for being "of the type that litters our environs."[29] Others were cleaned out, too. Vladimir Beliavskii remembered sitting in a mess hall with a colleague named Ivanov. A blond woman walked in and looked their way. Ivanov, noticing the woman, turned to Beliavskii and announced he would be leaving the job now. Indeed, the next day Ivanov disappeared. Ivanov turned out to have really been Schultz, of German heritage, and so not fit for top-secret work. The blond woman evidently had denounced Schultz to Tkachenko.[30]

Plant managers also rarely hired Tatars and Bashkirs, Muslim religious minorities who made up the majority of the population in the towns and villages surrounding the plant, because they could not pass the stringent security requirements. Those who were hired later charged that plant management discriminated against them in allocating housing and jobs.[31] Generally, Tkachenko and his staff interpreted loyalty and trustworthiness in national terms—as largely Russian and sometimes Ukrainian.[32]

Security measures were continual and pedagogical, a form of public theater meant to instruct employees and threaten them into following the rules and working selflessly. People heard the stories of "Abram in the zone" and of colleagues who disappeared into paddy wagons, and they passed these stories around. There was no city newspaper or radio station. Rumors stood in for media, and they taught people that it was just a short hop from the regime zone to the Gulag zone.

Workers put up with the restrictions and fear in part because they were locked in, but also because they were rewarded for their work with better rations. This was the first step toward plutopia. Tkachenko instituted a bonus system in which manual laborers (prisoners and soldiers) received 50 grams of vodka for finishing the day's quota. Soon after the regime zone fences went up, Tsarevskii set up a special cafe for scientists and supervisors. The cafe served candy, fruit, wine, and meat, all without a ration card, as much as a person could eat.[33] Regular workers had no shortages of bread or scarce, highly caloric food—sausage, caviar, and chocolate. Chocolate and caviar, in fact, were easier to come by in the zone than vegetables and milk were. Workers slipped out of the closed zone so frequently to trade their valuable chocolate for milk and carrots that locals took to calling the people of the regime zone "the chocolate people."[34] Word got around about the extra rations, and locals did their best to get a job on the site. The "special regime zone" that imprisoned residents also became a haven, one of the few places in the region where a working person could live, albeit behind barbed wire, without fear of hunger.

Outside the closed city, in neighboring industrial settlements bearing utilitarian names such as Asbestos and Labor, workers finished their shifts, waited to buy gray macaroni, and then disappeared, stooping, into dugouts. In Cheliabinsk in 1948, two hundred to three hundred people lined up daily for bread before dawn, and the line remained until 3:00 a.m. the next day. Seven years later, little had changed. Sallow children walked miles to attend the second or third shifts of their primary schools.[35] Hunger, illness, and crime haunted the towns and cities surrounding the well-fed Base 10.

Thanks to the closed city, life only got worse outside the zone. In the spring of 1948, Tkachenko ordered the creation of a fifteen-mile buffer zone around the plant. He then commanded that this new zone be cleansed of "undesirables: ex-cons, exiles, and people who had lived under foreign occupation."[36] Since the Urals were a major destination for deported kulaks, ethnic Germans, and repatriated citizens, a full 3 percent of the population in the sparsely populated buffer zone made the deportation list.[37] After the deportations, Tkachenko wanted to make sure that no outsiders wandered into the zone and learned something. He decreed that residents of the buffer zone register all houseguests with the police. People caught harboring unwarranted guests would be charged with violation of state secrets.

To stop the flow of people entering the zone on official business, Tkachenko issued one more command, ordering the closure of the region's cultural institutions: a mining institute, a teacher training institute, a nursing academy, several orphanages, retirement homes, resorts, and sanatoria in Kyshtym and neighboring towns and cities. Tkachenko also banned city and village governments from holding sporting or cultural events.[38]

When citizens admired Stalin and the Soviet government, it was often because the state had brought progress to the countryside in the form of educational institutions, hospitals, libraries, and theaters. Despite the hardships Stalinism imposed, many of Stalin's followers saw the payoff to be enlightenment and opportunity. As a consequence, Tkachenko's order to shut down education and culture in the buffer zone was so antithetical to the Stalinist notion of progress that a local party leader gathered the courage to write a letter of protest:

"How can we keep grade schools open if we can't prepare teachers to teach in them? What do we do with the children in the orphanage? The mining school trains workers for the factories in Kyshtym. Who will educate them? The nursing academy prepares nurses to work inside the regime zone. . . . At the TB sanatorium there are *160 patients.* Where will they go?"[39]

Depriving Kyshtym of its hard-won cultural and educational institutions made it, in Soviet parlance, no longer a "city" but a big, stinking, muddy village. Tkachenko answered this letter by showing up in person in Kyshtym to chew out the city party committee for allowing a choir to enter the city and perform right in the town square in broad daylight, against all orders.[40] As Tkachenko lectured and glowered, he underscored how the regime zone and its accompanying buffer zone had become a dividing line between the elect and the forsaken, between the emergent beneficiaries of the expanding military-industrial complex and those who paid its extravagant bills. In the early thirties, Soviet leaders had zoned into existence an unequal system of distribution by mandating that only people with passports and registration cards could live in the better-supplied showcase cities—Moscow, Leningrad, and a few republic capitals.[41] Collective farmers, who were not issued passports, bankrolled Soviet industrialization and urbanization with their wheat and sugar beets, with their crushing toil and pressing poverty. This was part of the plan. In order to keep farmers in place, working and producing, they were deprived of passports and the chance to move legally to a better-supplied city where a villager could get an education and a good wage. In 1948, Tkachenko took advantage of these established invisible lines that zoned prosperity from poverty. In subsequent decades, the towns and villages encircling the plant would supply the regime zone with workers, food, and machinery to support an increasingly affluent community of plant operators and their police wardens, while those in the buffer zone lived with less, much less.

This unequal exchange created a great deal of resentment lasting into the twenty-first century. In 2007, Elena Viatkina, the editor of Ozersk's newspaper, invited me to meet with some important cultural figures of the closed city, held at the same resort in Kyshtym that was first used in the forties to house arriving recruits. In 2007, the resort still belonged to Ozersk and was frequented by vacationing plant workers.

The resort amounted to ten long, low Bauhaus-style buildings along a small, weedy lake. Workers in hospital smocks smoked lazily in the sunshine. Vacationers strolled arm in arm. Older couples practiced a tango in the dance pavilion to the clang of an accordion. At the lake were little sheet-metal changing stalls, a tennis court, and a small beach. The tennis court was missing its net, but a middle-aged couple, pressed into tight Speedos, played anyway. The pair moved slowly toward the ball, the ball slowly moving toward them. Nearly twenty years had passed since the disunion of the Soviet Union. I felt, though, on that summer day, as if I were back in the USSR, in the time of pre-collapse innocence and congenial Soviet-style languor.

This meeting at the resort had taken my friend in Cheliabinsk a month of networking and phone calls to set up. I had looked forward to it, figuring the important cultural figures would enlighten me about the closed city, which I could not visit. On arriving, Viatkina led me into a conference room set for tea. She placed before me a creamy-looking cake. "No calories," she said, nodding encouragingly. The important cultural figures neither nodded nor smiled. They showed me some glossy brochures of marble monuments and monumental apartment buildings that looked unsurprisingly like those found in most Soviet-era cities. With little enthusiasm, they told me about Ozersk's theater and music programs. They told me their city was a wonderful place to grow up: "You never had to lock your doors." "Everyone worked hard." "It was safe." "The old-timers lived to ripe old ages." I had heard all that before.

We ate the cake, and that was it. The important cultural figures gathered to go, just as it dawned on me that I had been given the runaround. I was never to get closer to the closed city than this meeting with minor officialdom, and I had learned nothing significant.

Just then an older man trundled in, a janitor with keys and a mop. He looked to be Bashkir. I asked him what he thought of Ozersk. He shrugged, meaning, *Not much.* One of the important cultural figures briskly corrected him, saying that relations between Ozersk and Kyshtym had always been good, "excellent," in fact.[42]

Viatkina, obviously feeling sorry for me, invited me for a swim in the resort's lake. Viatkina assured me this lake was "clean," meaning free of radiation, but thinking of her "no calorie" cake, I did the breast stroke to keep my head above the cool, brown water. Later Viatkina saw me off at the bus station, where, alone, she spoke more openly. She made a point of saying she was not from the closed city but from Cheliabinsk. Her husband grew up in Ozersk, which is why she lives there now.

Like most residents, Viatkina described the first time she entered the closed city. It was 1987, and she was eight months pregnant and living in Cheliabinsk when her husband had a serious car accident while visiting his parents in Ozersk. With difficulty, using connections, she got a pass to enter the closed city to visit him in the hospital. But she had to figure out how to get there. She was told to go to the train station in Cheliabinsk and find a bus stop on Postal Street, behind a warehouse. After nightfall, Viatkina located the street, but she saw no bus stop. She asked people walking by which stop was to Ozersk. When others heard her utter the name of the secret city, they steered sharply away from her. They turned their backs on the tear-streaked woman heavy with child, violating a powerful

Russian social code to help strangers in need. Exhausted, Viatkina wandered in the dark until she came across an unmarked building, and in front of it, an unmarked bus, engine idling. Already wiser, Viatkina asked if the bus was going to "the city." The driver nodded. Viatkina boarded, sat down, looked at the other passengers—well dressed, well fed, calmly reading—and burst into tears.

16

Empire of Calamity

In June 1948, Igor Kurchatov sat in the control room of the first Soviet production reactor, A, or "Annushka," as it was lovingly called. Finally, one year behind
schedule, on June 10, Kurchatov pulled the switch that lifted the control rods out
of the reactor face.[1] The men cheered at the darting wattage indicators. For the
scientists, the dials illuminated the path to the Soviet "nuclear shield." For the
rest of the world, when they learned of it, these were the first lights of the costly
and risky Soviet-American arms race. For posterity, the hum of Annushka's turbines let loose a geyser of radioactive isotopes, the product of nuclear technology
hitched to endemic Soviet poverty.

On June 19, the reactor was fully loaded and Kurchatov, later admitting he
was in a hurry, gave the order to operate at full power.[2] That evening Kurchatov
called Beria to report that Annushka was fully operational, but the call was premature. Annushka ran for less than twenty-four hours before an operator noticed
that water pouring from the reactor was thirty times more radioactive than the
permissible level. Apparently, cooling water levels had dropped too low in several canals, causing uranium fuel slugs to overheat, rupture, and spurt radioactive steam. Fearing an explosion, Kurchatov dropped the control rods back into
the graphite reactor. He called Beria to tell him the bad news. Tersely Beria asked
how long it would take to start back up.

For the next three weeks the scientists fretted over how to fix the ruptured
slugs, working around the clock while the irradiated uranium emitted harmful
gamma rays.[3] Beria was not concerned about workers' health. In general, the
leaders of the atomic project displayed a cavalier attitude toward the dangers of
radiation. General Zaveniagin sat on a stool in the reactor hall in his street clothes
and ate a mandarin out of his pocket. The plant director, Boris Muzrukov, stood
alongside him, aware of the danger but afraid to leave the general's side. Later
dosimetry readings of Muzrukov's house registered an exposure ten times
greater than the permissible norm.[4] Exposing oneself to radioactive contamination was part of the unwritten code at the plant. In June 1948, Tkachenko
penned this worried denunciation of Kurchatov: "Academic I. V. Kurchatov at

times ignores all rules of safety. He personally goes into the premises where the activity is exceptionally higher than the acceptable norm. Comrade E. P. Slavskii behaves even more carelessly."[5] Kurchatov, Tkachenko continued, descended into the reactor chamber even as the alarm was signaling radiation 150 times higher than permissible levels, his bodyguards unable to stop him.

In mid-July Kurchatov again had the reactor going at full power, though the problem of fractured fuel slugs remained.[6] Ten days later more slugs in the reactor blistered and burst, provoking another crisis and more telegrams to Moscow. This time, however, Kurchatov kept the ailing reactor running, which continued to leak radioactive isotopes. The men called the cracked and radiating fuel cells "goats."[7] With this household slang they assimilated the "emergency situation" and domesticated its dangers into the daily working order. Kurchatov ran the hazardous and polluting reactor until January 1949, when Soviet scientists estimated they had enough plutonium for exactly one bomb. Only then did Kurchatov shut Annushka down. Engineers calculated they needed a year to dismantle the broken reactor and repair it. Beria gave them two months.

The staff had to decide how to get to the damaged fuel cells out of the reactor. If the reactor was working normally, they would have dropped the irradiated slugs into a pool below the reactor, where the slugs would cool. But the Soviet Union's entire stockpile of uranium was loaded into Annushka. If operators dropped all the slugs in the pool, they would lose the good slugs with the bad and have no fuel to produce more plutonium for a second and third bomb. Rather than waste the precious uranium, Beria and Vannikov ordered workers to unload the reactor by hand, sorting the cracked slugs from the undamaged ones to be reloaded into the reactor.[8]

It is hard to imagine what it meant to gather the courage to enter the central hall of a reactor and pull irradiated slugs from the reactor face by hand. Everyone took their turn—prisoners, deportees, soldiers, operators, supervisors, and scientists. Even Kurchatov snatched a gas mask and ran in. Returning from the reactor face, the men gulped down a cleansing glass of vodka while fighting back a dizzy, sick feeling.[9]

In the first thirty-four days of 1949, Kurchatov and his staff unloaded and reloaded thirty-nine thousand irradiated uranium slugs. Hundreds of men suffered nausea and nosebleeds followed by intense pain and knee-buckling fatigue. At the time, the official tolerance dose for a year was 30 rem. Cleaning up Annushka, workers received doses from 100 to 400 rem.[10] Four hundred rem is enough to produce early "radiation aging," which leads to chronic exhaustion, painful joints, and crumbling bones, and concludes in cancers and diseases of the heart and liver.

After sorting, the first batch of irradiated uranium was cooled underwater. Plant engineers knew that it was best to cool the slugs for 120 days to reduce by a thousand times radioactive iodine and other short-lived, harmful isotopes. However, the plant leaders were in a hurry and abbreviated the cooling period to thirty days, forcing them to process "green," or highly radioactive, fuel.[11] They built five-hundred-foot smokestacks to channel the radioactive gases high into the atmosphere, hoping that this would diffuse the contaminants safely over a large territory. The winds directed the contaminants in an arrow-shaped path in a mostly eastward direction over fields, pastures, lakes, swamps, and streams.[12]

After cooling, the irradiated slugs were ready for processing and went to Area B to be dissolved in nitric acid in order to distill the resulting toxic cocktail into plutonium. But the new processing plant, Factory No. 25, was not yet ready to begin processing. The engineers were still working out the plant's design and equipping it. Faina Kuznetsova, an original lab technician, remembered how security officers pressured her supervisor to finish faster. They took his pass and kept him at the factory, under guard, until it was ready. "What could he do alone?" Kuznetsova recalled. "Of course, we all stayed to help." For twelve days and nights the staff remained at the factory, racing to prepare the shop.[13]

In the late forties, Soviet biophysicists believed that workers in chemical processing were safe because plutonium and its long-lasting by-products do not give off gamma rays, but far weaker alpha and beta radiation, which cannot penetrate skin. By contrast, they knew that reactors were dangerous, issuing gamma rays strong enough to penetrate skin and directly irradiate a person's vital organs. The relationship between gamma rays and health was direct: after a strong gamma dose, a person immediately felt unwell, and Soviet researchers learned that larger doses caused lab mice, rats, and dogs to drop dead.[14] It took several years for Soviet researchers to grasp the harmful effects of ingesting radioactive substances, and before they did they hired mostly young women, right out of high school, to staff the chemical processing plant.[15]

To women, who had worked in conventional chemical plants and had learned to interpret danger as fire, smoke, and noxious smells, the chemical processing plants appeared safe. Nor were there the usual hazards of a factory worker's life—finger-chopping lathes, bone-crunching cranes, or swinging blades. Meanwhile, the young women made good workers. They were usually single. They had come of age in a wartime labor climate where to arrive at work just twenty minutes late was a crime. They were disciplined, accurate, and responsible.[16] More important, they were available.

In December 1948, the new, specially built Factory No. 25 was ready for plutonium processing. Soviet designers had sought to hide the factory from aerial detection, so rather than copy the design of the ship-sized T plant at Hanford, they built the plant vertically to make a smaller footprint. As a consequence, the processing chambers were stacked on top of one another, with pipes flushing radioactive solutions and vents with radioactive gases running up walls and across ceilings. These design features meant that if leaks or spills occurred in one area, radioactive substances could drip down to workstations below, expansively spreading contamination.

There were a lot of spills, starting with the day the plant opened, when a crowd of scientists, security, and military men gathered into the final chamber to see the first plutonium solution emerge from the vast still. At the appointed time nothing filtered out. They waited longer, the scientists nervously discussing the technical processes. Zoya Zverkova, the shift supervisor, was checking and rechecking the instruments, the generals behind her menacing. Everyone knew in those years that failure was the result not of accident or miscalculation but of enemies and saboteurs.

Finally someone noticed yellow pulp dripping from a vent in the ceiling onto the men. Investigating, they found that the plutonium solution had bubbled into a foam and been sucked into the factory's ventilation system. In search of plutonium, workers climbed to the roof and scraped up the precious residues. The scientists made changes, ran the process again, and were happy to see precipitates drip into the filter. Checking its composition, however, they found the solution held no plutonium.[17] Finally, a third bath landed plutonium in the still. The first two failed attempts, however, meant that plutonium was everywhere else: inside chambers, vents, equipment, vessels, and the control room, and even on the rubber galoshes of the generals.[18]

Running through the plant was a large, cement-lined "canyon" in which radioactive solutions were to move from chamber to chamber along remotely controlled conveyor belts. When the plant was built, the canyon was sealed off with "stones," thick safety doors, which were supposed to remain in place permanently after start-up. Soviet engineers, however, did not know how to produce metals that could withstand the heat and corrosive qualities of radioactive solutions. They plated beakers, cups, and equipment with gold, silver, and platinum, hoping they would hold up to radioactive toxins. But the precious metals, as well as rubber stoppers and gaskets, corroded before the powerful heat, toxic chemicals, and alpha particles of the radioactive solutions.[19] A month after start-up, a pipe containing plutonium solution cracked and leaked onto guards at a door. Soon leaks sprang up throughout the plant.[20]

Many spills occurred inside the sealed canyon. Since the spills contained valuable plutonium, the bosses demanded that staff go in and sop up the solutions. Entering the highly radioactive canyon violated the most basic safety regulations, but workers rolled the stones aside and descended into the canyon anyway. Once the stones were pushed aside, they stayed that way.

"Everyone went into the canyon many times," Faina Kuznetsova remembered. "It seems strange now, but no one had planned for cleaning up spills. There was no method to safely collect spilled solutions. We had only washcloths, buckets, and sometimes rubber gloves. We mopped up the spills and poured them into big glass bottles. It was a very expensive compound and we were expected to recover every drop. Our spills weren't too big, from fifty to one hundred liters, but there were spills in the earlier stages of processing that lost as much as two to three tons of solution. To collect those spills with washcloths was impossible. Those were real disasters."[21]

I. Dvoryankin described entering the canyon: "We worked without any protection other than gas masks. One by one we climbed down to the canyons. When blood began pouring from our noses, we pulled on the ropes to be brought up. We received extremely high doses of radiation, but thanks to our work, the plant was not stopped."[22]

Why so many spills? Kuznetsova blamed the haste and the strict regime of secrecy and fear. Security officers closely watched the young workers and kept track of the valuable gold-plated tools and final product. Young, inexperienced technicians caught violating the rules or making mistakes were transferred to the zone's labor camps for two to five years of hard labor.[23] "When we were hired to work at Maiak," Kuznetsova related, "we were not warned about radioactivity. We didn't even know what it was. That is why we handled the radioactive solutions. We were afraid only of the KGB [the successor to the NKVD]. Everything was done under the personal control of L. P. Beria and his envoys and they would convict for any blunder. And so fear pushed people to take steps that led to accidents. On top of that we worked with very expensive equipment and chemicals. They kept a close watch over the machinery, the gold and silver lab vessels. They cared more," Kuznetsova remembered bitterly, "about that equipment and the final product than they did about people."[24]

Beria had taken note of how a common worker, David Greenglass, had reproduced technical documents at Los Alamos and passed them to his Soviet handlers. He did not want Soviet employees walking off with plans and formulas in similar fashion. As a consequence, he decreed that there could be no charts or manuals to guide workers, nothing that could be copied down and stolen. While training for operations, employees were required to memorize the complicated network of plumbing, electrical schemas, and machinery in their sector. They also had to commit to memory the procedures of their workday. "People were in a constant state of stress," Kuznetsova remarked, "fearful lest they forget something important. And frequently they did forget, especially in the first period."[25]

There were other reasons, too, for the frequency of accidents. Soviet propagandists used to plaster a slogan on the side of buildings: "Cadres make the difference." The slogan points to the inherent problems that came from building the first Soviet plutonium plant with incarcerated workers, illiterate guards, chemists who got their degrees from culinary schools, and engineers who used sledgehammers to repair expensive machinery.[26] The official history of the Maiak plutonium plant records three accidents during four decades of operation.[27] In reality, however, accidents trailed Soviet plutonium production like a loyal dog, from the first days of operation.

After the process of separating plutonium from uranium, the plutonium solution was delivered to Area V, where workers transformed plutonium in liquid form into metal ingots, and finally into the achingly desired softball-sized orbs of weapons-grade plutonium for a bomb core. In February 1949, the first flasks of plutonium concentrate were ready for final processing into metal, but the specially designed chemical-metallurgical plant was still under construction. Rather than delay, plant managers ordered construction workers to make over a couple of old navy warehouses in a nearby village to serve as a makeshift metallurgical plant.[28]

These shops, Nos. 4 and 9, looked just like any other chemical laboratory, with wooden tables, glass cabinets, beakers, and stainless-steel sinks. At the workshops, the employees, mostly young women, processed radioactive solutions by hand in vented cabinets or simply on tables. For lack of stools, lab technicians rested on wooden bins containing radioactive waste. They poured solutions from vat to beaker, beaker to test tube. They stirred the coagulated goo in platinum cups. They ground radioactive powders on high counters. They walked with solutions down hallways to burners and ovens to calcify, roast, and dry them. They carried buckets of radioactive waste down the same halls, past toilets, canteens, and offices.

These employees had known work since they were children, and they toiled as at any factory or farm. It was one young man's job to carry glass flasks of just-radiated solutions from Factory No. 25 to Workshop No. 4. He put the flasks of plutonium in a bucket and slopped them over. Waste brigade teams lugged barrels of radioactive solutions to the forest not far from the plant, and, treating it like any other garbage, burned the coagulated gels. They stood over the fires, raked the coals, and tossed the ashes into shallow pits.[29] Blue-collar workers did not know they were working with radioactive solutions. They knew the elements only by coded number. The young women were given only the most basic instructions to stir, heat, and pour. The pale blue production manual was locked in a vault, and only supervisors with special permission had access to it.[30] Many workers, however, guessed from rumors that they were working on an atomic bomb.[31]

As the lab technicians started to become more anxious about the products they handled, the bosses disabused them of their fears. Kuz'ma Chernyshov, head of Workshop No. 9, told his staff they had nothing to worry about. To reassure them, he would hold up a flask and ask, "Want a lick?" He repeated his joke so often that employees started to call him "Wanna-Lick." Another boss would push his staff to work faster, telling them, "Uncle Sam won't wait. Hurry up!" Even if they had been fully aware of the dangers, the young workers probably would have carried on anyway, out of patriotic duty. The bosses told their workers that the country was still at war: "People died at the front," they said. "This is also the front."[32]

At every step along the plutonium production line, Soviet workers, with their supervisors leading the way, were exposed externally and internally to radioactive and other toxic contaminants. Many of the intricate devices and state-of-art technology built by rushed, tired, and harried workers on shoestring budgets failed, broke, or never really worked from the start. The automated trolleys transporting radioactive solutions to the chemical processing plant jammed, and workers had to climb down under the radioactive tracks to fix them. Plumbing filled with radioactive effluent clogged in narrow, twisting passages, which regularly had to be opened by crews with steel rods pushing the deadly solutions along.[33] Because of design flaws, staff often had to put their heads into glove boxes and directly inhale toxic substances. Eventually the employees dispensed with the boxes and evaporated plutonium in the open.[34] Workers climbed into vents in search of plutonium dust. Filters got clogged and had to be cleaned out by hand. Rubber stoppers fell apart and choked pipes, which plumbers had to cut open, clean out, and weld back together. Radioactive waste and radioactive equipment were carelessly dispensed or left in rooms where people worked. In some of these rooms, radioactivity reached 100 microroentgens a second, meaning that without any extra accidents, staff received ten times the already generous acceptable limit at the time.[35] In the first year and a half, 85 percent of all workers received more than the permissible dose. Exposures got so bad that in May 1949 a plant doctor, A. P. Egorova, boldly sent a letter to Beria complaining about the "underestimation of the leaders of the Object to the fact that workers are getting irradiated."[36]

Fighting these technological snags, the young, fleetingly trained employees rushed to make their deadlines. As they did, a beaker fell with a crash on the stone floor, a bucket was kicked over, a hand slipped into a solution, a valve was left on, or two barrels placed too close together went critical and exploded.[37] The staff tended to refer to these unclassified releases in shadow fashion, as "spills" (*utechki*), "crumbles" (*possypi*), "dispersals" (*vybrosy*), "hotbeds" (*otchagi*), or "slaps" (*khlopki*). Many of these events went unmonitored, unmeasured, and unreported, in order to stave off the security men with their black vehicles and pitiless investigations. These unnoted incidents were the usual fare of the dangerous reality of factory work, but as these episodes occurred in the world's second plutonium plant, they gradually became part of a hazardous and invisible geography.

As the young workers met for meals in their work clothes, hands unwashed, they laughed and talked. Most employees went home in their work garments, spreading radioactive contaminants as they went. Workers at Factory No. 25 lived in the emerging plutonium city. Operators of Factory No. 20 resided in their own settlement, called Tatysh, nine miles from the main settlement. They lived just a ten-minute walk from work, in sight of the ugly yellow plumes of the plant's smokestacks.

Larisa Sokhina remembered Tatysh fondly for its abundance and beauty. "The village was surrounded by lakes and a forest, which was full of mushrooms and berries. The industrial Lake Kyzyltash had the most fish. The fisherman loved that lake, though they could not fish there for long."[38] Unlike many smog-filled industrial cities of the Urals, the fifteen-mile buffer zone around the city, created for security, dished up an accidental nature preserve. "The impression that we were depressed, dispirited and frightened," Sokhina recalled, "is wrong. We were mostly young people at the plant. We were energetic, joyful, and full of life." After miserable years of war, the young employees played volleyball and basketball and arranged skiing competitions, hiking trips, and cookouts. They formed a wind orchestra and had parties and dances, spinning on Saturday nights into a welcome oblivion. Cut off from their kin, the young employees remade family in the most immediate, nuclear way: couples married in modest little weddings, and as the young brides became pregnant, they kept working, their bellies swelling over the lab tables.

Plutonium plant workers after a secret church wedding, 1948. Courtesy of OGAChO.

This panorama—the village next to the plant processing "green," highly radio-active fuel, the makeshift labs with no safety features, the workers walking home in radiated clothing spreading a path of contagion, commercial fishing in Lake Kyzyltash as it became a radioactive reservoir, the birch leaves in the village shimmering with gamma rays—was a calamity spawned of ignorance, haste, and a sense of mission that spared no room for individual safety. Saving money, plant leaders cut the budget for radiation monitoring, and so few grasped the dangers surrounding them.[39] The young employees in their hamlet surrounded by the snow-muffled silence of a Russian forest had no idea that they had already stepped downwind of their fate.

17

"A Few Good Men" in Pursuit of America's Permanent War Economy

In 1946, Paul Nissen, editor of Richland's *Villager*, wrote that Richland was a "nervous, wondering community." Residents were nervous not because they had discovered that they had been living alongside a plant producing the world's most dangerous and volatile material or that they were vulnerable to enemy attack or a plant explosion. Their fears were of a more ordinary variety: anxiety about their jobs and the survival of their city. With the war over, Nissen wrote, Richland's "purpose for being [was] suddenly shot out from under it and [people] worried about what, when, how and if the blow would fall that would make it another ghost town."[1]

Indeed, the future did not look bright for the shiny new atomic city. After the victory celebrations, the number of Hanford employees dropped by half, workers were dismantling neighborhoods of prefab houses, and many local businesses were closing.[2] At the plant, engineers worried about the health of the first reactors. The piles had contracted a condition called "graphite creep," caused by swelling graphite blocks that clogged the uranium fuel chambers. Powerful chemicals colluded with radioactive isotopes to corrode pipes and pop open the canned uranium. AEC officials worried the reactors might fail "any day" and they would have to shutter the plant.[3]

In the American national imagination, ghost towns are associated with the West because many western towns, built in boom-time haste, went belly up as mining and farming ventures failed. Nissen emphasized the green lawns and whispering sprinklers, the pretty suburbia in the desert, in order to link Richland's future with plutonium, because plutonium production made for a very special economic relationship. Richland existed solely on one product. The government monopolized plutonium's production as well as consumption.[4] Government officials in Washington, D.C., made decisions about plutonium on a planned basis, without reference to market forces, behind closed doors, for reasons kept from the public. As political scientist Rodney Carlisle put it, the whole process was similar to the command economy in the Soviet Union.[5] In 1946, Nissen was well aware that without plutonium's injections of federal money—which kept the water pumping, the rents paid, and the $69 million in paychecks flowing—Richland's local economy would roll off with the tumbleweed.[6]

Nissen wasn't the only one fretting. Ralph Myrick grew up in Richland in the forties. Myrick's family came from a hardscrabble coal town in Texas, unlovingly named Gamerco after the coal company that owned the town and everything in it—the little shotgun shacks, the mines, the company store, the slag heaps, and the trucks chugging by. His parents worried when their boy came home bruised and scraped from battles with Mexican boys on the other side of town. When Myrick's parents learned of jobs in eastern Washington on a government project, they were only too happy to escape, strapping their possessions on top of an old Ford and rolling north like a snapshot out of *Grapes of Wrath*.

Myrick's father found work in the electrical switch house at the plutonium plant. He got paid well and the work was steady. Most of his life he had had to worry about layoffs and falling behind on the bills, but in Richland, for the first time, he could save. Myrick's father was assigned a two-bedroom prefabricated plywood house in Richland for $35 a month, utilities, maintenance, and furniture included. Myrick's mother cried when she first entered their new house. She

Richland prefab. Courtesy of Department of Energy.

had never lived anywhere so new and clean, with appliances and plumbing, in a pleasant town with good schools filled with kids from nice families.

Myrick remembered that his father worried a lot: that the plant would close after the war, that the need for plutonium would be satiated, that he or his children would do something wrong and he would lose his job. Everyone knew that if an employee was fired from the plant, he had a month to move out of Richland.[7] Myrick's father had never finished high school, and he knew that nowhere else could he provide for his family so well with his skills and education.

Myrick need not have fretted, because there were people in high places campaigning for a permanent American war economy that would sustain Richland and communities like it for the rest of the century. In 1944, Charles E. Wilson, the charismatic CEO of General Electric, spoke to top military leaders in the frescoed banquet hall of the Waldorf-Astoria.[8] Hinting at the development of nuclear weapons, Wilson said, "The power to destroy will be greatly magnified," and added that it would come quickly, with little notice. Wilson called for a new program of postwar weapons research and development, a program that was "once and for all, a *continuing* program, and not the creature of an emergency."[9]

Wilson, who was drawing his GE salary while serving as the head of the War Production Board, spoke as much from the perspective of a corporate executive as that of a government functionary. He lectured the military men that "the leaders of industry are as much the leaders of their country as are the generals, the admirals, the legislators, and the chiefs of state." Referring to the Senate's Nye Committee, which investigated the tremendous profits of weapons manufacturers such as DuPont during World War I, Wilson warned: "Industry must not be hampered by political witch hunts, or thrown to the fanatical isolationist fringe and tagged with a 'merchants-of-death' label." To encourage industry, Wilson said, businessmen needed financial guarantees "supported by congress, by regularly scheduled and continuing appropriations."[10]

Wilson was the kind of charismatic leader who could get his followers to march around a flagpole singing "Onward Christian Soldiers."[11] Amidst the chime of porcelain and the whispers of waiters, Wilson envisioned a globally ambitious future of postwar American muscle backed up by corporate big science. Navy brigadier general W. F. Tompkins was so taken by Wilson's vision that he obtained copies of the speech and sent one to Vannevar Bush, head of the Office of Scientific Research and Development. Tompkins wrote Bush that they should have a town-hall-style meeting to toss around ideas for just such a council. Bush replied that the federal government already had the National Academy of Sciences, the National Advisory Committee for Aeronautics, the OSRD, and other government agencies connecting science and industry with the military. Tompkins persisted. He sent Bush a second copy of Wilson's speech. Bush finally agreed to a meeting, at which the men decided to form a committee along the lines Wilson had suggested. They called it the Wilson Committee, and Brigadier General Tompkins proposed that Wilson chair it.[12]

Working swiftly, by August 1944 the Wilson Committee proposed a new Research Board for National Security that would oversee large federally funded labs where corporate contractors would carry out weapons research while retaining profits from patents and commercial applications derived from the publicly funded research.[13] The clause on retention of patents made government contracts potentially very profitable for American corporations.

Wilson worried, however, that secret weapons work offered to educated employees "very few attractions."[14] Wilson decided they would need to actively cultivate "good men" with "ample leavening." Money, status, and continuity were to serve as the leavening agents: "pay scales must amply compensate for work less appealing than normal scientific or technical posts." Offering a sample of "highly attractive" pay scales averaging twice that of university salaries, the report stated that these higher salaries were only for "young men of high caliber and promise." "Lesser individuals" were to be paid "with less differential for attractiveness."[15] The Wilson Committee's report dwelled obsessively on incentives for attracting and keeping "good men," distinguishing them from "lesser men" so much that the larger goal—national defense—got lost. The rhetoric of "good men" also masked the welfare qualities of Wilson's plan for permanent federal subsidies to finance corporate research.

Crafting a blueprint for the military-industrial complex was not a bad day's work for Charles E. Wilson, who left government service to return to GE soon after he chaired the Wilson Committee.[16] In 1946, GE took over the management of the troubled Hanford plutonium plant in hopes of generating patents for a future civilian nuclear power industry. GE executives oversaw the plant's vast expansion, and in subsequent years, GE's portfolio rocketed skyward.[17] Wilson cycled back into government service in 1950 at President Harry Truman's request to chair the Office of Defense Mobilization, where Wilson led negotiations to hand over yet more military-industrial research and development contracts to private corporations, immensely expanding U.S. military budgets.

Wilson did more than merely service his own and his shareholders' fortunes. Wilson's words echoed across the silver-plated service of the Waldorf-Astoria into a gold-plated future. He reflected a revolution in thinking about the role of business and science in national defense that profoundly altered American history.[18] Wilson's vision helped transform Richland from a temporary wartime project into a permanent metropolis flourishing at the crossroads of technology, corporate management, and government funding. In the subsequent decade, corporate executives in Richland rewarded its salaried employees so generously and built up the temporary "village" so solidly that Richland acquired among its residents a permanence not originally intended by its founders.

Richland was also saved from abandonment with a little help from the Soviets. In 1947, officials of the newly formed Atomic Energy Commission suddenly discovered that they had no stockpile of nuclear weapons, just as Soviet political forces were consolidating power in East Germany, Poland, and Czechoslovakia—the very places that, AEC officials knew, contained uranium deposits. Panicking, AEC leaders resolved to double plutonium production in five years.[19] Five years was a crucial interval because it was the shortest American projection of how long it would take the Soviets to produce a bomb. Exhorting the need for "wartime urgency," AEC chairman David Lilienthal allocated $85 million for GE to build three replacement reactors, a new processing plant, and one thousand new ranch houses in Richland.[20] The resulting construction boom rescued the Richland economy.

Planners estimated they would need between fifteen thousand and twenty-four thousand construction workers, none of whom who would be allowed to live in Richland. To accommodate temporary workers, GE planners built a short-term construction camp on a sandy strip of land just outside Richland, called North Richland. They moved former barracks from the defunct Hanford Camp and the convict Columbia Camp there and set up racially segregated pre-fab houses, barracks, trailer parks, and a cramped Quonset-hut school running on shifts.[21] North Richland housed single men, migrant construction families, and racial minorities. North Richland was also home to taverns, brawls, and crime, all carefully cordoned off from the nuclear families in Richland by the GE-managed police force, the Richland Patrol.[22]

AEC officials believed that commission projects should be handed over to well-chosen individuals, "good men" with broad powers to act with "complete responsibility and authority." Without "total decentralization," they theorized, "only chaos, confusion and lack of accomplishment can result."[23] And so AEC officials gave GE executives wide powers of discretion and independence in the expansion project. GE executives in New York also believed in decentralization and gave their Hanford managers a great deal of authority and autonomy.

That proved to be a mistake. GE engineers were untested in the field of "nucleonics." They had never supervised construction projects so large and complicated.[24] When offered design help, GE executives rejected it, but they were slow to draw their own designs and begin construction planning. Within a few months, GE managers at Hanford were behind schedule, and AEC officials began to worry openly about GE's competence and reliability.[25]

The picture darkened in January 1949, when GE accountants sent a curt note to AEC officials announcing a tripling in the cost of the new 234-5 processing plant, from $6.7 million to an astonishing $25 million. A junior high school in Richland, originally estimated to cost an already generous $1.7 million, had doubled to $3.3 million.[26] No American public school had ever cost so much.[27]

Flabbergasted, AEC officials asked to look at the books and realized that GE accountants had allocated $17 million of the extra costs to "overhead," "contingencies," and "project engineering costs."[28]

Shoddy accounting, cost overruns, and poor management were, for the corporation, great business. GE had a cost-plus contract with the federal government, which meant the company earned a negotiated percentage on the total cost of contracts. The higher the final tab mounted, the greater the profits.[29] Reviewing the GE construction program, AEC secretary Roy Snapp reported that GE charged $41 million of the $85 million project to "overhead and distributives," nearly a 50 percent return on the company's no-risk, investment-free endeavor. Meanwhile, because of AEC's policy of extreme decentralization, AEC officials in Richland stood idly by as the bills mounted, unable, they said, to control GE managers.[30]

After the press got wind of the story, a congressional inquiry into AEC mismanagement of funds was started, and Congress instituted a cap on AEC spending. In the midst of this gathering storm, GE was saved again by the Soviets, who on the Kazakh steppe were making the final arrangements to test the first Soviet atomic bomb. When the American press reported the test in September 1949, GE's extravagance was quickly forgiven. Senator Brien McMahon argued that big budgets were the price America had to pay for security: "The fact that an overrun exists does not disturb me at all," he said. "There should be overruns, I think, in this business."[31] In October, American analysts predicted that the Soviets would make two bombs a month in their secret plutonium plant in the Urals, and President Truman asked the AEC to double its atomic bomb production.[32] The Senate dropped spending curbs imposed on the AEC. The AEC in turn granted GE an additional $25 million for construction in Hanford and Richland, more than enough to pay for the cost overruns.

The local, business-minded *Tri-City Herald* cheered, calculating the numbers of new jobs and the paychecks that would enrich the region.[33] Headlines trumpeted Richland's largest school budget on record, twice the per capita national average, and counted up the number of new houses, new paychecks, and new businesses.[34] Just as Charles E. Wilson had hitched his corporation's fortunes to permanent defense work, local business leaders and politicians grasped the possibilities of plutonium as the engine of regional growth for this long-sidelined corner of Washington State.[35] Congressmen Hal Holmes and Henry Jackson and Senator Warren Magnuson eagerly used Hanford to justify a continuing current of federal dollars to their state. They were not alone. Across the West, the urbanizing frenzy of the postwar period was pushed along by defense dollars.[36] In nuclear affairs, the West became what historian Patricia Limerick called "the center of gravity."[37]

Few regions won this contest as handily as eastern Washington. In 1950, Republican Hal Holmes, a fiscal conservative, boasted that Washington State's

Fourth Congressional District received more federal grants than any other district in the country.[38] The money came in not just to manufacture plutonium but to sustain an entire infrastructure designed to shore up the community that made plutonium. The items on the regional wish list, for which local boosters had long lobbied, were suddenly granted when justified by the assembling conflict with communism. In the fifties, the federal government built a series of dams (for flood control and electricity for arms plants), "national defense" highways and bridges (for emergency evacuation), and regional army and naval bases (to defend the plant in case of attack), and it implemented large-scale irrigation projects and agricultural subsidies (for national self-sufficiency in case of war).[39] All local politicians, businessmen, and job seekers had to do to get a share of this incredible public generosity was overlook the 570-square-mile nuclear reserve, fenced off and guarded—to forget about, in the parlance of contemporary B-movies, the Godzilla lurking in the desert.

It wasn't a hard sell—nuclear weapons for nuclear-family prosperity. In the early fifties, as news flowed in of spy scandals and global skirmishes with Soviet forces, few Americans questioned the need for atomic bombs.[40] Richland, an especially vulnerable Soviet target, celebrated the atomic bomb.[41] Rather than approach atomic weaponry with dread, people welcomed the bomb as part of the community. Richland residents bowled at the Atomic Lanes, snacked at Fission Chips, and strolled under the icon of whizzing neutrons at the Uptown Shopping Center. Richland teenagers adopted the mushroom cloud as the mascot for the Columbia High Bombers. At pep rallies, they danced around a three-foot-high green and gold rocket, while city leaders at civic celebrations joyfully detonated mock "Little Boys."[42] Once the smoke cleared, the adults went back to their work, peacefully producing weapons of mass destruction.

What was harder to sell was the notion of government subsidies and regulation dictating the terms of daily life in the conservative political climate of both the interior West and the inner corridors of the GE Corporation at the height of anticommunism. Local newspaper editorials railed against communist-style big government, high taxes, social welfare programs, and government control. Like a mantra, accounts of the liberating qualities of private property and free enterprise resounded through the city's coffee shops, community meetings, and editorials. GE managers attended obligatory annual training sessions where they took in the message of a "better business climate" and "right-to-work" laws while extolling the virtues of American democracy, defined as a free enterprise system, and freedom from oppressive taxes and intrusive government programs.[43]

To square the contradiction of small-government political conservatism with the fact that eastern Washington was quickly becoming the nation's largest per capita recipient of federal funds, GE managers used the blinding glare of the Cold War. Faced with the sinister dawn of global communism, watching the

Atomic Frontier Days, XM rocket float. Courtesy of Department of Energy.

high-stepping Russians lead the Koreans, Chinese, Hungarians, and Poles on a mission to take over the world, Americans began to see the costs of making bombs as a patriotic sacrifice. Subsidies for Richland's middle-class comforts were justified by calling the comfortable bedroom community a "Critical Defense Area."[44] The *Tri-City Herald* glorified plutonium work, comparing Hanford operators to their courageous and daring Indian-fighting ancestors.[45] This trope of the frontier concealed the region's growing dependence on federal subsidies and corporate oversight.

Dependency came with a price. Even in the flush years, people in Richland never ceased to worry that federal largesse would evaporate and their pretty city would follow the path of other western ghost towns. After GE's first building boom, terminations followed. Workers decamped to the construction project on McNary Dam or sought to hire on with the Bureau of Land Management constructing irrigation networks. In 1952, AEC commissioned GE to build two more reactors and another processing plant, PUREX, and again thousands of workers streamed in looking for jobs, only to be laid off eighteen months later. For the rest of the century, Richland and its neighboring communities suffered from the boom-and-bust swings of large-scale government projects followed by periods of downsizing and worry.[46]

And that was the blister point. When a massive government project was not under way, there was little else to sustain the bloated regional workforce on the sparse ecological and economic terrain of the Columbia Basin. In addition to nationalism and a "patriotic consensus," the fear of unemployment, recession, termination, and economic stagnation pushed business leaders, political representatives, and their constituents in eastern Washington to embrace the bomb and all of the infrastructure—roads, bridges, dams, schools, airports, levees, and irrigation networks—that could appear to support national defense.[47] Perhaps that is why in Richland people feted the bomb with reckless abandon and an insensitivity that shocked outsiders. The desire to keep the government-stimulated communities alive led residents to blithely exchange the possible dangers of radioactive contamination for the certainties of growing prosperity, bankrolled by an expanding federal government, which, as they grew more dependent on it, they politically derided.

18

Stalin's Rocket Engine: Rewarding the Plutonium People

In Moscow in 2007, at a small exhibit on the sixtieth anniversary of the Soviet atom bomb, rare footage of the 1949 Soviet test played against the wall. In the declassified footage, the explosion did not have the infamous stocky column of the first American bomb, Trinity, blossoming skyward above a bouquet of radioactive gas. Instead the Soviet bomb lifted thin, twisting fingers of Kazakh soil, like a fist opening to the sky, sending the earth hurtling to the heavens, transcending time.

The first Soviet test occurred in great secrecy, but it is hard to hide a nuclear explosion. Kazakh farmers stepped out of their huts and watched the blast, enjoying the pyrotechnics, the curious heat, and the powerful winds. Scientists in Europe noticed the seismic activity. U.S. Air Force pilots circling the Soviet perimeter picked up telltale radioactive debris in filters placed in their planes for such an eventuality.[1] In Ozersk, however, no one announced the first test or congratulated the plutonium workers. In general, Moscow propagandists left the veneration of the mushroom cloud to the Americans and instead trumpeted the "peaceful atom." In the early fifties, Soviet physicists began designing the world's first "civilian" reactor, Kurchatov's pet project. Journalists emphasized the vast difference between the creative violence of capitalism and the pacific technologies of socialism.[2] Soviet citizens believed their bomb was not a weapon of destruction but a "nuclear shield" against capitalist aggression. In Ozersk, residents saw the colossal effort to build nuclear weapons as an act of personal sacrifice. They were the front line defending the globe from the horrors of nuclear apocalypse.

After the first test, nothing changed in the status of the handful of closed nuclear cities in the Soviet Union. They remained cloistered and unmarked. Ozersk existed as a phantom city with fifty thousand phantom residents. Yet an important feature of the crash bomb-making program was the promised rewards. Soviet culture had a very public quality in which citizens outwardly demonstrated their dedication to the nation and the government publicly rewarded them for their service. People wore their medals with immense pride. A person gained a lot of respect, as well as a seat in the front of the bus or a spot at the head of the line, because of a ribbon on a lapel. The question in Ozersk was, how did one reward phantom heroes?

Beria had a prepared a list of medals and cash for the leading scientists on the project. But what could a Soviet citizen in a closed city buy with cash? And what good was a classified medal that could not be displayed on one's chest? Nor could scientists publish their discoveries or engineers patent their inventions. One scientist suggested to Beria that cars and dachas would be better prizes than cash.[3] Beria agreed to have special dachas built in the closed zones for leading scientists, and he added other perks, too.[4] The scientists' children could attend any Soviet university and travel for free on public transportation throughout the Soviet Union. But these, too, were hollow gifts for people locked into a zone.

Many residents felt the reward should be liberation. With the bomb finished, young employees figured they had done their service and they deserved, after several years of captivity, to leave the ramshackle forest outpost for life in a city with services and opportunities.[5] The dream of upwardly mobile Russians was to get a registration card for Moscow, the glistening, glimmering seat of power, culture, and enlightenment.[6]

Social mobility in the Soviet Union was spatial because rights, freedoms, and opportunities were attached to birthplace and residence. A successful person levitated from village to settlement, settlement to city, and city to provincial capital to finally arrive in Moscow. Ozersk, in 1949 a claptrap settlement called Base 10, was an aspirational cul de sac, and a lot of people wanted to leave.[7] It became clear, however, that there were to be no mass departures. Construction of two new reactors was in full swing, and leaders in Moscow were asking for more plutonium and spending billions of rubles—a quarter of the national income—on defense.[8] The plutonium settlement was there to stay, and so were its residents.

There had to be another way to reward residents and keep them happy. In 1946, Stalin first suggested to Kurchatov that he could offer a comfortable life in exchange for a bomb. "Our state has suffered very much," Stalin reportedly told Kurchatov, "yet it is surely possible to guarantee that several thousand people can live very well, and several thousand people better than very well, with their own dachas, and with their own cars."[9]

This promise of a good life for a few sounds counterrevolutionary. Bolsheviks overturned the existing order in the revolution of 1917 in order to do away with the privileges and entitlement of the Russian elite and build a classless society. Stalin's anointment of a few good men with dachas, cars, and affluence signaled an intervention in the goals of the revolution. Stalin was celebrated in part for this reason—because he normalized the radically egalitarian revolution into a meritocracy. Literary critic Vera Dunham called this shift the "Big Deal," an unspoken compact with the middle classes to grant them a peaceful, prosperous private life, enriched by individual consumption, cultural pursuits, educational opportunities, and upward mobility in exchange for loyalty and submission.[10]

Stalin had in mind rewarding top scientists and the most important managers. To continue manufacturing plutonium, he had imagined a modest garrison of two thousand soldiers.[11] Kurchatov, however, radically altered this plan in requesting a new special city as a reward for plutonium. In 1948, he gave a speech in Base 10:

> And to spite them [our enemies abroad] [a town] will be founded. In time your town and mine will have everything—kindergartens, fine shops, a theater and, if you like, a symphony orchestra! And then in thirty years' time your children, born here, will take into their own hands everything that we have made, and our successes will pale before their successes. The scope of our work will pale before the scope of theirs. And if in that time not one uranium bomb explodes over the heads of people, you and I can be happy![12]

Kurchatov's speech served as the founding proclamation for transforming the closed settlement into plutopia. He was offering nothing less than a nuclear Big Deal: middle-class urban affluence for working-class operators in exchange for the risks of plutonium production. As Vera Dunham observed, in the Big Deal of the 1930s the Soviet working classes got left out of the arrangement, remaindered to dead-end jobs, their children, too, detoured from education to farms, factories, or apprenticeships that were akin to indentured servitude. In the late 1940s Kurchatov stunningly reconfigured the Big Deal. He declared that the benefits of the new city would go to everyone, so the working class could live like the middle class.

But why buy off the working class? Trained managers and specialists were scarce and difficult to replace. Regular employees, on the other hand, were expendable. There were plenty of critical defense industries—rocketry, aeronautics, armaments—where only the management were rewarded.[13] So why enrich plebian plant operators?

In 1948, G. Safranov, the general procurator of the USSR, announced that the Soviet Union was nearing the completion of a classless society. Most Soviet citizens could judge for themselves the veracity of the official statements, but cut off as they were from the rest of Soviet society, Ozersk's residents did not recognize official perfidy, or at least they pretended not to. In 1949, when economic planners announced that the country's postwar supply problem had ended, a party member in Ozersk complained, "Now that the country is nearing an abundance of consumer goods, the unsatisfactory situation of trade and supply in our city is unendurable." Then, as if on a chessboard, the speaker made the next move: "The trade question includes a political character; it creates an unhealthy political mood among the workers at the base."[14] Someone was bound to utter it—the prospect of unhappy, agitated plutonium operators. Like a club arrested in mid-air, the speaker ever so lightly dropped the threat of labor unrest in the security-sensitive, highly volatile plant.

Nor was the threat idle. In Soviet society, workers, underpaid and overworked, were far from submissive.[15] Soviet managers watched their subordinates anxiously, the dark faces emerging from factory gates, the spitting, muttering men alongside seething, quarrelling women, people who had labored with hardly a pause since childhood, whose muscles were as taut as their nerves. The postwar years were an especially jittery time. Working people noticed the redistribution of wealth to the managers who issued fines, boosted their daily targets, and sped up the belt. They were bitter that they had sacrificed for the war but were still going hungry. Some wrote enraged letters to leaders complaining of the local elite, how they commandeered cars and apartments, how their wives dispatched their chauffeurs for supplies and never waited in line themselves. They described starvation wages, suffocating taxes, and their children "chased out to work at age nine." Some directed their hatred at the "American-English jackals," others against Jews ("American hirelings"). Many people despised "the bosses" closer at hand.[16]

These smoldering grievances ignited randomly. In nearby Cheliabinsk in the spring of 1948, thousands of workers abandoned their shifts to stand in line for bread. They worried because thousands of farmers had filtered into the city to try to buy back the grain they had grown but were not allowed to keep. The country and city folk got into fistfights. Brawls broke up mealtime in cafeterias. In the sharp March breeze, rumors circulated. Some said the shortages were because Russian grain was going to Hungary, Romania, and Finland. Others said it was just another sign of how "the bosses only care about themselves."[17] Just outside the closed city, a band of hungry farmers armed with shovels, knives, and pikes attacked a collective farm. They killed a few party leaders, stole sacks of food, and disappeared.[18] "Banditism," meaning crimes against the state, skyrocketed in the postwar years, quenched only with a lot of extra policing, including random searches of forty-five million homes.[19] Officially, the working classes were celebrated in the workers' state, yet in rough urban ghettos and smoldering villages, workers were often feared for their volatility and treated like the enemy.

Imagine such spasms of working-class wrath in the halls of a reactor or a radio-chemical plant burdened with highly radioactive waste. Imagine the damage a few angry workers could do. The military men who ran the plant did not have to imagine. They witnessed firsthand the combustible qualities of single men and women unhinged from their larger communities—they saw it daily in the plant's unruly Gulag camps and soldiers' garrisons. Their answer to the threat of battle-hardened

Family of plant workers, Christmas 1959. Courtesy of OGAChO.

vets and convicts producing weapons of mass destruction was to exchange them for something far safer, far more secure and stable—nuclear families.

In 1948, Kurchatov's promises of fine shops and theater sounded like fantasy, but within a year the gritty colony indeed boasted a jazz orchestra, a choir, drama and movie theaters, and new brick apartment buildings, which in Soviet visual culture signaled "urban." There was also a store that sold furs and boots. A person, however, had to put on Wellingtons to wade through the mud to get to the fine shop (in a former barracks), and the rows of the movie theater sloped backward, so viewers in the last rows had to stand to see the screen. The new theater produced plays that were "pale, formulaic and lacking artistry."[20] The new apartment buildings leaked, the walls slanted, mold spread under peeling parquet floors, the plumbing coughed, and the electricity sparked. A few new buildings were so poorly constructed that they were dangerous to enter.[21]

Just after the plant delivered the first bomb cores, a steady murmur rose among residents about their deplorable living conditions: "The question of housing and residential construction," a party member complained, "has reached crisis proportions."[22] Residents grumbled that the settlement had expanded willy-nilly, with no master plan to accommodate the growing population. Crooked little lanes of shacks and prefab pine cabins took the course of a tipsy driver. A swamp on the edge of town dished up great clouds of mosquitoes and pools of excrement. Sewer, electric, and water lines serviced only a few

Ozersk's first movie theater. Courtesy of OGAChO.

buildings. In the exchange of plutonium for prosperity, city leaders would have to do better.

City leaders tactically chose the months after the first nuclear test to request more building funds. In December 1949, Beria's First Main Department signed off on government subsidies to make over the grim, prisonlike Base 10 into what they planned would be a first-class "socialist city."[23] Soviet generals did not think to call Ozersk a "village," as DuPont executives had named Richland. In Soviet parlance, "village" had no romantic connotations. It meant mud, ignorance, and poverty. Soviet leaders sought instead to build a new, modern city—paved, electrified, enlightened, clean, and convenient. Creating a socialist city in the middle of nowhere, however, took time.

And in the meantime, "the chocolate people" became increasingly vocal. There were not enough preschools and kindergartens for the rapidly multiplying babies. The primary schools were short on teachers and textbooks. A quarter of students were failing basic subjects such as math and Russian.[24] With no high school, education ended at the seventh grade.[25] The city had no park or playground and little entertainment, nor enough dentists, tailors, bakers, shoemakers, butchers, and tailors. With a perpetual labor shortage, Ozersk had half as many workers as were needed for service and janitorial jobs. At the newly opened hospital, nurses and doctors had to make beds and clean up after surgeries themselves.[26] Six maintenance men raced to keep order in the city's public buildings, but forty people were needed. Public spaces, courtyards, and paths abounded with trash, broken glass, and rotting food. Construction debris, piles of earth, uncovered trenches, and discarded building materials were scattered everywhere, hazards especially on moonless nights.

Short on child care, working parents had to leave their children home alone. Elsewhere in Soviet Russia, extended-family members watched children, but grandparents were generally not allowed in Base 10. After school, kids and teens hung out, thought up pranks, got hold of beer, and caused trouble. Sometimes it took days for overworked parents to notice their kids were missing or hospitalized.[27] Toddlers, too, were at large. They fell into ditches, ran from dogs, or wandered lost in the swamp.[28] One small boy, Yuri Khors, became the poster child for unattended children. Roaming alone, he broke a shoulder falling into a trench and got a finger amputated by a heavy swinging door.[29] One woman from Ozersk told me in passing that in the early fifties a truck ran over and killed her four-year-old son. I asked how that had happened.

"He was out playing in the street, and the driver didn't see him."

"Out alone?" I asked, with sensibilities from the other side of the twentieth century.

"Well, yes. He was four by then. You know, already grown [*uzh bol'shoi*]."

Ozersk and the other closed nuclear cities were the first communities dedicated to nuclear families in the Soviet Union. The problem was that the Soviet nuclear family walked onstage before its cue. An extended Russian family looked after itself, while nuclear families are impossibly needy. The new cities lacked the services, appliances, and shopping conveniences that would substitute for extended families. The city's housing and supply crisis fused with rising expectations about the state's proper role in stepping in as next of kin to aid the isolated nuclear family—what Ozersk residents called "the growing demands of the Soviet people."[30] The needs of nuclear families were both physical and emotional. In demanding more goods and services, people sought to buy their way out of the dependence and loneliness of life orphaned in the zone.

To make their case for better services, parents and educators pushed children to the front as the target of concern. They worried that unaccompanied minors were budding hoodlums, the more disquieting as this was the first generation of Russian working-class kids freed from work, their childhood suddenly prolonged an extra decade. Children got in trouble, they said, because they had nothing to do. It was the unsupervised kids, a teacher argued, who committed "crimes unworthy of our Soviet society."[31] Parents and educators demanded after-school programs, sports, dance and music education, nurseries, playing fields, gyms, swimming pools, and more and better schools, all to produce, they said, good citizens who could eventually make the "product." In the wake of these demands, a new child-centered society emerged in triumphant, spectacular, entitled anxiety.

Responding to these demands was no small feat because the city had to be self-sustaining. Federal funds poured in to build a bread factory, a heating plant, a dairy, a meat-processing plant, a department store, cafés, restaurants—open not just to the bosses but to anyone. Residents volunteered for a city beautification program planting saplings that would also serve as cover from snooping American planes. A stadium rose under a conifer roof. Construction workers pounded away on several new apartment buildings, schools, preschools, and health clinics. The breadth of progress was audible, sensible, in the pulse and steady beat of hammers, the rumble of the city's first bulldozers, the excitement of the crowd dashing out to see the city's first slab of pavement.[32] Masons fashioned a marble embankment along the lake, a real promenade so that city dwellers could take an evening stroll. Engineers drained the swamp to make a city park. The city got an orchestra and operetta. First-class actors and musicians arrived from Moscow and Leningrad (though the theater's director, Tkachenko complained, was a "provincial").[33] Graduates from the best universities showed up to teach in the schools.

Far more important than medallions and prizes for a few, this crash building program enhanced the lives of all residents. Over the next decades, the scientists and engineers who were trained in Moscow, Leningrad, and Cheliabinsk, along with the working-class residents who had long dreamed of leaving the mud and dust of village life, came to live in a way commensurate with their importance in Soviet society. Kurchatov's city of culture and affluence would come to mean a great deal for the people who lamented their exile in the Urals, who placed their health in harm's way, and who agreed to a future of indeterminate indenture. The nuclear deal wasn't just a cheap bribe, an exchange of housing space and sausage for risky work. Kurchatov's city came to stand in for the larger society and the national project residents were working to defend: the construction of socialism. It took a decade, but as Kurchatov's vision materialized, the residents of the plutonium city would finally achieve socialism, if secretly, off the map, and only in one city.

19

Big Brother in the American Heartland

Once Richland looked like it was going to stay, its very permanence revealed to outside observers a troubling tumor on the American landscape. Seymour Korman of the *Chicago Tribune* traveled to Richland in 1949 to investigate the construction of a $3 million junior high school. He wrote a scathing article describing Richland as a "police state," with frightened residents whispering to Korman in his hotel room for fear of being overheard.[1] *Time* joined the chorus, depicting the mayor railing against the GE corporation's "benevolent dictatorship."[2] A visiting team of sociologists called Richland a "mutant" community.[3]

The critics had a point. GE managers and AEC planners did wield an impressive, Big Brother–like control over residents' lives that uncomfortably echoed McCarthyite images of the Soviet enemy. The U.S. government, acting through the AEC, possessed all the land in Richland except the cemetery. It owned the houses and shops, the streets, the hospital, the schools, the parks, the telephone and electrical system, and the police and fire facilities. Acting on the AEC's behalf, GE lawyers managed the city with largely unchecked powers.[4] GE community relations managers, for example, assigned houses or took them away at will. It helped to have influence in the Employment Office to get a job and in the Housing Office to rent a house in Richland. These connections paid off. A Richland rental amounted to a big bonus. Residents paid rents well below market value for modern houses with furniture, appliances, maintenance, and utilities included.[5] Residents had only to call Village Services when they needed help with unpacking dishes, laying rugs, or babysitting.

Richland had no free enterprise. Corporate architects drew up the city's master economic plan without public input and classified it as secret. Like DuPont managers before them, GE accountants calculated per capita demand for goods and services and then chose businesses by type, vetting them for financial and political security.[6] Once in business, proprietors had to submit monthly statements to GE accountants and hand over a percentage of profits.[7]

Businessmen feared complaining about their contracts, lest their leases be withdrawn.[8]

Richland had no free press. Paul Nissen, a former army censor, ran the *Villager*, founded by DuPont and continued by GE. Nissen later said he knew the kind of news the "top-brass" wanted and he gave them a veto over any story, promising not to "run around madly screaming about 'freedom of the press.'"[9] In exchange for a docile fourth estate, GE provided Nissen with free office space, materials, printing, and supplies.[10]

Every resident in Richland received a free paper, whether or not they subscribed. Project officials pressed commercial contractors, beholden to GE, to buy advertising space. "They had no choice," Nissen wrote. If they resisted, "they heard about it."[11] As one merchant put it: "I need advertising like I need another hole in my head, but I don't want anyone to get the impression that I'm uncooperative." So much advertising revenue poured in, Nissen soon lowered the ad rates.[12] After Nissen covered his expenses, the paper's profits funded the community organization, Villagers, Inc., which paid for sports, recreation, and entertainment programs exclusively for Richland residents.

It was a lovely arrangement. The highly subsidized Villagers, Inc. appeared to be a grassroots, voluntary organization, as corporate leaders insisted, but members did not have to sweat over bake sales. The community's programs were paid for by prosperous merchants granted a monopoly to do business in what Nissen claimed was "one of the richest spots in the country." Evidently merchants didn't mind footing the bill. As one man joked to Nissen, "I do so much business that it keeps me round-shouldered carrying the money down to the bank."[13] Nissen admitted that as editor, he didn't have to work too hard. The city's corporate managers took care of the paper's ad accounts and books, and corporate public relations staffers wrote most of the paper's "news."

The community organization had other uses as well. GE placed their employees on the board of Villagers, Inc. to keep an eye on the community. One informant in Richland told me confidentially that the head of recreational programs and his staff reported to Richland's corporate-managed police force about rumors, suspicious activity, and seditious remarks. I asked Annette Hereford about it; she worked in Richland recreation programs for much of her career. She angrily denied the charge, rejecting this Orwellian depiction of her hometown.[14] Yet from the security files I learned that corporate police kept a close watch on Richland's unions.[15] And that comes as no surprise. American corporations had long used private police and hired informants to control factories and company towns.[16] Officers of the Richland Patrol were GE employees.[17] Residents were warned continually about the severe penalties for spreading state secrets. Corporate bulletins encouraged employees to join community groups but cautioned citizens against belonging to "political" organizations, especially those engaged in civil liberties and international causes, for any well-meaning group could be a communist front. Roving FBI agents managed site security. The security division spent nearly a quarter of its budget to spy on employees and residents. That budget had line items for "technical surveillance" (listening devices) and "physical surveillance" (tailing).[18] Agents routinely knocked on doors to ask residents what they knew about their neighbors and if they had noticed any suspicious behavior. That made people nervous. Rumors went around that when a wife asked too many invasive questions the FBI removed that family overnight.

Visiting journalists, congressmen, and sociologists puzzled over Richland. With no democratic institutions, free press, free market, or private property, what about Richland was American? "Is it socialism," a team of sociologists asked, "or fascism?"[19] Concerns that Richland had qualities akin to Soviet totalitarianism caused grave anxieties—and not only about national security. Fascism and totalitarianism were on the minds of many people worried about the New Deal expansion of the federal government. The popular columnist Westbrook Pegler warned Americans that the enlargement of the executive branch threatened to create a fascist state in America.[20] Economist Friedrich Hayek theorized that state planning led societies directly down the road to serfdom. State planning, Hayek argued, was where Nazi Germany took a detour.[21] American conservatives regularly used the charge of socialism to push back or shut down government-funded programs. Tarring Richland as socialist constituted a somber threat to the city's subsidized affluence. In response, city leaders worked hard to reconfigure definitions of citizenship and democracy to justify state largesse for their select community.

In 1949, under fire from critics, GE managers set up an Advisory Community Council and school board as token representative institutions. At the first meeting, the council passed a bold resolution stating that as American citizens, Richland residents had the right to incorporation, self-governance, and free markets. GE managers wasted no time squelching this winged battle for democracy by calling the council members in for a talk. Afterward, GE lawyers sat in on council meetings, and no issue was too trivial for corporate oversight. Even dog walking and parking ordinances had to be approved by GE management to become law.[22]

Since most council members were GE employees, few people felt they could resist GE managers when the latter overrode the council, fired the superintendent of schools, or removed a rebellious school board member.[23] One member called the Advisory Community Council a "puppet group." As a consequence, local elections were lackluster. Candidates often ran unopposed, and public meetings were poorly attended. When asked about their political apathy, residents responded, "Why bother to vote for those people? They can't do anything anyway."[24]

Why did Americans put up with these restrictions on their civil rights and democratic freedoms? Paul Nissen explained it this way: "The question is as simple as, 'Do you like your present job?' 'Don't you want to live in that nice house and this pretty little city longer?' 'Are you likely to want to do anything, at least publicly, which will put you in a position where you will have to do any fast explaining either to GE or the AEC?'"[25]

Working to squelch the drift of negative public opinion about subsidies, AEC and GE managers proposed a major rent hike in 1949. They claimed the increase was to prepare Richland residents for eventual privatization and incorporation, though at the time they had no intention of giving up control of the federal city.[26] Instead, rent increases were for public consumption, a performance of fiscal austerity. It was good public theater especially because locals ideologically distrusted publicly subsidized housing. In neighboring Pasco and Kennewick, citizens loudly rejected low-rent public housing as "the first step down the road to complete socialism."[27] In Richland, AEC officials tried to bill the rent hikes as a step for residents in shouldering the burden of paying the real costs of their housing and community services, which, AEC officials admitted, were "exorbitant."[28]

Richland residents, depicted in the press as disenfranchised, frightened, and timid, proved to be anything but when faced with a threat to their financial security. After the rent hikes were announced, they organized petitions, wrote letters to the editor, lobbied their senators, and showed up for meetings to heckle and boo the big brass.[29] They suggested that if the rent hikes went through, they might just walk off the job like construction workers had in North Richland. Union workers tossed off phrases such as "agitation" and "unrest among employees." The suggestion of a strike or "loss of enthusiasm in work performance" at the plutonium plant got managers very nervous. They begged the gathered crowd not to take any rash action. Finally AEC officials backed down, postponing the rent increases "indefinitely."[30] With their collective protest, several thousand Richland residents had the rent hikes rescinded, delayed, or reduced for years.[31]

The debates and protests about the rent hikes in Richland drew the attention of neighboring towns. For many the fact that Hanford workers earned 15 percent more on average, had subsidized rent, and paid no local taxes felt like an injustice. A self-identified "farmer-tax-payer" wrote the *Tri-City Herald*: "They [in Richland] get top wages, as we all know. They probably think they are doing us a big favor by working for GE. Well, I figure we are doing them a bigger favor by raising their fruit and vegetables, plus taking their load of taxes."[32]

The suggestion that Richland residents did not pull their civic weight triggered an explosion of responses to the paper; long, literary conflagrations in which writers argued that without Richland, neighboring Pasco and Kennewick would be nothing but dustbowl ghost towns with "that farmer co-piloting a horse and buggy." Richland writers asserted they deserved low rents because of the sacrifices they made living in this "jerk water town." They wrote that they had earned subsidies because they won the war and another war might be coming. Most tellingly, all the writers assumed that the farmer would move to Richland if he had a choice, but he didn't because he lacked "the mental aptitude to obtain a Q clearance," or, as another quipped, "maybe he's just plain too darn old."[33]

This exchange shows the power of zoning space by class, race, and occupation. The zones designating Richland for permanent employees and other areas for temporary and low-paid workers transformed territory in ways that appeared to naturalize difference, as if people outside of Richland were inherently "wooden-headed" and those inside were gifted and bright. It was this sense of superiority that stuck in the craw of Richland's neighbors.

Indeed, Richland residents embraced a self-image of a community of scientists and engineers, educated, urban, and cosmopolitan, contrasting with the blue-collar farming communities surrounding Richland. Yet in the fifties, most Hanford employees were working-class, had no more than a high school degree, and were doing shift work.[34] Despite this demographic reality, Richland residents, GE propagandists, and later historians assumed that Richland was, in fact, middle-class.

Amidst the early Cold War invective against the communist workers' state, the American working class acquired a reputation as a political and security threat. GE, long a "right-to-work" company, mechanically distrusted unions. The company worked with FBI agents to smear union activists as Red, treacherous, and disloyal.[35] Union-bashing could be useful. When a second cost overrun scandal threatened GE in 1952, Hanford managers rushed to blame million-dollar delays on union "foot-dragging" and "feather-bedding" to dispel a picture of corporate padding and profiteering from government contracts.[36]

Making the working class disappear was another, even more effective way to battle working people and organized labor. Calling working-class Richland middle-class, setting rents so that this group could live like the middle class, and passing regulations mandating a neat, orderly middle-class appearance all contributed toward making the working class of Richland disappear, a mirage that lasts to this day.

The only problem was that a lot of people still acted like common workers, so GE executives tried to change behavior, too. They instructed managers and supervisors to become community leaders in local organizations so as to spread the GE message and mores. Corporate handbooks instructed hourly employees not to "chew the fat" on the job, but to "listen and be quiet." They reminded workers to bathe frequently, watch for dandruff and halitosis, and, on greeting someone, say a dignified "How do you do?" rather than bellow "Glad to meetcha!"[37] The employee handbook, in short, did not hide a disdain for the common man.

This sentiment created a fault line running through Richland, one alluded to by former shift workers. One man remembered that his parents had a chance to move to a bigger rental house in a neighborhood where management lived, but his mother refused to go: "She didn't want to live with those people." A former union steward sighed and laughed as he characterized the GE management: "They were so arrogant." A retired electrician recalled the dances the salaried staff had in the management club. When I asked if he had ever attended one, he chuckled and said he had gone there a few times—to fix the electrical system. But these are just cracks in a larger memory of a united, wholesome American community. Most people remember the social harmony, the community-mindedness, how "everyone was the same."[38]

Historian Jack Metzgar argues that the American national memory has been created and shaped by middle-class professionals who appropriated the working classes, spoke for them, and subsumed them in an amorphous classless American society.[39] In Richland, as working people began identifying with their middle-class bosses and their aspirations for education and social mobility, they were less likely to regard the working people outside Richland as kin, to mobilize as a class, to resist or question, or to strike over workplace safety or health issues. Workers went on strike at other AEC sites, and GE executives were bedeviled by striking workers in the late forties and fifties, but GE had no strikes in Richland, just as it had no spies during the war or whistle-blowers in subsequent decades. A sociologist conducting a poll in Richland in 1952 found to his wonderment a "dominant universe of contentment."[40] Residents thought their city safer, nicer, and more prosperous than surrounding cities. They liked the city's planned streets and orderly corporate management. They liked its homogeneity. One resident remembered how comforting it was that everyone's parents "made about the same amount of money and so we all lived in the same quality of housing, ate the same quality of food and were able to dress about the same."[41] Most residents wanted nothing to change in their classless society except perhaps to have more places to shop.

Richland residents were a mobile, car-owning community. In Richland, a car helped to span the long distances between sprawling lawns and wide streets under a hot sun or sharp wind. Residents wanted to drive everywhere, but parking and traffic were problems. In 1948, the Richland Chamber of Commerce lobbied GE and won a new shopping center with ample parking, one of the first postwar strip malls in the nation.[42] The new Uptown Shopping Center replaced the old downtown as the community gathering place. Downtown merchants grew upset about the competing shopping center and demanded that GE managers buzz-saw the city's prewar town square with its rare shade trees to make way for a parking lot next to downtown stores because "people won't walk to shop."[43] GE managers at first resisted, appalled, but citizens organized a petition, and within a few months the town green, the traditional open space for democracy, was paved over.[44]

The new parking downtown and at the Uptown Shopping Center served other purposes, too. The large carton-shaped, cement-block buildings of the mall doubled as bomb shelters. The asphalt acreage surrounding them made for a firebreak in case of a spreading inferno. The encircling five-lane arteries served for emergency evacuation. In the fifties, civil defense considerations slipped imperceptibly into American architecture.[45]

There were miles of open space and streams surrounding Richland, but residents wanted parks, pools, and playgrounds for organized, supervised play. As in many American towns at the time, residents were concerned about juvenile delinquency, especially teens who had time to drag-race, drink, and dance in disturbing ways.[46] They believed Richland had a particular problem with delinquents because the city had few grandparents to watch children while parents worked swing shifts. In compensation, parents demanded and received a child-centered community with "superior" schools, plus science and nature clubs, dance classes, and after-school recreational programs.[47] To combat unruly children, city leaders censored city shops selling comic books and established curfews, and members of the Richland Patrol visited Sunday-school classes. Ill-behaved children reported to their ministers for counseling, and really bad youth were sent out of the city to a home for juvenile delinquents.[48]

Still, critics complained that the federal-corporate control of Richland wasn't "normal."[49] As home ownership spread across the United States in the fifties, the fact that well-paid GE employees enjoyed low-rent public housing was irksome, even suspect. For many Americans, democracy and home ownership belonged together. As William J. Levitt, the founder of Levittown, put it, "No man who owns his own house can be a communist. He has too much to do."[50] Congress in the early fifties pressured AEC officials to incorporate Richland and sell resident-employees their houses, so as to make it a "normal" community. AEC officials resisted, however. If residents owned their houses, AEC and GE managers would no longer be able to control who lived in Richland. Nor could they monitor workers as easily. When employees tracked home radioactive contamination, GE management could commandeer a rental house, clear it, clean it, and move the family elsewhere, all in less than a day.[51] Instead of incorporation, AEC managers announced a plan for a progressive "disposal" or privatization of Richland over many years.[52]

Residents again balked at this threat to their financial security. In polls in 1952 and 1955, voters rejected "privatization," "disposal," and "self-government" by margins of three to one.[53] Renting, they argued, was cheaper, and they feared that if they bought their houses and the plant closed, there would be no jobs and no market for their real estate investments. They voted, in essence, for the conveniences and especially the economy of government ownership and corporate management over private property, a free market, and local democracy. In turning down incorporation, residents also took a pass on the right to self-government, free speech, free assembly, and an uncensored press. No one said it because it would have breached security regulations, but by living in Richland, people also gave up rights to their bodies, placing urine samples on the front stoop in the morning and undergoing compulsory annual medical exams.

Richlanders' resistance to self-governance was puzzling. Why would people make plutonium to defend American democracy and capitalism but oppose it in their own community? One man explained it this way: "We speak up about rent hikes because we are Americans."[54]

This statement is, at first glance, perplexing. What is American about demanding federal subsidies for private, single-family housing? In the backdrop of these contests over Richland's rent subsidies pulsed the frantic scramble for more nuclear weapons alongside a dawning realization that American cities, lit up in magazine depictions with the lurid light of a nuclear fireball, were no longer safe.[55] In Richland, the Red-hunter Al Canwell warned audiences that communists were everywhere, probably in their city, too; they had to be alert.[56] Against this tableau, the battles in Richland over rents and shopping centers seem trivial, but to dismiss them is to miss the point.

As the ideological match between the United States and Soviet Union sharpened, American polemicists latched onto Soviet poverty and shortages as evidence that communism was flawed. Data showing the superior consumption of average Americans drove this point home. As American workers followed the debate, they came to see a rising standard of living as an important right for citizens. This message meshed with a long-standing conservative argument about free enterprise and consumer freedoms as the cornerstone of democracy. Richland epitomized this ideology, what historian Lizabeth Cohen calls the "Consumers' Republic," where activist-consumers gave way to consumer-citizens who were told their civic duty and democratic freedoms lay in shopping. In Richland, then, it was logical for residents to seek to cash in their physical security for financial security, to exchange their civil rights for consumer rights, their freedom of speech for the freedom to pursue prosperity.[57]

In this regard, Richland residents were not exceptional. Across the nation, white men won federal subsidies via the GI bill and FHA loans. Increasingly segregated suburbs pumped with federal aid for roads, schools, and infrastructure multiplied across the landscape, while inner cities fell into decline. As consumption came to define American freedoms, Richland and its poorer neighboring towns exemplified the inequities, spatial exclusions, and hierarchies of postwar America. In Richland, rising up over rents was important because in their role as consumers residents of Richland could win back a bit of their lost voice and power. Disenfranchised, under surveillance, and silenced, Richland residents could safely express their opinions only about issues such as parking, dog walking, and shopping. These disputes, though seemingly insignificant, mattered a great deal because they simulated the motions of American democracy, for which people in Richland were putting their lives on the line.

20

Neighbors

C. J. Mitchell came to the Tri-Cities from east Texas in 1947, following the rumors of good jobs. Mitchell left the segregated South to find that in eastern Washington Jim Crow had followed him up north. Mitchell's first job was putting up barracks in North Richland for construction workers. As a black man, Mitchell could only find a place to live in the Pasco ghetto, a collision of shacks and trailers between the rail lines and Second Street, an area shot through with mud-packed yards and heaps of garbage city workers failed to collect. No shade or lawns graced the settlement. The wind kicked up at night, slamming doors. A few toilets and spigots serviced the community, one for every eighty or so people. Mitchell stayed in a pup tent outside the trailer his uncles rented for $100 a month, three times more than a two-bedroom house cost in Richland. Mitchell worked construction, building new ranch houses in Richland, where he could not live.[1]

The Hanford plant produced very different destinies for those who lived near it. As Richland gained in federal support, neighboring communities often lost out. Federal allocations to the West created a terrain where, like class, race was etched onto the landscape, invisibly, but powerfully and with lasting effect.

When he founded his newspaper in Pasco in 1947, Glenn Lee, the ambitious publisher of the *Tri-City Herald*, promoted an image of three united communities, but the newspaper's moniker never really took. People in the farm town of Kennewick, the railroad town of Pasco, and the atomic city of Richland resented, feared, and at times hated one another. The high schools' sports rivalries were legion. Tensions got bad enough that the GE community relations division started in Richland a "Hello Neighbor Day" to try to warm up people in nearby towns.[2]

But a day of music and games could not patch up the main source of friction—that Richland used neighboring communities as dumping grounds for transient workers who flowed in like jetsam with each major construction boom at Hanford. The migrant workers drank, fought, spat, and swore; they were dirty and stank, committed crimes, and moved on, or at least that was how locals saw them. Race played a role in these perceptions. About 70 percent of laborers employed by the GE construction subcontractor Atkinson Jones were African American.[3]

To keep black workers from settling in Kennewick, city leaders established a curfew, banning African Americans after dark. Sherriff Ward Rupp was serious about the curfew. Once, after catching a black man in Kennewick on the wrong side of sundown, he tied him to a post and called the Pasco police to collect him.[4] In Richland, city leaders did not have to resort to Deep South tactics. They had a more genteel and impersonal solution. They explained that Richland was open to all races. No person of color lived in Richland because GE and the AEC had no employees who ranked high enough to qualify for housing there. It was a matter of education and rank, not racism, they said.[5]

With Kennewick and Richland restricted, most of the area's two thousand black residents lived in Pasco. In 1948, after the Supreme Court struck down racial housing covenants, the American Civil Liberties Union (ACLU) sent a couple of investigators to Pasco. The investigators documented the conditions of the black community living on a "weed-ridden, dust-choked" five acres on the eastern margins of the city. They described the reeking outhouses and ice-coated spigots. They documented how city leaders failed to connect water and sewer lines because they wanted the black community to come up with $5,000 to help pay for it. The investigators took note of the high rents and snapped photos of the miserable circus trailers and shacks.[6]

Boy at water spigot. Photo by James T. Wiley Jr.

Shanties in the Pasco ghetto, 1948. Photo by James T. Wiley Jr.

Black residents told the inspectors that they would like to build their own houses, but they could not get government-sponsored FHA loans because they could only live on Pasco's east side, which was too financially risky to qualify. Blacks could live in North Richland, where space for a trailer rented for $4 a week, but to live there, one had to have a trailer. To get a trailer required financing . . . and so it went. Pasco had a school dedicated to black children, and the newly opened public indoor pool allowed anyone to swim. But there was no dentist who would serve African Americans, and an African American dentist from Seattle couldn't get office space in Pasco. Most Pasco restaurants, bars, hotels, and boardinghouses posted Whites Only signs. With no hotels, people passing through were forced to spend the night pacing the streets, lying in doorways, or sleeping in parked cars.[7]

Zoned out, the minority community made do its own way. People needing temporary quarters slept on the pews of the Baptist church for a dollar a night or at the trailer of a woman named Queenie. In some shacks, women cooked dinners and packed lunches. After dusk, smoke, music, and the clink of bottles sallied forth from trailers gussied up with bare bulbs to serve as taverns.

The unregulated quality of Pasco's east side made it a magnet for drink and the devil. Police didn't much frequent East Pasco, and it had an anything-goes quality. On weekends, workers poured into the red-light district of taverns, gambling dens, and bordellos. Crime followed the revelers: drunken driving, fights, theft, domestic violence, and verbal assaults.[8] C. J. Mitchell, scarcely sixteen when he lived there, remembers being scared most of the time. City leaders blamed Pasco's vice and crime on minorities. Some residents talked of breaking from East Pasco, making it a separate town. Others mentioned starting a chapter of the Ku Klux Klan. A few community leaders sought to end Jim Crow and replace the ghetto with safe and clean public housing.[9] The *Tri-City Herald* publisher, Glenn Lee, had a different plan. He waged a campaign to shut down Pasco's trailer parks for sanitation violations in an attempt to clear out people he variously termed "gamblers," "procurers," and "fugitives." Lee collaborated with the police chief to conduct nighttime round-ups of black residents for vagrancy, illegal cohabitation, "investigation," and other vaguely defined crimes.[10] As the jailhouse filled, the local prosecutor, William Gaffney, refused to prosecute citizens arrested in the sheriff's mass raids, claiming there was no evidence of wrongdoing. Lee retaliated, directing his editorial rage against Gaffney. He charged him with neglect and incompetence and eventually ran Gaffney out of Pasco.[11]

Lee may have been racist, but most of all he was a successful businessman. He saw that white residents were moving out of Pasco as black residents moved in, and he was working desperately to secure his real estate investments in Pasco before the city became tagged as blighted and real estate bottomed out. The designation of some territories as financially insecure grew out of secret "Residential Security Maps" that depicted urban territories according to the security of investment. The Federal Housing Authority manual instructed that once undesirable elements—ethnic minorities and especially African Americans— "invaded" neighborhoods, those areas fell into a red zone and declined in value.[12] A cautious proprietor such as Lee naturally fought against such a money-losing invasion.

The naturalizing qualities of spatial practices made it seem that culture and genetics, not poverty and lack of sanitation, explained why Pasco's blacks and local Indians were "dirty" and "smelled," why they "lacked ambition" and appeared content to live in squalor.[13] The slippery qualities of spatial arrangements also made inequities hard to pin down and contest. Citizens can fight a law or dismantle a regulation, but how would they go to battle against a customary boundary or financial security zones on unpublished maps?

How, for instance, could Pasco African Americans counter Kennewick sheriff Rupp's announcement to the ACLU investigator: "Let me tell you, if anybody in this town ever sells property to a nigger, he's liable to be run out of town"?[14]

When investigators asked AEC Hanford manager David Shaw why the AEC and GE hired no permanent African American employees, he replied just as unapologetically: "The fact that there are no Negroes permanently employed here speaks for itself, doesn't it?"[15]

Even the influential and powerful publisher Lee could not fight the racialized zones set up with the arrival of Hanford in the Tri-Cities. While Richland was knighted a "Critical Defense Area," Pasco gained the financial designations of "blighted" and "risky." Lee eventually moved his paper to Kennewick.

Because of the low visibility of spatial arrangements, white Americans in the Tri-Cities did not see themselves as racist. When polled, most whites, the same people who walked by the No Dogs or Negroes signs in shop windows, said they thought blacks were treated fairly. Seventy percent of local whites said they agreed with fair employment and housing laws, but at the same time nearly half said they would not want to work with or live next to a black person.[16] Respondents cheered laws that guaranteed equality yet supported zoning that made equality impossible. It was a bewildering state of affairs. In a 1963 hearing, the Washington State Board Against Discrimination found no illegal discrimination in Kennewick. "But we heard again and again," said board chairman Ken MacDonald, "that there is something in the air in Kennewick, and that Negroes know they can't come and live here."[17] With the principles of equal opportunity affirmed again and again in U.S. law in 1944, 1948, 1954, 1964, 1965, and 1968, white Americans could continue to believe that those who languished in ghettos, in prisons, or on the dole did so solely because of their own moral shortcomings. In popular perception, the more visible and heroic battles for civil rights eclipsed the federal government's role in creating largely white suburbs at the expense of "blighted" urban ghettos.

As they orbited each other in the Tri-Cities area, Pasco was the nadir and Richland, insulated from crime, poverty, unemployment and race problems, was the zenith. Pasco suffered from deepening crises of poverty and racial tensions, as people who had choices left and people who didn't stayed. Over the decades, Pasco remained a place for minorities. It remained poorer and darker than Kennewick and Richland, which continued to be predominantly white and middle-class into the twenty-first century.[18]

It was in part the Soviets who saved C. J. Mitchell from spending his life as a day laborer living in a ghetto. Soviet propagandists helped African Americans fight against the most overt institutions of American racism. The Soviet press dwelled so obsessively on Jim Crow and the Ku Klux Klan that civil rights became something American officialdom could no longer overlook. In direct response to Soviet challenges, the U.S. State Department sent abroad black entertainers to illustrate American tolerance, but it was embarrassing to promote African Americans abroad only to have them ejected from public eateries at home, as happened to the singer Hazel Scott, who was refused service at the Pasco bus depot diner. Scott, married to Congressman Adam Clayton Powell, successfully sued the Pasco proprietors, generating national press and bad publicity for Hanford and the Tri-Cities.[19] In response, Glenn Lee attacked Paul Robeson for working with the Soviets to divide America to make it weaker.[20]

Nonetheless, Soviet charges that Americans were racists stung. On the government-funded project, AEC and GE managers had to address charges of discrimination and the fact of lily-white Richland. Pursuing the case in the early fifties, the Seattle Urban League and the NAACP broke the unspoken color bar at the plant. AEC and GE managers agreed to hire a number of black employees located by the NAACP, but activists criticized this as a "quota system."[21] Yet thanks to this concession, in 1955 C. J. Mitchell landed a job at Hanford, working as an operator in fuel preparation. It was a good job, giving Mitchell and his family the opportunity to eventually rent a three-bedroom house in Richland.

Mitchell described to me the day they moved in. As the van pulled up in front of their new house, Mitchell looked up to see a white woman walking quickly toward him along the sidewalk. Mitchell froze, a young man from east Texas carrying the imprint of generations of insult and invective, sick with the thought of an ugly scene playing out before his kids. Then Mitchell noticed that the woman had in her hand a plate of cookies. Mitchell is still grateful for that gesture. The woman's name was Polly Cadd; though she is no longer living, he wanted me to make sure I had her name written down in my notebook.

21

The Vodka Society

For security reasons public processions were banned in Ozersk, but every day the city had a parade anyway. Nikolai Rabotnov recalled how as a boy he would stop to watch:

> In the morning the construction zones filled up. Infinite columns of prisoners chaperoned by guards with automatic rifles passed through the city along Prospect Stalin and turned onto Prospect Beria toward barbed-wire cages. At times they arrived earlier and were planted inside the barbed wire in several rows, faces to the street. Then I had to pass them. I especially disliked that. With shame I remember now that picture, which I used to watch every day with complete indifference.[1]

The indifference of the schoolboy to the suffering of the chain gang speaks to the dailiness of prisoners in Ozersk in the fifties and the banality of their gaunt faces. In Ozersk, the inmates were more common than newspapers, more dependable than the sale of vegetables, more punctual than the bus service. Arriving soldiers would announce the start of a new civic construction project, whether a school, apartment building, or theater, by marching in and putting up a coop of barbed wire called a "zone." In their wake prisoners would follow to dig, build, plaster, and paint. When the crews finished the project, soldiers would return to remove the fencing to another site. "Zone," Rabotnov remembers, "was the most oft-repeated word in the city. A fence with barbed wire was as common on the city landscape as houses and trees."[2]

In Ozersk, convicts, conscripts, and ex-cons built the emergent closed city. Forced labor came at a price, however, one that hounded plutopia for over a decade. Ozersk was designed to be sealed off from the economic hardships, crime, and uncertainties of provincial Soviet life, but inmates brought with them the violence and brutality of the Gulag along with the misery and resentments of the underclass.

It was an uncomfortable arrangement. Free and unfree citizens were sup-posed to be separated through zoning and regulation, but city dwellers lived in dangerous, intimate proximity to their service classes. This arrangement bound well-paid plant operators to prisoners, soldiers, and ex-cons by means of a des-perate inequality. After prisoners served their time, security officers often required ex-cons to remain working in the zone, detailed to civilian service jobs in the growing city. Ex-cons, in fact, became the public face of the socialist city. Free residents disliked them greatly. They said the ex-con waiters, clerks, and janitors behaved crudely, cheated them, and stole state property.[3]

General Tkachenko, city prosecutor Kuz'menko, and police chief Soloviev blamed the city's high crime rate on prisoners and ex-cons. Kuz'menko rolled out crime statistics showing that convicts and ex-cons committed the most se-rious crimes—theft, murder, rape, and assault. Worse, Kuz'menko argued, the criminal element negatively influenced the city's youth, which was the cause for the doubling of crime among blue-collar workers under age twenty-five.[4]

There was, for example, Skriabina, a woman who took up with an ex-con, married him, and got pregnant. After that, she stopped attending meetings of the Young Communist League and ceased to pay her dues.[5] General Tkachenko called her relationship "fraternizing," and he had his agents watch to make sure employees did not get too friendly with prisoners, ex-cons, or soldiers. Female employees were a particular problem. Because of the demographic void caused by the deaths of so many young Soviet men in the war, single women took up with parolees and prisoners. For that, however, they could get fired.[6]

Tkachenko promised to solve the problem by doing what he knew best. "We are taking measures," Tkachenko told a gathering of communists in 1951, "to purge our city of all formerly sentenced and all this kind of criminal element. We just sent off two echelons [of ex-cons and prisoners]. By the end of the year we will banish all the Germans from our city."[7] Two months later, Ervin Polle, the son of two doctors, woke one morning to find his parents packing. Because they were ethnic Germans, the family was shipped in June 1951 to Kolyma, the most notorious Gulag territory in the USSR.[8] In the subsequent three years, Tkachen-ko's forces deported from Ozersk to the Far East a total of twelve thousand exiles, ex-cons, and convicts, but Tkachenko's security forces did not entirely purge the city of former inmates or ethnic Germans.[9] Their labor and training were too precious in the isolated, understaffed city.[10]

It was handy to blame ex-cons for the city's high level of crime, but after Tkachenko's cleansing, crime continued unabated. The police chief, Soloviev, postulated vaguely that the Americans were behind it: "You might believe that the workers of our city are isolated from possible provocations of this or that enemy influence. Some comrades placate themselves by saying that we have a 'special regime' here, where everyone has been carefully selected and checked,

Ervin Polle with family members, Ozersk, 1950. Courtesy of Ervin Polle.

and so we don't need enlightenment work. But to make that argument is a serious political mistake. It is known that the American imperialists try to inculcate their bourgeois ideology into the consciousness of our Soviet people."[11] Soloviev proposed a "promotional campaign" to combat the "temporary feeling" of the settlement and "to teach people that the city is not alien, that it is a wonderful city, a wonderful factory." The security boss Tkachenko, less given to persuasion, cut Soloviev off. "We need to send a message," Tkachenko announced, "that there is no place in our city for disorganizers and drunks."[12]

By 1950, inhabitants of Ozersk annually drank more than thirteen quarts of vodka and wine per capita, more than double the national average.[13] Party officials calculated that residents spent three times more money on alcohol than they did on entertainment and culture.[14] In the first months of 1951, nearly all of the three hundred violations for missing work were due to drinking.[15] People drank before work, during, and after. People drank and drove. Teenagers drank. Children drank. Parents drank with their kids. Citizens drank at home, in cafes, in parks, on the street, on buses, at the lake, and in the town square. Employees went on benders that lasted for days.[16] Time and again, residents had to stand up at party meetings and own up to their binges, such as one Comrade Sokolov, an officer in the counterespionage force, who on various occasions, while drunk, barged in on his commander at home at midnight, broke into a woman's apartment, and caused a row in a shop, cursing violently before children. He could remember none of these events, Sokolov said, "because I was so greatly intoxicated."[17]

"Drinking was a relic of capitalism," one official declared, "it emerged with the exploiting classes."[18] Yet alcohol was more than a relic in Ozersk. It factored into the pay scale and was part of the health plan. In 1947, Tkachenko came up with the policy of rewarding soldiers, prisoners, and workers with vodka for reaching their daily targets.[19] Plant doctors believed that three substances, much loved and hard to obtain in Stalinist Russia, worked to cleanse radioactive isotopes from the body: chocolate, red meat, and vodka. After a person had been exposed on the job, he or she received vodka for medicinal purposes. Wives hated this policy.[20] Their husbands would leave the plant red in the face, primed for more, and head for one of the many pubs and kiosks that surrounded the bus stop nearest the plant. When party bosses tabulated the cases of public drunkenness, workers at the most contaminating radiochemical and metallurgical factories had the most violations for abuse of alcohol.[21] In most Soviet cities, it was difficult to get vodka, but in Ozersk, vodka, beer, and even imported wine and cognac were sold freely. After hours, salespeople sold booze from shop windows.

A woman who worked as a lab tech in the plant's central laboratory in the fifties showed me a rare picture taken of her female colleagues sitting around lab tables in white coats and head scarves. Petruva pointed to one woman and told me that she used to steal rubbing alcohol from the lab and drink it. Then she used the special Russian verb *spit'sia*, which means "to drink oneself to destitution." That is what her colleague did, Petruva said, her eyes widening at the memory of it.[22]

Drinking in Russia had long been a working-class and village pastime. The city elite spent a lot of time and money building up services for more cultured diversion—a symphony, a choir, drama and opera theaters—but the rows of seats remained half empty. A party member complained, "Only the upper and middle classes go to the theater. Average workers don't attend our theaters. They are not interested." But to leave young people to their own devices, the cultural authority argued, was not an alternative. "People say the young people need to show their own initiative, start their own clubs. I'll tell you what kind of initiative they have. On payday, they—independently—start to drink!"[23]

To combat drunkenness, police set up a drunk tank, which filled on week-ends.[24] Bosses garnished wages of workers who showed up drunk or missed work on a binge. Party members were reprimanded and then lost their party cards for excessive drinking. Teenagers who routinely abused alcohol and caused trouble were shipped out of the city to juvenile detention, never to return. Despite these measures, the problem persisted. The most enduring topic at city meetings over the decades was alcoholism.[25] Of the dozen elderly women I interviewed who lived in Ozersk from the first years, all but one had broken marriages, caused, they said, by their husbands' drinking. Somehow Ozersk, the affluent city of specially chosen plant operators, cultivated a vodka society.

There were lots of reasons to drink. Foremost, residents drank because they could. In other Soviet towns, obtaining booze was difficult. War veterans and sick and injured plant operators used alcohol to self-medicate, to anesthetize themselves against the pain of illness.[26] Judging by the number of discussions on the need for more sports programs and entertainment, vodka helped relieve the boredom and loneliness of people who spent every weekend and holiday locked in a zone far from their native realms. Finally, drinking served as a form of rebellion against the bosses who told them to work faster in dangerous con-ditions and against party leaders who instructed them to read books and go to the theater.

Soviet propagandists described Soviet society as classless, but when com-munists were talking about alcohol abuse they were also talking about class. Sharing closed quarters with plant operators, ex-cons, and soldiers, educated elites were appalled at how the other half lived. One night inspectors made a spot check of a women's dorm. The inspectors' disgust lay in the details: the men in the women's beds; the women inebriated; the floor a carpet of rotting potatoes, vomit, spit, and cigarette butts; the table cluttered with wine bottles and dirty dishes.[27] Plant operators should behave better, they said, in a manner "deserving of socialist society."[28]

For the Soviet elite, part of the hardship of living in the closed city was being locked in with the lower orders, people fresh from villages and urban ghettos. The city leadership could have cut off the sale of alcohol, easy enough in a closed city, but they decided against restricting this important consumer freedom. Rather, party members worked for years to improve their blue-collar neighbors. For party members, a classless society meant a middle-class society, and with it the pur-poseful elimination of working-class habits, manners of expression, and values.

In memory, these divisions and evictions were forgotten. Former residents describe a close-knit community of upstanding, select citizens. Over the years, neighbors claimed to no longer remember who had come to the city as a hired employee and who came as convict. But after the 1990s revelations of Stalinist repression, the memory of living among Gulag convicts began to haunt former residents of Ozersk.[29] Writing in 2000 about his childhood in Ozersk, Rabotnov wondered: "How could we carry on peacefully, happily rejoicing in our comfortable life, while living in a huge concentration camp? How come the horrors of the Gulag did not cast a shadow over our lives?"[30]

In the 1990s, many Russians came to reimagine their past by appropriating the image of convicts in order to place themselves in the role of victims rather than perpetrators of the crimes of Stalinism. In retrospect, the swelling number of convicts in padded jackets suggested to former residents their own incarceration in the closed city. One writer recalled how the convicts from inside their zone used to taunt him: "We are here for robbery. What are you in for?"[31]

But when they lived there, the elite of Ozersk did not see themselves as having any kinship with convicts. Ozersk's carefully filtered residents had no cause for empathy with Gulag inmates because they were selected as people with no relation to the Gulag and no relations in it.[32] Soviet ideology asserting that the USSR was the world's freest and most democratic country reinforced the notion of political purity, especially as it played out in a closed zone where residents had no chance to check official assertions against reality. In short, the spatial hierarchies that segregated and eliminated also inspired confidence in the Soviet polity. Rabotnov recalled his youthful conviction that he lived in the best country in the world: "I was sure that within our barbed labyrinth, I inhaled the air of freedom!"

Part Three

THE PLUTONIUM DISASTERS

22

Managing a Risk Society

Dr. Herbert Parker, a man of above-average abilities who landed far from his native England on the burnished hills of the Columbia Basin, faced an impossible mission in the late 1940s. As the head of the Health Physics Division, it was his job to manage the risks the Hanford plant posed to public health and environmental welfare. Parker knew what to do: monitor the plant and environs to maintain safe, "permissible" levels of radiation to minimize risk. He relied on a model of risk management from the pre-nuclear era, when the hazards were charted and could be controlled by a system of communal-corporate responsibility. Parker struggled to reconcile this pre-nuclear concept of rational safety management against what he and his colleagues were beginning to recognize was a vast chasm of unknowability about the unintended consequences of radioactive contamination. Wedged between two epochs, scientists such as Parker in this new field of health physics were caught like salmon swimming up the newly dammed Columbia River, reeling between the confines of conventional risk management and the emerging wall of diminishing control and persistent confusion of nuclear safety.

What left Parker calm before his impossible mission was that like most educated men of his time, he believed in abstractions. He placed his faith in science and progress to come up with solutions to the increasingly hazardous coexistence of humanity and radioactive isotopes.[1] In the end, progress failed Parker. The harder Parker and his colleagues worked to produce technologies to protect life, the greater the peril in which those technologies placed humans and their sustenance.[2] As head of health physics, Parker was given a great deal of responsibility, yet very little authority and power. He had every bit of the humanity and courage his job required, yet not enough of either. A person would have needed heroic qualities to violate social codes and secrecy regulations in order to stave off the approaching dangers. Parker sent up murmurs, but in the end he could not single-handedly stop a plutonium disaster in the making.

Let me be more specific and describe Parker's dilemmas. In 1945, Parker was very concerned about the spread of radioactive particles in the gases and effluents pouring from the Hanford plant.[3] Following the Japanese surrender, 7,000 curies of radioactivity gushed monthly from the plant's stacks because the plant was still processing green fuel on wartime express schedules. The excessive contamination mounted. Rainwater, Parker noted, had three times more radioactivity than permissible.[4] Worried, Parker sought to discover the effect of high concentrations of radioactive iodine on nearby livestock. But since AEC officials had assured local populations that the plant was safe, Parker feared that his monitors in lab coats wielding electronic gadgets would attract unwelcome attention.

Thinking creatively, Parker had his staff dress up as cowboys, and "furtively" wrangle neighboring ranchers' sheep to measure radioactive iodine.[5] The results were alarming. The animals' thyroids had more than a thousand times the permissible exposure. Scientists knew by then that high doses of radioiodine could damage the thyroid, leading to thyroid disease and cancer. Using these studies, Parker argued that the high concentrations from the stacks "were a potential hazard for food sources."[6]

Warnings about the food supply worked. Parker managed in 1946 to get plant engineers to double the cooling time of radioactive fuel to sixty days, and then in 1947 increase it to ninety days. In the subsequent months the levels of iodine contamination fell steadily.[7] Longer cooling times, however, were overridden by the plant's rocketing production. By 1948, four reactors were online, supported by two plants processing a heavy workload. With higher volume and more reactors, the radiation levels went up again. Monitors found hot spots fifty miles away at levels two and a half times the permissible dose.[8] Parker requested that GE managers increase the cooling time to 125 days to allow short-lived radioactive isotopes yet more time to decay.[9] Parker's GE bosses were new on the job, new to the nuclear business, and overwhelmed by the job of running the mammoth plant. In these first years, they often deferred to employees who had been on the job since the DuPont days. Parker got the cooling time he felt was safe, a real victory for public health.

Soon Parker faced another crisis. In 1948, monitors found on ground near the T plant "large" (milligram-sized) particles, fiercely radioactive. Searching downwind, they discovered many more particles, an average of one per square foot. Lab scientists determined they came from corroded ductwork inside the processing plant, where radioactive and chemical toxins ate through metals with superhero potency. Workers replaced a few large corroded ducts, but innumerable flakes continued to issue forth. Monitors tracked them as far as Spokane, a hundred miles away.[10] Parker estimated there were eight hundred million flakes, which, if sucked into workers' lungs or eaten on a french fry at Richland's Hi-Spot Drive-In, could lodge in soft organs and remain in the body for years, a tiny time bomb that Parker feared would produce cancer.[11] Parker presented the problem to his GE managers and congressional representatives. In the midst of his dense briefing on the problem, Senator Hickenlooper interrupted Parker. "Would you want," the senator asked, "your son working in the 200 area [chemical processing]?"

"No," Parker barked out emphatically.

When the risks were put in these personal terms, Hickenlooper and GE executives quickly grasped the essence of the dangers. On October 6, 1948, they issued an order to stop the dissolving operations until the particle problem was solved, for the safety of workers, they wrote, and to prevent the spread of radioactive particles across the Northwest.[12]

If this story ended there, it would have a happy ending. It would mean that once the wartime emergency was over, American officialdom heeded the warnings of their scientists and prioritized the mitigation of clear hazards to public health over nuclear weapons production. That conclusion would illustrate the effectiveness of Parker's Health Instruments Division's mission to monitor the environment to ensure safe levels, even if it required slowing down or even temporarily ceasing plutonium production.

But the story didn't end there. In October 1948, an AEC auditor noticed the longer, 125-day cooling times and upbraided GE executives for slowing plutonium deliveries to Los Alamos.[13] Two days after GE engineers shut down the plant because of the particle problem, the AEC Advisory Committee for Biology and Medicine met in Richland's Desert Inn, where Parker presented his data to them. The minutes read that after a "thorough" discussion, the scientists ruled there was no evidence to believe the particles presented "an unwarranted hazard." The scientists recommended the plant start dissolving again, while workers installed filters under the stacks. Advisory board members also iced out Parker and his monitoring program, reminding him that the Hanford lab was "primarily a production center, and research in biology and medicine should be directly applicable to local problems."[14]

Parker does not appear to have fought the decision to restart production hazardous to public health. To do that, he would have had to appeal above the advisory board—that is, he would have had to "go political," deploying his expert knowledge to make arguments about policy in the halls of government. Then as now, most scientists consider that to "go political" is to cease to have credibility as a detached, objective scientist. To do so in the chilling anticommunist climate of the late forties was risky. Parker no doubt had heard how scientists such as Leo Szilard and Harold Urey, who advocated international control of nuclear weapons, were marginalized within the AEC. Apparently Parker wasn't up for that kind of fight. After the rebuff at the Desert Inn, Parker kept to the prearranged schedule, escorting the visiting committee members to his house for dinner.[15]

Later, AEC officials criticized Parker for shutting down the plant and jeopardizing national security. Chastened, Parker sought to justify his decision, writing that he had not "dared" to contaminate workers and the surrounding populations.[16] Nor was Parker the only one concerned about radioactive waste. Reporting to the AEC in 1948 in a routine review, scientist Sidney Williams announced that "the disposal of contaminated waste in present quantities and by present methods if continued for decades, presents the gravest of problems."[17] Dr. Kenneth Scott, inspecting the waste disposal practices at Hanford, criticized inadequate, stopgap measures devised in the rush of war. Scott worried that as more reactors went online, discarding more scalding, contaminated water, the health of the Columbia River and with it the $10 million salmon industry faced risks that "may exceed an already narrow margin of safety."[18] Chronic exposures to radioactive waste, another scientist noted, could cause cancers, organ failure, or "a marked reduction in vigor."[19] As Parker worried over emissions, the National Committee on Radiation Protection reduced the permissible levels for human exposure to radioactive iodine by a factor of ten.[20] Even those numbers, however, were guesswork. "Safe tolerance," Scott wrote in 1949, "is more a matter of scientific opinion rather than fact."

With the plant restored to faster speeds, Parker could not but take away the message that, although the plant was no longer functioning on a wartime footing, production still trumped public health.[21] Further, he grasped that AEC officials were comfortable with doubt and uncertainty. Although aware that the Hanford plant was pouring forth historically unprecedented quantities and types of hazardous waste, AEC officials dedicated few funds to studying the medical effects. Parker's lab ran on a minimal budget, less than the subsidy the AEC gave to Richland's schools.[22] Consequently, Hanford scientists made little progress discovering the deposition of particles in lungs and digestive tracts, though it was a problem that plagued the plant for decades.[23] Parker was left to assume the lower scientific and safety standards of his superiors.

Perhaps for that reason, Parker didn't bother bringing up to the visiting scientists the fact that tunneling muskrats had undermined an earthen wall holding back a waste storage pond, sending sixteen million gallons of radioactive effluence rushing into the Columbia River. Soon after, in the same troubled 300 area, drinking-water wells located dangerously close to the collapsed radioactive pond were found to be contaminated, and truckers had to ship in clean water. Contamination was so widespread that workers in the 300 area, which included Parker's labs, had to start wearing full safety gear to go about their daily tasks.[24]

Nor could Parker stop, in 1949, one of the riskiest ventures of the Health Physics program. Nuclear risk is like gambling—once the first dice are thrown, it becomes easier to gamble again with higher stakes. Three months after the

Soviets tested their first atomic bomb, with military and AEC leaders in a panic, Hanford scientists risked running an experiment later known infamously as the "Green Run." Hanford health physicists conducted the test in collaboration with the U.S. Air Force in early December. The experiment called for processing a ton of twenty-day "green" fuel and tracking its distribution across the Columbia Basin.[25] This was exactly the kind of highly polluting procedure Parker had fought against the previous year. The experiment's purpose is not clear, but a scientist in Parker's division told a reporter in 1988 that they were trying to determine measurable levels of radioactive isotopes when processing scantily cooled fuel, which, they correctly guessed, was how the Soviets were processing in the Urals. If Air Force officers could find out how much short-lived radioactive iodine came out of a ton of green fuel, they could estimate from monitoring the air on the borders of the USSR how much plutonium the Soviets were making.[26]

There was nothing "green" about the Green Run. It was a jaundiced plume sailing over a russet landscape under an asphalt-gray sky. From the start a lot went wrong. Scientists expected the gases to contain 4,000 curies of radioactivity (from radioactive iodine), but as the gases left the stack the scientists measured a value of about 11,000 curies, a massive quantity topping all plant records. The researchers had waited for stable, dry weather, but the week served up "the worst possible meteorological conditions," as a scientist complained.[27] Soon after the experiment started, the wind picked up and sent stack effluent bouncing along the ground. Later, the temperature dropped by half, and rain brought down heavy concentrations of radioactive iodine over Spokane and Walla Walla. Meteorologists expected to be able to track the radioactive plumes along a predicted path, but the winds shifted, swirled, stagnated, and sharply changed directions. Pilots lost the Green Run's trail, only to find it again in unexpected places. Monitors recorded levels of radioactive iodine-131 on vegetation in Kennewick that were a thousand times greater than the permissible limit.[28] But the researchers weren't sure of their readings because their equipment got clogged up with contamination and gave false readings or no readings at all. Unexpectedly, the major portion of the toxic cloud passed south over Richland, where the scientists' families lived.[29] As radioactive eddies swirled over their houses, the researchers slipped imperceptibly from executors of their risky experiment to its victims.

The only good news emerging from the run was that the scientists predicted that with better weather conditions they would be able to track the stack gases for up to a thousand miles and so, presumably, be able to follow Soviet radioactive tailings. Basically, the test charted the path for a new field of nuclear surveillance. The Green Run was a first attempt for Hanford scientists to drill metaphorically through the earth to the closed city of Ozersk. And that is the strange irony. Most residents of Richland and Ozersk had no idea how much plutonium their plants were producing, but enemy scientists halfway around the globe came to know plutonium quantities in elaborate detail.

In the years following the Green Run, permissible doses declined, while the volume of radioactive waste increased terrifically. In 1951, with the plant dissolving more irradiated uranium than ever, emissions of radioactivity from iodine-131 climbed to an average 181 curies a day, while the safe target dropped to one curie a day. In 1955, operators were managing eight reactors and three processing plants. During World War II, the plant issued a peak of 400 curies a day into the Columbia River. From 1951 to 1953, retention basins dumped an average of 7,000 curies a day into the river. In 1959, discharge into the river peaked at 20,000 curies a day.[30]

Much of this waste was the product of economizing. AEC officials found it cheaper to dump radioactive waste than to deploy expensive technologies to contain it. In the fifties, the annual budget for waste management was $200,000 compared to the cost of operating Richland's $1.5 million annual school budget.[31] The meager budget went to deposit millions of gallons of effluent cheaply into cribs, trenches, tanks, ponds, and the river, with the most dangerous waste going into temporary underground containers.

Meanwhile, as GE built five new reactors from 1948 to 1955, they cut corners on design and basically reissued the original DuPont reactors, duplicating their risky features. The new reactors had no containment shells in case of an explosion and had a single-pass system where river water poured through the reactors once and exited into basins, from which the effluent, contaminated with radioactive isotopes, flowed into the Columbia River.[32] The designers considered building reactors that recycled river water and also contemplated larger holding tanks to allow the radioactive water to decay longer, but these plans were dismissed as too "elaborate," meaning expensive.[33]

As large quantities of waste amassed on plant grounds, scientists grew confident that, since no serious mishaps had yet occurred, the risks of contamination were manageable. In the late 1940s, for example, a Hanford radiobiologist, Karl Herde, carried out a study of pheasants near the plant. He wandered seventy miles from the chemical processing plant, shot pheasants, dissected them, and measured their thyroids for radioactive iodine. Herde was happy with what he found: in 1947, all the birds he trapped tested positive for radioiodine, but a year later, only a few of the birds had trace amounts. Herde concluded that "the extremely low levels found in the birds indicate that the present control of atmospheric contamination is adequately effective."[34] Considering Herde's unequivocal conclusions, I supposed that his study included hundreds of birds. I looked up the database for Herde's study.

Herde had shot ten male pheasants.

AEC reviewers reflected Herde's embroidered confidence about the manageability of radioactive waste. "Due to the extraordinary record of the health physics group, the risks [of radioactive waste] now are well understood."[35] Subsequent Hanford managers largely put aside worries about the radioactive waste. Decades later, when a growing environmental movement was questioning nuclear safety, reviewers discovered that the AEC had no commission-wide policy on waste, nor an office to oversee waste management. AEC officials had largely left the management of radioactive waste up to their contractors, and contractors left it to their divisions. Centrally, no one had a good idea how much radioactive waste there was, where it went, or how to contain it safely.[36] In sum, the cost-cutting design, management, and research decisions taken in the late forties and early fifties did not come cheap. They cost subsequent generations dearly in terms of hundred-billion-dollar cleanup funds and future health problems.

23

The Walking Wounded

Angelina Gus'kova got her start in radiation medicine as a young doctor in 1949 when she was assigned the lowly job of treating prisoners and soldiers in a Gulag medical unit at the Maiak plant. At the time, the settlement had no hospital, just a barracks clinic. One day in 1951, a dozen prisoners checked into the clinic with nausea and vomiting. Gus'kova treated them for food poisoning and sent them back to work. Later the convicts returned, complaining of weight loss, fevers, and internal bleeding. This time Gus'kova diagnosed acute radiation poisoning. Apparently the men had dug trenches in highly radioactive soil near Radiochemical Factory No. 25. Monitors went to the site to take a count. They estimated that three of the men had received about 600 rem, for most people a fatal dose.[1]

Gus'kova recalled that the prisoners received the best care: specially prescribed foods, vitamins, fresh bedding, blood transfusions, and anti-infection medicine. This was highly unusual treatment for Gulag inmates, but Gus'kova and her colleagues were interested in these patients as a first encounter with acute radiation poisoning. Of the three with the highest doses, one died, but the other two checked out after several months. Gus'kova remembers that accomplishment proudly. If she and her colleagues could figure out how to cure radiation poisoning, they could do their part to keep the plant running—a value for humanity and for national security.

It makes sense that Gus'kova first treated soldiers and prisoners. Plant supervisors assigned the most hazardous tasks to the most expendable and least knowledgeable workers.[2] When they had a really dangerous job, they asked for prisoners to volunteer, in exchange for time off their sentences. Usually it was the lifers who signed up to fix rails under leaking pipes or clean up spills. At Plant No. 25, prisoners volunteered for "special apparatus teams." The teams got a bonus for removing filters clogged with plutonium precipitates. The men didn't last long on the job. "I saw people doing that work who were coughing up blood," Inna Ramahova recalled. "Every two to three months the people changed, and I never saw them again."[3] Men who burned and buried radioactive waste grew noticeably weak and pale, and then they disappeared.[4]

From the standpoint of production, the health crises commenced only when skilled employees fell ill. Taisa Gromova worked in Factory No. 25. She was often first to work, the most enthusiastic and hardworking. Like all other employees in production, she had been given a clean bill of health before employment.[5] In 1950, Gromova began to complain of headaches, sharp pains in her bones, and constant weariness. She lost weight. Her gait slowed. While her friends danced and swam, she watched from a bench. In 1953, Gromova wheezed heavily and started to show signs of heart disease. Plant doctors diagnosed tuberculosis and sent her to a TB ward, where doctors found no TB and released her. Soon others fell ill, including Shalygina, Simanenko, Nagina, Modenova, Klochkova, Gribkova, Dronova—all chemical engineers from Factory No. 25, all in their early twenties.[6] People started to take note of the deathly pale girls from No. 25 who sat silently in the cafeteria weakly chewing black bread, the young women who suddenly looked so old.

When plant doctors examined the women, they were puzzled. The young workers' doses of gamma rays from external sources were not alarmingly high. Why were they sick?

There was a secret lab just down the road from the Maiak plant that might have answered this question. In 1946, General Zaveniagin set up a biophysics research institute, Lab B, and staffed it with German scientists brought in from occupied Germany. The German scientists experimented with radioactive waste from the Maiak plant, injecting it into soils and lab animals and mixing it with aerosols. The lab's director, Nikolai Timofeev-Risovskii, was most interested in the effects of radioactive isotopes on biological forms. He hoped he could use radioactive waste to grow bigger, better plants. Other scientists were trying to invent solvents to cleanse biological forms of radioactive isotopes.[7] Because the German scientists were prisoners, they were not allowed into the plutonium plant to test radiation levels.

In late 1949, security boss Ivan Tkachenko inspected the lab, about twenty miles from Ozersk. He found the German scientists settled in comfortable housing and well-equipped labs, but, Tkachenko complained, they produced few results. The problem, Tkachenko wrote to Beria, was that the lab was staffed entirely with prisoners, "all sentenced for anti-Soviet activities. They cannot do proper science because they are cut off. They have no idea of the latest discoveries and are completely isolated from work in their fields."[8]

It's strange to read the Gulag general's critique of science behind barbed wire. Tkachenko recommended appointing a free, civilian lab director who could maintain connections with the outside world and initiate the exchange of information on which science thrives. By that time, however, it was late. Reactor A was leaking wildly, sending radioactive waste into Lake Kyzyltash, where a large fishing concern harvested and processed tons of whitefish for commercial sale. Prisoners and soldiers were toiling in soils saturated with fission products, and the girls at the radiochemical plant were ingesting and inhaling radioactive dusts and vapors on a daily basis. All of this information was off-limits to the imprisoned biophysicists.

In other words, the state-of-the-art lab staffed with Western scientists was an opportunity lost. Soviet research on the biological impact of radioactive isotopes made little progress from 1946 to 1953. Meanwhile, of the ten thousand pages of documents purloined from the Manhattan Project, Soviet agents did not get their hands on the medical studies laying out the dangers of ingesting radioactive isotopes.[9] With monitoring protocols in place that concentrated on doses from external sources, Gus'kova and her colleagues did not consider the hazards of the minute amounts of radioactive isotopes clinging to dust particles as they cascaded down the esophagus to the soft tissue of lungs or followed pathways from skin abrasions into the bloodstream to lodge in vital organs.

Among the female workers of Factory No. 25, Gromova died first, at age thirty. An autopsy showed she had 230 times more plutonium in her bloodstream than "the acceptable norm." Eight colleagues soon followed her to the city cemetery.[10] Vladimir Beliavskii remembered the cemetery near the quarry, "even in those [early] days," he wrote, "the cemetery quickly outgrew its boundaries."[11]

There were only a handful of trained experts in nuclear chemistry in all of the Soviet Union. The sickly young chemical engineers were irreplaceable, and their illnesses sounded a shrill, though classified, alarm. Moscow leaders earmarked funds for two new hospitals, several clinics, dosimetry equipment, payroll for an expanded medical staff, and a branch of the Moscow Institute of Biophysics in Ozersk.[12] Soon the town had more doctors per capita than any other city in the Urals.[13]

With medical facilities, a lab, and more staff, Gus'kova and her colleagues set to work. Rather than discover the harmful effects of ingestion of radioactive isotopes through experimental research on mice and rabbits, they uncovered the link between radioactive particles and health effects as they treated their patients. In 1950, the doctors came up with a new disease, diagnosed so far only in the Russian Urals—chronic radiation syndrome (CRS), caused by extended exposure to low doses of radioactive isotopes. One memoirist described the terrible ache of CRS as a pain that made him "want to crawl up the walls."[14] The doctors learned to predict the onset of this mysterious new illness by changes in the blood, often signaled in severe anemia.[15] The doctors took to drawing blood from employees in production jobs every few months. Exposed employees might look and feel fine, but changes in blood cells could be sharply telling. One clinician remembered looking at a blood smear of a woman exposed in a criticality accident, a self-sustaining neutron chain reaction. She was appalled to find, instead of a multitude of white blood cells, only one lone lymphocyte swimming on the glass.[16] When doctors recorded alarming cellular changes in a worker, they requested that the worker be removed from contaminated jobsites. No boss, under pressure to produce, wanted to hear that doctors were withdrawing scarce, trained staff. "We had some very trying conversations with the supervisors," Gus'kova recalled.[17]

Plant doctors argued that if young people became invalids within a few years at the plant, bosses would have trouble keeping an adequate supply of workers. They pointed to the fact that it was already difficult to attract qualified employees and scientists to work in production.[18] The doctors finally convinced leaders—the same men who had failed to ensure biological protection, who had cut the budget for radiation monitoring, who ordered shorter, dangerous cooling times—to remove valuable, trained personnel from radioactive environments before they fell gravely ill. To get recalled from a production job meant a sharp pay cut of up to 50 percent or, worse, termination. Terminated employees had to leave the well-supplied, increasingly affluent little city and go out into the "big world," as they called the grim landscape surrounding the enclave. Many people experienced termination as a form of deportation or forced resettlement, a reversal of the promise of social mobility.[19]

Removing workers before they suffered symptoms of chronic radiation syndrome, Gus'kova recounted, saved thousands of lives. In 1954 alone, medical staff had 805 workers reassigned from production work. Gus'kova asserts that of the 2,300 workers who were eventually diagnosed with CRS, only nineteen died of it within a decade of exposure.[20] A significant number of these patients, however, died in their thirties, forties, and fifties. Among the young women working in Factory No. 25, over half succumbed to cancers by the time they reached age fifty.[21]

For Gus'kova, the deaths were distressing but nonetheless indicated a victory of medical science. She and her colleagues learned how to diagnose CRS in a timely fashion and treat it so that most workers, she argued, fully recovered. Of the more than two thousand employees who received an astonishingly high 300 rem, half were still alive forty to fifty years later. In fact, she argues, this group of CRS patients lived longer on average than the Soviet population.[22] From these statistics Gus'kova argued that radiation, even at higher than permissible doses for prolonged periods, is survivable and that rogue radioactive isotopes are not dangerous when coupled with careful medical monitoring, good health care, and sound living conditions. That message was warmly embraced by the bosses in Moscow who were projecting a future of nuclear power. Gus'kova's career rose along with the success and appeal of her research results. At the end of the fifties she was promoted to Moscow; there, a rare woman in the field, she became the director of research at the Institute of Biophysics. After the Chernobyl disaster in 1986, she became the public face for discussions of radiation and health.

In front of TV cameras, Gus'kova worked to mollify a worried populace and convince them that radioactive contamination was containable.[23]

Gus'kova's statistics, however, raise some questions. The closed city of Ozersk had a very young population of relatively affluent residents who enjoyed excellent, universal health care. The city had one elderly person and no one who was poor, indigent, or chronically ill. In contrast, the surrounding postwar Soviet hinterland served up malnutrition, a host of infectious diseases, shoddy and scant medical services, people suffering psychological and physical illnesses from wartime trauma, and majorities living in poverty. These conditions made a mark on public health. In some places, 25 percent of all infants died of disease and malnutrition. The epidemiological picture of Ozersk residents, preselected for their good health and youth, should indeed have looked far better than Soviet averages.

Medical diagnoses obscure as much as they reveal about complex medical realities. With the CRS diagnosis, plant doctors could designate some as sick and others as unharmed by the plant. In later decades, circumscribing the number of people diagnosed with work-related illnesses meant saving millions of rubles in workers' compensation. There was real economic and political pressure, in other words, to come up with medical data that described only minor health impacts for a limited number of workers.[24] What I am arguing is that although Gus'kova's statistics sound like scientifically verified facts, they mean very little. Gus'kova's tabulation of CRS patients hardly reflects the sum of people who were exposed at the plant, fell ill, or died untimely deaths. Those numbers are lost to history. What we can do now is only glimpse a snapshot of the larger picture.

Gus'kova's records enumerate only paid staff assigned to main production who were medically monitored and remained in Ozersk after termination from production—less than 10 percent of all employees. Employees with CRS who were later fired and evicted from Ozersk fell off the plant's medical radar. These people's medical histories joined the statistical tide of the larger Soviet population, which chronicles rising cancer rates and shorter life spans.[25] Gus'kova also did not include unmonitored workers, prisoners, and soldiers.[26] About a hundred thousand of these "jumpers" passed through the Maiak plant in the first ten years of operation.[27] Nor do Gus'kova's records account for the guards who stood watch in front of dirty labs or in breezes blowing from the stacks. They fail to include cooks, plumbers, electricians, cleaning staff, and clerks working in the contaminated plant premises, or of construction workers who built new reactors and plants on radiated territory.[28]

What happened to these men and women is a subject of speculation. Soldiers served a few years and were demobilized. It took a year or two for the symptoms of CRS to appear. It took a dozen years to develop circulation disorders or tumors. There is no way of knowing how many soldiers succumbed to exposures from the plant. CRS symptoms resembled that of a long list of infectious diseases and illnesses that also could be caused by malnutrition, stress, and exhaustion. Chronically infirm prisoners were eventually classified as "invalids" and sent to camps outside the zone.[29] If these prisoners died from or survived radiation poisoning, no one recorded it.

The new disease, the sick young people, and the cause of plant employees' deaths were classified as state secrets. After medical exams, workers were not told their doses or diagnosis. But the prisoners with bleeding noses after a shift, the pale young women haunting the grocery stores, the growing number of doctors and medical staff converging in the new city—all were secrets in plain sight. Residents began to understand the dangers of their factory. A rumor went around that men at the plant were sterile and that husbands paid soldiers to impregnate their wives.[30] One salesgirl, referred to as "blabbermouth," went home to Magnitogorsk to visit a sick family member in January 1951. There she recounted that Ozersk was a virtual prison. She said she was offered higher-paying work at the plant but turned it down. "To go work there [in the plutonium plant] is to bury oneself," she reportedly said within earshot of an informant. General Tkachenko had the woman arrested for "spreading state secrets."[31] Nonetheless, her sentiments were troubling.

Sick workers and rumors of dangerous work conditions were bad for morale. Tkachenko had to admit that people shied away from workshops with reputations as "dirty." Bosses noted that many people stopped paying their party dues and ceased attending obligatory meetings once they arrived in the closed city. They did so because party members, when told to report for work in a dangerous sector, had to obey, while those who were not party members could refuse.[32]

Soldiers and prisoners became important not only for the economic health of the community but also for its physical health. As the plant spilled more and more waste, they served as an expendable labor force on contaminated ground. One of the services of modern states is not only to redistribute wealth but also to reapportion risk.[33] In Ozersk, temporary workers shouldered the lion's share of risk, and that helped make the dangerous, poorly designed, and leaking plant palatable to the permanent crew of elite, trained workers who ran it. The temporary workers in their garrisons and camps insulated plutonium workers from the risks and health consequences of the plutonium disaster they were making.

24

Two Autopsies

In 1952, the AEC won the first of many awards for its record of exceptional safety.[1] Years later, the official account is that the Hanford plant had no accidental fatalities due to radiation in its forty years of operation.

I started to doubt this accounting when I came across two autopsies for one corpse. On June 9, 1952, Ernest Johnson, a maintenance foreman, left work early, complaining of a burning sensation and a sore throat. He went home, lay down on the couch, and died. Johnson's wife, Marie, called Richland's GE-managed Kadlec Hospital. Dr. William Russell arrived within twenty minutes. Russell later conducted the first autopsy, which stated that the forty-eight-year-old employee died of an aneurysm.[2]

Mrs. Johnson was suspicious about her husband's death. The mortician pointed out an unexplained burn mark on her husband's arm.[3] Meanwhile, Johnson's coworkers dropped by and whispered to Marie something about Johnson "getting dosed." Also, the FBI tailed Marie around Richland. Marie Johnson brought her husband's body home to Chicago for the funeral; afterward, she dropped the body off at the office of the Cook County coroner, Dr. Thomas Carter, who conducted a second autopsy. Carter stated the death was from an aneurysm caused by contact with radioactive substances, which explained the burn marks. He wrote to Marie Johnson, "I am positive his death was due to exposure to radiation, and that you will have no trouble in proving your claim for insurance or compensation." Carter then added a troubling clause: the autopsy, he wrote, was incomplete because "some of the important evidence [on the body] had been removed."[4]

Marie Johnson sent Carter's autopsy to GE seeking compensation for her husband's death. Alarmed, Dr. Philip Fuqua, second in command of Hanford's Health Physics Department, flew to Chicago to ask Carter to withdraw his conclusion. Fuqua suggested to Carter that, given the conflicting opinions, they could call in experts in radiation medicine to review the case.[5] This was a fix. Fuqua's proposed list of experts—Robert Stone, Shields Warren, and S. T. Cantril—were all hard-boiled AEC insiders. Previously Stone and Warren had approved the experimental exposure of thousands of U.S. servicemen in nuclear field tests. Fuqua knew these doctors could be counted on to give the kind of medical testimony the AEC would like.[6]

With Fuqua in his office, Carter relented, but a week later he changed his mind, refused to retract his original conclusion, and called Marie Johnson to tell her GE was pressuring him.[7] Switching tactics, GE lawyers then requested that the Washington State Labor Board redact information about radiation exposure in the record of Johnson's second autopsy and deny Marie Johnson's claim. Washington State officials complied and, even more helpfully, informed GE lawyers of the content of Johnson's trusting phone calls to the Labor Board.[8]

This all sounds suspicious—the mysterious death, the conflicting autopsies, Fuqua's rushed flight to Chicago, the missing body parts. I asked a former Hanford health physicist about Johnson's case. He pointed out that harvesting organs was standard practice at the time and may have had less to do with hiding evidence than with collecting material for future research. The health physicist also disparaged the coroner's second autopsy. "What did family doctors know about radiation?" he asked rhetorically. "All a family member had to do was mention he worked at a nuclear installation and doctors would jump to conclusions."[9] But the Cook County coroner was a trained pathologist in a major city, not a family doctor.

Sixty years after Johnson's death, I pondered who was right. Did Johnson die of radiation-related injuries or from an unrelated condition? I ran searches in archival databases to try to exhume Johnson's fate from the tens of thousands of declassified Hanford documents. I never caught sight of Johnson again, but I did find some clues. A June 1952 Hanford monthly report described a "Class I" radiation incident on the day Johnson died, in the area where he worked, and involving men whose job description met Johnson's, but this incident did not end in a reported fatality, and the account of the event appears to be innocuous, having to do with workers unknowingly handling radioactive barrels.[10] As the men were unmonitored, the dose they received was not recorded.[11]

That is all. I could not be sure this radiation incident involved Johnson. According to regulations, there should have been other records about Johnson, documenting his early departure from work and his untimely death, but those papers no longer exist, or maybe they were never written in the first place. I did find evidence that some accidents failed to make it into the official record. On a day, for instance, when two million gallons of radioactive effluent leaked into the Columbia River, a monitor scratched in a notepad: "We will leave this as an informal."[12] For another accident describing the discharge of an unknown quantity of high-level waste, an official wrote, "We do not consider this a reportable incident."[13] In December 1955, AEC officials in Washington noted, in just one sentence, that a Hanford worker had died an accidental death. Checking the operation reports, I found no sign of this fatality or of any injury at the plant that month.[14] Perhaps these accident records were among the majority of documents purged routinely from AEC files after six and fifteen years. Or, when GE lawyers were worrying about Marie Johnson's phone calls in late 1952, Johnson's documents might have been more purposely "sanitized," as they called it. As I puzzled over this case, it plagued me, this reminder of the oblivion that swirls in the vortex of memory.

Searching through accident reports to try to distill the kinds of events that might have killed Johnson, I was astonished at the wealth of ways to get injured. Johnson could have been splashed with contaminated water overflowing a storage basin. He might have been hit with a neutron shower from under the reactor.[15] He could have breathed in burning uranium oxide when venting fans failed.[16] He might have been tasked to climb a crane to work near a pipe flowing with ragingly hot radioactive solutions. He could have passed too close to leaking waste tank pipes.[17] He could have gotten a radioactive shard in his arm as he repaired control rods. Radioactive solutions might have burned him while he was changing clogged pipes, unjamming ruptured slugs, or dodging a high-velocity radioactive spray from the reactor's rear face.[18]

There were a lot of things that could go disastrously wrong in a typical day at the Hanford Plant. AEC experts coolly asserted that accidents could be prevented by design, regulations, and worker training. However, sociologist Charles Perrow, studying the Three Mile Island accident, points out that the complexity of high-risk nuclear industries "outruns all controls." In these elaborate systems, he argues, even a benign problem, such as failure of an electrical source, can lead to catastrophe. Perrow calls these events "normal accidents" because they cannot be prevented, as they cannot be anticipated.[19] Moreover, a nuclear power plant ages fast, prey to highly corrosive chemical and radioactive toxins. "The plant is wearing out," Parker wrote when the Hanford factory was just four years old, "and this gives some difficult maintenance work."[20]

There wasn't much time for repairs. In the fifties, operators were under pressure to keep the line going in order to stay ahead of the Russians. Workers were under the gun to churn out plutonium as quickly and cheaply as possible. From 1947 to 1951, the plant tripled production; after President Dwight Eisenhower made nuclear weapons the cornerstone of his defense strategy in 1955, production rose 30–40 percent in each of the subsequent three years.[21] Unfortunately, as production doubled, the volume of effluent tripled. Running at higher power caused slugs to rupture at a rate of ten to twenty a month. When slugs cracked open, operators had to shut down, dislodge the slugs, fix malfunctioning equipment, and start back up, with a loss of valuable production time. To make up for lost time, they again rushed.[22] Hurrying, inspectors had little time to canvass old materials and test equipment. Waste lines collapsed, irradiated paint peeled off, pumps failed, rubber gaskets dried out and leaked.[23] Ordinary wear and tear on the plutonium plant translated into minor and major radiation incidents. Reading the monthly reports, one notices that the factory seemed to be more than a little out of control. In 1955, the year Eisenhower announced his special "nuclear defense materials initiative," the disorder reached a crescendo.

In November a fuel element ignited and the blast spread contaminated particles across five square miles.[24] In December the basin of the F reactor cracked and leaked 1.7 million gallons of effluent a day into the Columbia River. Yet no one rushed to fix this radioactive geyser. A. R. McGuire wrote that the radiological sciences department was aware of the leak but had not filed a report. "I rather suspect that they prefer not to go on record with any statement now," McGuire wrote. "I'm inclined to wait, since their answer, if forced at this time, might require basin repairs."[25]

McGuire was only too happy to overlook the effluent pouring into the river because he had much bigger problems. I came across these lines buried in a three-hundred-page report: "On December 22 a wind storm with gusts up to 80 miles per hour caused extensive damage to the 100 Area [the reactors]. Approximately, 35,000 square feet of roof were damaged to the extent of the concrete blocks being flown from the roof in some cases."[26]

I had to read that passage several times to make sure I had it right. In its brevity, the statement leaves a lot to the imagination: the wind's deafening roar, the air luminous with sound, and through it, like a shot, the sustained crash of behemoth concrete blocks ripping from rebar moorings and rising with a terrifying snap to fly off the reactors. As the blocks bounced into walls, fencing, and retention basins, they shattered and the pieces rained down on trucks and fleeing personnel. Afterward, contaminated steam and water gushed while workers carrying radiation monitors crawled over cement shards and operators rushed to scramble the reactors.

It took workmen six months to fix the reactors' roofs.[27] All the while, the reactors kept going. In the days and weeks after the storm the three damaged reactors with "missing roofs" functioned exposed to wind, snow, and rain. In this heady wartime climate with no war, the exposure of a few workers, the irradiation of hands and faces, were just metaphorical eggs broken to make the proverbial omelet.

As production increased, and with it radioactive effluence, construction projects on the reservation continued apace. In 1951, J. Hofmaster of the United Brotherhood of Carpenters wrote to Washington congressman Henry Jackson to tell him in strict confidence of the hazardous conditions for construction workers at the plant. The carpenters toiling on the new reactors, which designers had located very close to the old reactors to save money, worked in prevailing winds that went directly "from stacks to earth."[28] They worked through periods when radiation monitors recorded snowflake-sized radioactive particles skating from stacks and radioactive iodine levels spiking. The exposures could be very high.[29] Construction workers had no training to work in radiation zones, so they did not recognize the dangers when they borrowed a contaminated hose, which would set a Geiger counter clicking at 1,000 counts a minute, to hook up to a hydrant for water.[30] The workers had no protective clothing or personal monitors. They passed the sanitation inspection line while the monitors stayed seated. With no water to clean up before lunch, they ate with contaminated hands. A couple of men developed inexplicable sores on their arms and face. Hanford doctors told them not to worry, "that they were not cancerous."[31]

Because the presence of construction workers was temporary and the workers were employed by subcontractors, they did not qualify for safety regulations prescribed for GE employees at the plant, and so their exposures were not accounted for and their health was not monitored. Once the job was done, the men moved on, taking with them possible health effects. At least one construction worker wrote the Hanford management in the 1970s, describing a mysterious exposure and charging that afterward his stomach "pulferated," his lungs perforated, and his vision blurred. Since leaving the plant, he wrote, he had had half his stomach removed and a lifetime of medical complaints. He claimed that five others in his crew were already dead. Nothing came of the inquiry. An official wrote back to say they had no record of the man's employment.[32]

At the plant, workers came to describe accidents in their own way. They told each other on the bus ride home how they got "crapped up," "burnt out," or "jetted," or how they were exposed to a "cutie pie," "big sucker," "skimmer," "pig," "doorstop," "trombone," or "totem pole." The language was opaque to all but the initiated. One worker told me the story of a guy who got home late. When his wife asked what delayed him, he replied, "I was working in decon and the pig slipped out of my hand and crapped up my whites. The RM took a while to release my arm." His wife sputtered, "If you wanted to stop off for a few beers, you didn't have to lie about it!"[33]

Especially vulnerable were people who worked outside, exposed to airborne dust, unmarked hot spots, burial grounds, leaking effluent, and the downward drafts from stacks issuing an oily yellow fog that ate away secretaries' nylons. Not just operators were exposed. Radioactive isotopes, following the paths of wind, water, and feet, were tracked all around the reservation. Truckers hauled sloshing loads. Guards were prey to a gentle breeze and a patter of light rain; over time the mattresses in the guardhouses accumulated so much radioactivity that they had to be thrown away.[34]

Health physicists referred to fission products as something distinct from the environment and from human biology. They used verbs such as "contain," "decontaminate," and "clean up," though it was known that there was no way to neutralize radioactive isotopes. They could only hope to move fission products from one place to the next and wait until the isotopes decayed. As the firmament became saturated with fission products, monitors had an even harder time distinguishing Hanford's by-products from atmospheric contamination floating in from Nevada, the Pacific testing grounds, Kazakhstan, and the Russian Arctic, where Soviets carried out mega-tonnage tests. In the late fifties, scientists ran employees through whole-body counters to measure internal levels of radioactivity. The counts were high, the monitors determined, because the subjects' hair was sullied with fission products that inflated the measurements. When the researchers had the subjects wash their hair, the count only increased because of the radioactivity of the local water supply, drawn from the effluent-rich Columbia River.[35] And so it went. Radioactive isotopes, so readily combining with biological forms, had no discrete boundaries. In time, they were no longer distinct from the local environment, from scientists' bodies, or from human evolution.

Radiation monitors insisted on regulations and caution because the fission products were so capricious and tenacious and because monitoring was not an exact science.[36] Workers remember that safety regulations drove them crazy.[37] The training, reports, drills, monitoring, and alarms were repetitive, numbing features of the workday.[38] Employees, however, rarely knew the extent of the radioactive fields surrounding them.[39] They wore personal counters, usually in the front breast pocket. If a body got exposed from the back, feet, or hands, the monitor recorded a muted reading. Furthermore, the personal counters recorded only gamma rays, not internal exposure to beta and alpha radiation from radioactive particles that had made it into the body. While analysts took urine samples to estimate internal exposures, that too was often guesswork.[40] When workers received high doses, they were quietly removed from dangerous work and reassigned to what Hanford veterans call a "lifetime job," meaning an employee got paid whether or not he or she ever worked again, but the employee was not officially granted worker's compensation, which would have entered the record.[41]

In 1958, AEC executives reported to Congress on the safety record of the nuclear industry. According to AEC officials, there had been only one radiation-related injury at the Hanford plant among eighteen thousand workers toiling on the nuclear frontier from 1944 to 1958. It is hard to square the hundreds of accident reports describing radiation incidents with this rosy picture from AEC headquarters. I am not suggesting that AEC executives purposely misinformed lawmakers on the Joint Committee on Atomic Energy. Rather, the data they presented were an honest reflection of what they knew. Information passed through many filters as it went up the chain of command, from workshop to supervisor, from monitor to department chief, from GE in Hanford to AEC in Washington. As facts left the factory floor, they were streamlined, with details filtered out; some events were designated "informals"; major individual health episodes were computed down to a statistical average. The result was that in Washington, D.C., an image emerged of a well-run, orderly, increasingly efficient plutonium plant mastering nuclear technology from the ground up. This is a hopeful narrative, a naturally American story. For obvious reasons AEC executives were drawn to furnish it to congressional leaders, who were happy to hear it.

Yet this narrative demanded a good bit of monitoring, containment, and constant surveillance to suppress extra autopsies, bury major radiation events in long reports, and deny compensation to other widows who followed Marie Johnson to GE's offices to make claims in subsequent years.[42] This repression of fact and detail was not born of a conspiracy to deceive the American people. On the contrary, the process of reporting was more habitual than that. It originated in goodwill and optimism, loyalty oaths and security regulations, and it thrived alongside a belief that American know-how and technology would work it all out for the best, as if Voltaire's Pangloss had set out on a nuclear odyssey.

25

Wahluke Slope: Into Harm's Way

Lt. Col. Fritz Matthias chose the Hanford site because it was sparsely populated. Plutonium, however, brought an economic boom to the region in the form of jobs, a spreading electrical grid, more towns as markets for farm produce, and more highways and bridges to move that produce. In addition, the federal government bankrolled in eastern Washington the nation's largest irrigation network, the Columbia Basin Project, which planned to use power from Columbia River dams to pump river water to a million acres of dry land and make it green. With the new prosperity of the Hanford area, the irrigation project made even more sense, but there was one hitch. AEC officials had enclosed 173,000 acres, called the Wahluke Slope, near the reservation as a secondary "control zone." A handful of ranchers owned this land and farmed it, but AEC officials did not want the Columbia Basin Project to irrigate in the control zone and place new farms at risk of harm from the plutonium plant.[1]

AEC officials worried especially about a reactor explosion. In the early fifties, Hanford managers built more reactors and ran them at power levels higher than their design capacity, in order to prevent the graphite cores from swelling. As they did, AEC consultants worried that the risk of a runaway explosion ranged from negligible to "almost fantastic proportions."[2] In case of accident, they postulated, they might need to evacuate an area from twenty to a hundred miles in diameter.[3] The threat of a noisy, radiation-showering catastrophe most worried AEC officials, in part because it was the kind of event most "detectable" to the larger public.[4]

The Wahluke Slope, however, if it went unirrigated, represented a great financial loss for a couple of big landowners, among them Leon Bailie and N. D. Olson, who had bought up tens of thousands of acres on the slope during the Depression at fire sale prices. They expected to cash in on their investment once their dry land became irrigated at public expense. The Columbia Basin Project was originally conceived not as a windfall for speculators such as Bailie and Olson but to aid small independent farmers, seen as the ideological cornerstone of American democracy. Like many federal public works programs in the West, however, the social welfare goals of the Columbia Basin Project were vulnerable to powerful special interests and big business.[5] Leon Bailie was one such businessman who stood to lose a lot of money if the AEC withheld land he owned from the irrigation project. Bailie organized a posse of local businessmen, called it a "grassroots movement," and enlisted the help of junior congressman Henry "Scoop" Jackson.[6] Jackson won an audience with the AEC chairman, David Lilienthal, when Lilienthal visited Richland in 1949. The transcripts of the meeting show how the notions of risk and security were parried and eluded.

Lilienthal told the gathered businessmen and ranchers that because of security regulations he could not tell the men much, "or risk going to jail," but, he said, a panel of high-level scientists had all agreed that the territory northeast of the plant should remain a secondary buffer zone until nuclear safety was improved in the future. Not content with this explanation, Olson, one of three farmers living in the secondary zone, asked Lilienthal if his farm was safe.

> Personally, I would be interested in remarks about the security of the people living in this additional controlled area, and I would say that if it wasn't safe to irrigate, it wouldn't be safe to live in at the present time. It has been reported that is one of the reasons they are holding this from irrigation. I think if my security isn't safe, I don't want to live there. . . . I guess I am about one of the closest people to the plant of anyone in the project.[7]

Lilienthal evaded Olson's question, but later another farmer asked about the risks: "The thing that concerns you is the number of people on the project? You couldn't protect them?"

Lilienthal replied, "We don't think that additional people should go into that area until we know more about the safety angle."

A third farmer followed up: "Are you afraid of escaped fumes? Radiation? Plant blow-up?"

"That is not the real problem," said Lilienthal. "In the immediate vicinity we might have some concern about that, but in talking about the secondary area, we are only talking about what we only firmly believe to be a remote possibility of disaster."

Like Peter in the Garden of Gethsemane, three times farmers asked Lilienthal about the dangers of the plant, and three times he denied AEC fears of a catastrophic accident contaminating the basin for a hundred miles around. He did not inform the men that as plant officials ran more reactors at higher power, the reactors were harder to control, and if they blew, they would do so more powerfully. He did not tell them that by saving money doubling up reactors, plant designers had also doubled the potential exposure in an accident. Nor did he mention the lack of containment around reactors or the hot, burping waste storage tanks. He did not tell them that plant scientists were secretly testing their sheep on the slope to measure radioactive iodine-131.[8]

The Wahluke Slope became another episode where AEC officials threw caution to the winds. They reasoned in classified reports that the boundaries of the secondary zone were "arbitrary." With unforeseen releases of fairly large amounts of radioactive iodine and with more facilities going online, the hazards were only increasing, but that was true for all communities within a fifty-mile radius because the winds distributed the hazards "capriciously."[9] In other words, the secondary zone was not uniquely at risk. Richland and Kennewick, officials rationalized, would be no safer in a disaster.[10] In fact, studies showed that the prevailing winds usually skirted the disputed Wahluke Slope, blowing instead toward Mesa and Pasco.[11] Yet to admit to this generalized risk would have undermined AEC assertions of safety. Secretly worrying, AEC officials contended publicly that any potential disaster would be localized within the fenced-off confines of the nuclear reservation, as if fences could contain radioactive isotopes.

Businessmen and ranchers took from official AEC statements the assurance that their risk was no greater than that of all Americans facing a possible nuclear attack. They had no idea that AEC consultants repeatedly advised against opening the slope, so they kept up the pressure for the release of the secondary zone until they won, in 1953, 87,000 acres and, in 1958, another 105,800 acres freed for irrigation, farming, and habitation.[12] This was a victory for special interests but a serious compromise for AEC risk brokers, who assessed dangers on behalf of others who they admitted were "completely ignorant" of the hazards.[13] AEC officials had no emergency plans in place in the event of a catastrophe.[14] All they could do was pray it would never happen.

Who would get the strategically sensitive, newly irrigated farmland once it was released? Officials at the Bureau of Land Management decided the land should go to American veterans who passed financial and political background security checks. They held lotteries and gave the vetted veterans a chance to buy Columbia Basin Project land. Most of the land, however, was retained by large landowners for agribusiness ventures or sold at great profit.[15] Leon Bailie grew rich from the greatly increased value of his land.[16]

The newcomers settled their young families on the high bluffs east and north of the plant. At first the families struggled financially. They lived in trailers and huts and ate what they grew, saving money on groceries.[17] A lot of small farmers went bankrupt, but for those with capital and patience the venture paid off. With channeled water soaking the crops under the fierce desert sun, with hard work, and good terms from government-backed loans, the families began to prosper like no other farmers before them in the Inland Empire. They built new houses, bought cars, and sent their children to local schools.[18] With cheap credit and savings from subsidized water, the farmers bought fertilizers and pesticides, which they poured on their crops, and they watched astonished as the crops grew, this green revolution brought to rural America by the ingenuity of corporate scientists and the largesse of federal agencies.

Releasing the buffer zones around the plant to farming families was a gamble. As farms, towns, schools, and families clustered around the plant, the potential consequences of a large-scale radioactive accident grew exponentially, but these risks were not apparent because they came masked in a familiar narrative of American prosperity.

In fact, the Columbia Basin Project's spatial reorganization of the landscape helped push calculations of risk to the margins. Within the nuclear reservation control zones, the land was brown, rocky, depopulated, and barren. The reservation, surrounded by cyclone fencing and punctuated with nuclear hazard signs, indeed spelled danger. But just outside the control zone, new irrigation networks created a prosperous, lush landscape of crops and cozy hamlets, the very image of health and prosperity. AEC claims that the plant's threats were safely contained within the reservation made visible sense of the two polarized landscapes. As years went by and no reactor blew, that fact too increased the confidence of plutonium plant employees and their neighbors, who were allowed to remain oblivious to the steady release of radioactive isotopes into the surrounding air, ground, and water.

Confidence in American know-how and progress enabled complacency in other ways as well. When Herbert Parker first worried about radioactive waste, he operated from a belief that the same scientists who had so quickly produced atomic weaponry would make similar breathtaking discoveries about how to safely contain radioactive effluence. But the quick bursting creativity of the thirties faltered in the postwar years among scientists sequestered in classified labs.[19] During the Cold War, engineers did not figure out how to safely and permanently store radioactive waste.[20] They have yet to find a solution.

Hanford scientists and their superiors could not admit to having no contingency plans, no way to safeguard the populace, including the new farmer-veterans settling in the lee of the plant. To do so would have meant shutting down, and since ceasing plutonium production would be a national security risk, that was not an option. Instead, public figures and corporate scientists carried on as if nuclear technology had not just changed the calculations of risk management irrevocably, rendering the notion of public safety meaningless when it came to nuclear affairs.

26

Quiet Flows the Techa

Maps of the region around the Maiak plutonium plant were rare before the war, and after June 1947 they were classified secret. A hand-drawn map on onionskin in the Cheliabinsk archive is the only historic depiction I have seen of the area around the plant. Across the map flows the Techa River, a soft breeze of blue pencil. The river winds along the paper from Lake Irtiash to Lake Kyzyltash, where, emerging, it splits in two; after the two branches meet up again, it wanders over to Metlino Pond before heading eastward off the map.[1]

The map is fascinating because it arrests a moment in the transition of this remote territory from small-scale farming and fishing to a nuclear production zone with global significance. The map dates from May 1947, just before the district was hurled into the postindustrial nuclear age without ever having stopped for industrialization. This was a stunning transformation, perhaps unique in the history of industrial development. Looking at the area covered by the map over the course of the subsequent decade, say from the view of a spy satellite, an analyst would notice an increasing naturalization of the nuclear landscape as populations receded, evacuated to make room first for the plutonium plant and later to clear contaminated territories. Time-lapse photos would show hamlets fading away, fields turning into forest, roads overtaken by brush and swamp—nature winning handily, or at least a ghostly postnuclear version of it. After five decades of nuclear production and unregulated dumping of radioactive waste, the lake region surrounding the plutonium plant appears nearly as untouched as the first scouts found it in 1945. Yet in 1990, to stand on the windswept, reedy shores of Lake Karachai for an hour was to get a fatal dose.[2] The difference from 1947 is that the landscape, still beautiful to behold, is now dangerous to traverse.

It is not clear who first decided to measure levels of radioactivity in Metlino, a pretty village on a pond a few miles down the Techa River from the Maiak plant. Maybe word got to plant scientists of barren livestock and domestic fowl dropping in the paddock on neighboring collective farms.[3] Perhaps it was the ten-pound pikes with ulcers on their spines and eyeballs clouded with cataracts blindly jabbing at the riverbank.[4] Or maybe news spread that Metlino's infants

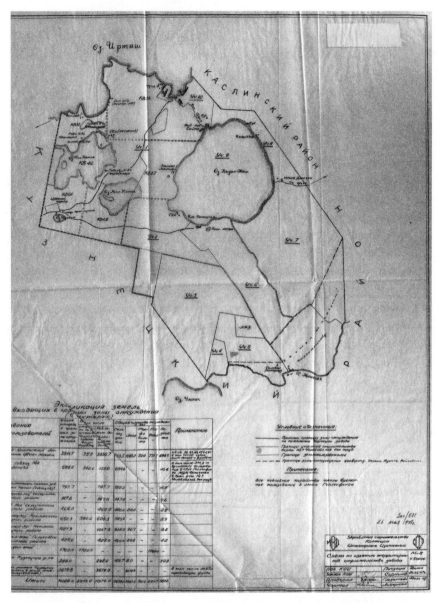

Map of land holding, Ministry of Internal Affairs of the Kuznetskii region. Courtesy of
OGAChO.

had an unusually high number of birth defects.[5] Perhaps in the lull of 1951, the
first break after five frantic years of construction and production, Soviet scientists finally had time to wonder about the effects of the plant's waste pouring
forth into the surrounding farms and forests.[6]

Maiak's gaseous waste was pumped unfiltered up through smokestacks and
was easily forgotten. Maiak's liquid waste was rated (high, medium, and low)

according to the amount of radioactive isotopes it contained. The high-level waste was and is very dangerous. A Dixie cup of such waste in a conference hall would kill everyone in it.[7] What should be done with such hazardous waste? Amidst the great belief in the inexorable progress of science, Soviet engineers assumed that they would soon invent a solution to the plant's refuse, possibly even recover the spent fuel for useful purposes, such as agricultural fertilizer.[8]

Following the American example, Soviet engineers settled on temporary solutions to dispose of radioactive waste. Soviet engineers built a handful of underground storage containers to store high-level radioactive effluent until a more permanent solution could be found. Low- and medium-level waste they dumped directly into the Techa River.[9] Soviet scientists knew that the liquids they were dumping exceeded permissible concentrations for an open hydraulic system, but they also knew that the Americans dumped radioactive waste in the Columbia River at Hanford.[10] The Columbia, however, was a swift, high-volume mountain river with a rocky bottom, and it raced to the Pacific Ocean. The Techa River was slow-moving and turgid.[11]

The medium- and low-level emissions into the Techa quickly became a problem. In July 1949, after six months of dumping, plant director Boris Muzrukov wrote a memo stating that the Techa "is exceedingly polluted."[12] That problem, however, was soon overtaken by a bigger setback. In late 1949, Muzrukov got word that the expensive, underground containers of high-level waste had been filled to the brim. Caught unprepared, Muzrukov had no place to stash the hundreds of gallons of extremely radioactive effluence discharged from the factory each day. Muzrukov faced a decision: either stop plutonium production long enough to build more containers or continue operation and find someplace else to dump.[13]

There never really was a choice. At the start of 1950, Soviet and American leaders were embroiled in Korea and China, and there was a lot of talk about an "inevitable" nuclear clash with the United States.[14] In the menacing doom of the early Cold War years, ceasing production was out of the question. Dumping into the river then was one of those solutions that made sense at the time. The beauty of rivers is that they effortlessly carry garbage elsewhere. That is why rivers have been man's chosen waste repositories for centuries. Who would notice? With radioactive waste, impalpable and invisible, slipping a few million gallons into the Techa would cause no breach in plant security. In January 1950, Muzrukov ordered his supervisors to dump the plant's effluent, all of it—low-, medium-, and high-level, an average of 4,300 curies a day—under the ice into Lake Kyzyltash, from which it quietly flowed into the Techa River.[15]

Calling the Techa a "river" is a generalization. It is more a shape-shifting hydraulic system rolling over a vast, flat sponge. Along most of the river's pathway the ground was poorly drained. Groundwater bubbled just below the saturated surface, which quickly turned marshy when it rained or when the snow melted. The Techa meandered, hooking north or south and dissolving into a lake, pond, or swamp before continuing lazily on. In spring thaws the river spread out capaciously across corpulent flood plains, refreshing forest and field with the river bottom's mineral-rich sediment. The summers in the Urals tend to be dry. Because of the Techa's spreading fertility and sweet, clear water, most of the farms in the region were located along the river and its lakes and tributaries.

Once the plutonium plant's pipeline started to drain into the river system, the river was rapidly overtaken. From 1949 to 1951, a full 20 percent of the river's water consisted of plant effluent. For two years the river took the blow of about 7.8 million cubic yards of toxic chemicals containing 3.2 million curies of radiation.[16] This colossal volume combined with the boggy river system created a radioactive landscape.

Like pearls on a necklace, forty-one settlements lined the Techa River. In the villages dwelled 124,000 people; a majority were Tatars and Bashkirs, Muslim minorities speaking Turkish languages. These people had inhabited the territory, according to Tatar historians, for centuries.[17] Metlino was the first downstream community, a little over four miles from the plant. It was an older, solid village with a stone mill, a handsome old church, a butter factory, and twelve hundred residents, including Russian employees of the plant and farming Tatars and Bashkirs.[18] In 1946, NKVD bosses requisitioned the collective farms surrounding Metlino and staffed them with exiles to produce for the closed city, but officials allowed the original residents to remain to work for the plant.[19] Because workers and farmers growing for Ozersk lived in Metlino, the party leadership in Ozersk took an interest in the village.[20] In 1951, a small team of radiation monitors and doctors arrived in the village to take measurements. They found that Metlino Pond radiated at five rads an hour. In the adjacent village, the meter read three and a half rads.[21]

This was appalling news. With these levels, a person could get a lifetime external dose in less than a week. Worse, the scientists were shocked to find that the villagers had been ingesting radioactive waste, as they used the pond and river as a source of water for drinking, cooking, bathing, and watering crops and livestock. In the villagers' homes, every household object—beds, tables, chairs, pots, and especially teakettles—radiated at levels hundreds of times greater than background radiation. The bodies of villagers also sent the needles on the detectors scrambling. Some bodies were radioactive sources worthy of quarantine.[22]

Working quickly, plant managers sent in a brigade of men in uniform. The men said not a word. They silently gathered up the villagers' geese and ducks, shoved the screeching birds in a truck, and drove off. A few days later, the men returned and, again explaining nothing, swept away twenty-nine families living directly on the pond, the soldiers pushing the confused men, women, and children through the hatches of waiting helicopters. No one ever saw those people again. No one has since figured out what happened to them.[23]

The soldiers returned a third time and began to dig wells in Metlino. They told the remaining villagers that the state had banned the use of the pond and river. They didn't explain why. The wells the men hastily dug were shallow, and the water tasted of sulfur and salt. So the villagers returned to the pond with their buckets, slipping under the fence posts strung with wire and warning signs. To enforce the ban, plant managers deputized guards to keep villagers away. The children especially thought it was fun to dash under the fence and into the water when the deputy turned his back. Muzrukov ordered the factory's central lab to take regular samples, setting up the first environmental monitoring service on the river, but he did not request medical exams of the residents.[24]

It is a sign of how seriously plant medical authorities saw the unfolding catastrophe that they assigned to Anatolii Alexandrov, a leading Soviet physicist, the job of leading a commission just after a major flood in the summer of 1951 to investigate the contamination of nineteen villages downstream from Metlino.[25] It wasn't easy to get downriver. The teams of medical researchers and radiation monitors beat their way along unmarked paths and dirt roads to the riverside villages. They took samples of water, fish, birds, plants, and animals. Analysis showed radioactive deposits at almost every point along the river. In remote hamlets far from consumer networks, they found that villagers subsisted off their own produce. They hunted, fished, farmed and gathered in the forests. They drank, washed, and watered livestock from the radioactive Techa. As the scientists traveled and measured, they grasped the extent of the tragedy.

The river served as a powerful mirror reflecting gamma rays onto plants, animals, and humans. Villagers also ingested radioactive isotopes through their agricultural produce.[26] Blood samples showed that the villagers had received both external doses of radiation and internal doses from cesium-137, ruthenium-106, strontium-90, and iodine-131 that had been deposited in organs and bone marrow.[27] Villagers complained of pains in their joints and bones, vague illnesses, strange allergies, intense fatigue, disturbances of mood and sleep, weight loss, heart murmurs, and increased hypertension.[28] The villagers' blood counts were low, their immune systems weakened. Mothers had miscarriages and bore infants with birth defects at rates up to three times greater than normal.[29]

Alexandrov's commission made several important recommendations. The first was to cease dumping radioactive waste into the river. Wasting no time, engineers drew up a plan to dig a series of canals to divert waste to Lake Karachai, a swamp with no outlet, which they renamed Reservoir No. 9. They also planned to block the Upper Techa with a dam to stop the spread of contaminated sediment.[30] Contractors started to build more underground waste storage tanks for radioactive effluence. The commission further suggested a ban on taking water from the Techa. They proposed digging wells in twenty riverside villages and putting up fences to zone off the river.

Muzrukov, meanwhile, discontinued shipments to Ozersk warehouses of contaminated produce from the Metlino collective farm.[31] He ordered an investigation to find the parties guilty of dumping into the Techa River.[32] Investigators, however, made no charges because in 1949 Muzrukov himself had ordered the dumping to commence and every plant boss along the production line had participated. Apparently, dumping high-level radioactive waste into open waterways was not a crime. As the chief of waste engineering at the plant remembered, "Until the sixties, there were no limits [on dumping], and no reliable count of disposed waste. Whatever was considered necessary was dumped. No one feared any kind of sanctions."[33]

In accord with the commission's recommendations, Moscow issued orders to dig wells in twenty Techa River villages. These orders were reissued at the province level, and again in the regions and villages.[34] Case closed—a job apparently well done. Americans in the nuclear field at times envied Soviet leaders because they imagined that the Soviets, with a centralized authoritarian state, could solve big problems quickly, without the fuss and inefficiencies of democracy.[35] After the orders were issued, however, leaders in Ozersk and Moscow were not aware that villagers felt no urgency to carry out the strange orders to dig wells. The plant was a state secret; so was its radioactive waste. Farmers were told they should drink foul-tasting well water instead of sweet river water "for epidemiological reasons." Not surprisingly, the collective farm chairmen ignored the directive. So did the local party bosses, just as they disregarded many other decrees from the provincial capital, especially ones that made no sense. Officials in Cheliabinsk, meanwhile, were slow to fill out the paperwork needed to pay for the new wells. Once they did, village chairmen didn't draw down the funds because local farmers, who were taxed with digging the wells, were busy farming.[36] If they did get to digging, they started in November, after the harvest, when snow covered the earth, the frost line steadily deepening. Hacking into frozen earth, they made slow progress, and the wells were shallow.

Eighteen months passed. During that time, provincial officials carried out programs that placed more people in harm's way. From 1951 to 1953, sixteen hundred more people moved into two new geological research stations directly on the irradiated Upper Techa, near the plant. In the same years, several new irrigation projects got under way, with pumps and pipelines drawing from the Techa to water thousands of acres of pasture and cropland.[37]

Putting more people on irradiated land and irradiating more farmland were not part of a deliberate, malicious plan. Nuclear security dictated that one hand of government did not know what the other was doing. Along the river, agronomists improved crops by supplying (radioactive) water, geologists set up stations on the (irradiated) Techa to search for mineral deposits, and farmers delivered (contaminated) milk, butter, and grain to warehouses in Cheliabinsk; all were competently carrying out their jobs, but in so doing were spreading a path of contagion.[38]

Finally in late 1952, A. Burnazian, head of the State Service of Radiological Protection, sent a letter stating that none of the proposed eighty-nine wells on the Techa had been built.[39] The plant boss, Muzrukov, deputized an inspector, Grigorii Markov, and gave him the job of making the difficult trip on horseback to oversee the well digging.[40] Markov was appalled at what he found. The villages closest to the plant had no wells in use. Villagers were still drinking right from the river. In a pique, Markov wrote that local bosses "view the well-digging work as spectators, taking no part in it."[41] At the same time, no one could figure out why fowl were dying in alarming numbers, as were cows and goats. On some farms, all the animals had expired.[42]

In August 1953, medical researchers from Moscow and Ozersk met to try to figure out the boundaries of exposure and damage. Of 28,000 exposed villagers, doctors examined 578 people. They determined that 200 had clear cases of radiation poisoning and 54 had possible cases. In other words, nearly half of the small portion of exposed people who were examined were sick from radiation poisoning—an ominous finding for the unexamined 27,400. Researchers also determined that although the plant was no longer dumping into the river, villagers' internal doses were still going up.[43]

The committee recommended building yet more wells, twice as many, and repairing and deepening unused existing wells.[44] They also advocated opening clinics in the remote villages and adding extra beds to regional hospitals.[45] But these measures would not contain the damage, the scientists admitted. The symptoms of the population were too widespread, and the scientists had no technological means to clean up the river and the polluted landscape. The only way to avoid new illnesses and cure those who were sick was to evacuate the communities on the river and clear a five-mile buffer zone around it. The solution, they admitted, was "radical" but necessary.[46] Indeed, the recommendation meant moving sixteen villages and carving out of the southern Urals a 230-square-mile zone of exclusion.

This "radical" solution of emptying sixteen villages on the river was proposed not by a fringe environmental interest group but by loyal Soviet scientists who were not given to exaggeration. At the time, Soviet scientists generally underestimated the impact of radioactive isotopes on human health. Yet the commission's scientists, the world's first eyewitnesses to a nuclear landscape, recommended a drastic, public, and expensive evacuation. Officials far away in Moscow, however, balked at the 100-million-ruble price tag. They indeed found the solution radical and trimmed the list to ten villages.

After the orders were issued, the resettlement program progressed slowly. Initially, plant employees refused to report to work on the Techa out of fear of contamination.[47] When they did go, they had trouble finding the remote villages along unmarked trails through woods and swamps.[48] In the end it would take more than a decade to resettle ten villages away from the Techa. The lethargic reaction to this emergency occurred in large part because the Techa villagers existed far outside the orbit of plant managers and medical personnel in Ozersk. The two groups were divided by tightly controlled spatial zones as well as by class, ethnicity and long distances traversed slowly on foot or horseback. These cultural and spatial distances proved critical.

27

Resettlement

The first nongovernmental radiobiological inspection of the Techa Basin took place in 1990, thirty-nine years after plant managers ceased dumping highly radioactive waste into the river. When Soviet officials first reopened the area, which had been closed in 1954, a team of volunteer scientists tried to raise money for their trip to the Upper Techa region. It wasn't easy. In 1990, the Soviet economy had just fallen off a cliff. Prices were climbing, and inflation was running away with people's life savings. The team could find no donors, so the small group of radiobiologists borrowed equipment and a car, and pooled their own money; even so, they had trouble getting enough gas to drive the seventy miles to the zone. They had no tent and spent nights in the open, fending off mosquitoes. They had little food with them. In rural areas shop shelves were barren, so they negotiated with farmers to buy potatoes, which they suspected had been grown in contaminated soil. With no place to shower, the sweaty, bug-bitten scientists jumped into the irradiated Techa.

V. Litovskii, a member of the expedition, kept a diary.[1] He described how, as they took measurements on a floodplain along the Techa, a farmer stopped his horse-drawn cart. With him was his ten-year-old son, who appeared to be mentally disabled. Litovskii asked the farmer whether the field, a floodplain that should have been restricted, was in use. The farmer said the land was his. He explained he had gotten the pasture through personal connections at the district council. He used the rich pasture for fodder. Litovskii took a reading of the grass—820 microroentgens an hour, a level higher than near the blown Chernobyl power plant. Litovskii tried to explain the problem of radioactive feed and showed the farmer the shaking needle on his Geiger counter. The farmer tried to tell Litovskii how much work he had invested in cultivating the field, and then, getting angry, he showed Litovskii the sharp tines of his pitchfork. Litovskii realized he was having this discussion in the wrong place. "This was a conversation," he wrote in his diary, "for the halls of power."[2]

The independent expedition was possible because Mikhail Gorbachev had encouraged Soviets to build a civil society and participate in their government. The

Boy in the Techa River region. Courtesy of Robert Knoth.

1990 expedition was one of the first skeleton crews of that civil society in the nuclear archipelago of the Urals, a region long controlled by powerful security forces. This self-funded research effort was far from comprehensive—"pathetic," Litovskii admitted. The handful of scientists wrote up their findings in school notebooks and, having no maps, sketched their own by hand. Yet, despite the shortcomings, the independence of the small expedition was greatly valued because by 1990 Soviet officialdom had as much credibility as a used-car salesman.

The volunteer scientists presented their findings to the newly elected Soviet parliament. They reported the hair-raising readings they randomly encountered on sites along the river—at deltas, under bridges, on floodplains as far as sixty miles from the plant. They noted that the Asanovo Swamp, with an estimated 600,000 curies of radioactivity, from cesium and strontium isotopes, was open and accessible to the local population. They recounted the secret history of the years of uncontrolled dumping, in which an estimated twenty-eight thousand people in the Techa region had received doses from 3.5 to 200 rem, where 5 rem is the maximum annual dose for a nuclear worker. These people had the third-highest occurrence of leukemia after Nagasaki and Hiroshima, plus a notable excess of cancers of the intestine, liver, gallbladder, uterus, and cervix, and a general mortality that was 17 percent to 23 percent higher than their neighbors who did not live on the river.[3] The news was broadcast across the nation. Emerging national politicians such as Boris Yeltsin used the Techa catastrophe as one more block to build the mausoleum for the expiring Soviet state.

In the Urals, the headlines had a more personal import. People began to inspect their memories to determine whether their lives had touched the Techa and the radiated areas around the plant. For many people, the breaking stories about the secret plutonium plant and the radioactive contamination dropped like a curtain, cutting their lives into separate acts. Act I consisted of unlinked illnesses, fertility issues, and sick children. Act II featured the fear that all one's problems were due to contamination in an unfolding tragedy descending directly from the state's deliberate mass poisoning.[4]

In the summer of 2010 I was living in a primitive cottage on the outskirts of Kyshtym. The cottage was the closest I could get to the closed city of Ozersk in order to carry out interviews with former plant employees and residents along the Techa. When my cell phone rang, I was aggravated, having just broken a sandal strap while hauling a heavy yoke hung with buckets of water from the well. The call was from Nadezhda Kutepova, a human rights lawyer in Ozersk. She said she had set up a meeting for me with a woman in Kyshtym. I wrote down the address, taped together my sandal, and called a cab.

I showed up at Galina Ustinova's house at midday. She was waiting for me, her big, vicious German shepherd chained in the yard. She came out to meet me, her red lipstick freshly applied, her voice at high volume and speed. She asked me about my accent. I was surprised Kutepova hadn't told her I was American. I realized why when Ustinova sat down upon hearing that piece of information, dramatically fanning herself. Foreigners, banned from the region for security reasons from 1947 to 1990, are still rare in the southern Urals. For Ustinova, having an American in her kitchen must have felt like being handed a canister of microfilm in a back alley. Nevertheless, she put on the teapot, heated up potatoes, and bade me sit down.

Ustinova told me right off that she did not have an education. She said that many times, by way of apology, by way of disclaimer. She said she had always dreamed of becoming a schoolteacher, but she was a village girl, and village schools were not so great. She couldn't pass her math exams and so failed to get into the teachers college. Instead, she was sent for a year's training to become a film projectionist. She worked that job for twenty-seven years in Kyshtym's only cinema. "Maybe radiation made it so I couldn't do math?" Ustinova wondered.[5]

Ustinova was born in Nadyrev Most in 1954. Her parents were from Belorussia. After the war, her father had gotten a job on a geological survey team headquartered in the village on the banks of the Techa. They had moved to the Upper Techa in 1952, a year after radiobiologists had determined the river was dangerously contaminated.

Nineteen fifty-four wasn't a good year to be born on the Upper Techa. That was the year leaders in Moscow decided the territory was too hazardous for habitation and issued orders to remove Nadyrev Most and nine other villages from the map.[6] I showed Ustinova a photograph from the first 1990 expedition on the Upper Techa. The scientists had passed through Nadyrev Most, and under a bridge crossing the Techa they found an old sign that read, "Warning—radiation! 1500 microroentgens/hr."

"I had my first bath in that river!" Ustinova said. "I ate from it. It was my bread."

When I asked her to elaborate, Ustinova told me she remembered nothing of her first years in Nadyrev Most. Instead, she said, she would always ask her parents why they had moved so often—first from Belorussia to the Urals, and then from Nadyrev Most. Her parents' answers were so vague and incomplete that Ustinova repeatedly returned to the topic. Ustinova learned only in 1997, long after her parents had died, the reason they left her birthplace.

In 1956 soldiers showed up, unannounced, in Nadyrev Most. They read orders for immediate resettlement. They had done this already many times, and they were quick and efficient. While the soldiers rounded up the live-stock, a team of young agronomists, the educated elite of the Soviet country-side, went from house to house making inventories, tallying up the number of chickens, pigs, and goats outside, and the numbers of pots, pans, shirts, and shoes inside.[7] The lists are sad to read. Most families didn't have much to count—the entirety of a household translated in rubles to less than a monthly salary of a supervisor at the Maiak plant.[8] An officer gathered the heads of the household and told them they were moving to a new settlement, where they would be given a new house and be reimbursed for their belongings. The officer said if they spoke of this evacuation they would face twenty years' im-prisonment; adults, confused and agonized, were made to sign a piece of paper to that effect. Most people in the region knew of earlier deportations of kulaks or ethnic Germans, who lived in multitudes in the Urals. The villagers figured they too must have broken a law and were being punished, though they weren't sure for what.[9]

Then the soldiers did something unforgettable, as if the nice boys in ill-fitting uniforms were enemy forces in the grip of war. They gathered in a great heap a century's worth of possessions—wooden plows and harnesses, lathes, linens, bridal laces, woolens, quilts, most made by the hands of the villagers—and burned them on a communal pyre. Next, the conscripts led the livestock to a field and began to shoot them. Green, canvas-topped trucks lined the roads, engines idling. Children, excited to take their first ride in a vehicle, shouted to one another, but their mothers wailed and men cried over the sound of the bleating, frightened farm animals. In the field, soldiers fired round after round. Drivers gunned their engines, propelling the trucks over rutted, icy paths.[10] As the first trucks departed, the voices and sounds merged in a spontaneous requiem to the river community.

As the last residents departed, bulldozers waddled in behind them. They dug trenches and pushed in the cottages, burying them in mass graves. In a few days, Nadyrev Most had disappeared.

This was an emergency that was never treated as one. In August 1953, plant scientists recommended removing sixteen settlements on the Techa. In Moscow, leaders procrastinated, and then fourteen months later, in October 1954, issued resettlement orders for ten villages. Security forces from the Ministry of Internal Affairs were to carry out the evacuations, while the plant's Construction Division No. 247 had the assignment of building housing, schools, clinics, wells, and roads in the new settlements, and digging wells and putting up fencing in the villages that remained on the river. The orders spelled out clearly that this project was to be given top priority—"equal to work building reactors and processing plants at the main object."[11]

Yet Nadyrev Most was resettled in 1956, two years after the decree was issued from Moscow and a full five years after the river was discovered to be contaminated. The delays were caused by the extremely slow initial reaction to the disaster and then the relaxed pace of construction of new settlements. During the years that spanned awareness and action, villagers ingested more doses of harmful radioactive isotopes, and more children were born.[12] Teenagers, growing rapidly, absorbed the largest doses. Fetuses exposed in utero, infants, and children, like Ustinova, were most at risk from the damaging effects of radioactive isotopes.

Had the leadership in Moscow and construction bosses in Ozersk acted more swiftly, Ustinova's life would have turned out differently. Ustinova's parents were dead by the time she was twenty-five. By her mid-twenties, her blood pressure was so high she could barely get out of bed. Doctors diagnosed her with heart disease and prescribed medicines. When she turned thirty, doctors diagnosed thyroid disease. "I didn't understand," Ustinova said, pointing to her chest. "Look at me—I am big and I look healthy, but inside I am weak." Ustinova's first child was born prematurely, failed to thrive, and died within a year, at the same size as when it was born. Her second child, a daughter, started to have medical problems at age twenty-five. She underwent several operations and now is in a wheelchair. In the last decade, Ustinova had a cyst removed from her liver and suffered a stroke that temporarily paralyzed her left side. Recently doctors have found tumors on her thyroid. Her daughter has not married and has no children. Ustinova and her daughter are the last of their line.

Galina Ustinova took me to the resettlement village where she grew up, a pretty lakeside hamlet called Sludorudnik, populated almost entirely with evacuees from the Techa. The evacuee settlements are strange parodies of mass-produced Levittowns in the United States. Identical prefabricated houses of thin plywood sheets line up in rows. The schools are stamped from one blueprint, clinics from another. The streets are named in the same repetitive manner—instead of Maple, Pleasant, and Sunnydale, they are Marx, Lenin, and First of May. One evacuee, Dasha Arbuga, told me when they arrived in the settlement they were made to hand in their contaminated clothing and everyone was issued a new suit of clothes. "Imagine a town," Arbuga said, "where the girls all walk about in identical skirts, shirts, and shoes, their mothers in larger versions of the same, and where the boys match their fathers too, down to their buttons."[13]

People didn't like the new settlement, especially the way the Siberian winds blew through the hollow walls of the prefab houses. They kept asking to have their sturdy old log houses back. They didn't know their houses were buried in the earth in undulations visible today from the road in the evacuated territories.

Ustinova had the taxi stop in front of one of the houses in Sludorudnik. A man ran out to chain up his two snarling German shepherds. "Good, mean dogs," Ustinova said, nodding approvingly. Ustinova introduced me to two elderly women, Anna and Dusia, who were wearing village garb of long skirts and head scarves, in stark contrast to Ustinova's urban appearance. We sat down to tea. Anna and Dusia matter-of-factly described their resettlement in Sludorudnik. The new settlement was placed next to a quartz mine, and after the evacuation the former farmers became miners. Anna said a lot of miners got silicosis. She described her job on the conveyor belt, lifting rock, fifteen tons a day. I eyed the diminutive Anna, about five feet tall and 120 pounds, and wondered how that was possible. It seemed that Anna and the others had jumped out of the radioactive frying pan and into the fire of hard-rock mining, with its more mundane yet still treacherous hazards. Work in the twentieth century was dangerous for a lot of people; in the twenty-first century, it still is.

Anna and Dusia made a list of the neighbors who were no longer with them, having died before they reached fifty. Reciting last names in the plural form, they ran through the families whose genetic line had expired: "Krasovykh nyet, Kupchitskikh nyet, Kanavalovykh nyet, Belkanovykh nyet, Ivanovykh nyet . . ."[14]

Anna started on another list, the kind I had heard many times in the communities around the plutonium plants. Her brother got cancer. Her niece is an invalid. Her two daughters have thyroid problems. Anna's bones ache from inside. Her nephew had psychiatric problems and hung himself.

I stopped her there. What did psychiatric problems have to do with radiation? The women shook their heads at my ignorance. Radiation works on the mind, they explained. Some families had it worse than others. They spoke of a neighbor: four of her sons had committed suicide, and the lone surviving son was in prison for murder.

I took this discussion as one of those cases of radiation phobia, where people who have no training in medicine or physics chalk up every illness to radiation. Looking into it later, however, I learned that some Chernobyl researchers have concluded that radiation exposure damages the central nervous system, inducing neurological and neuropsychiatric disorders. Among Techa River populations, researchers found that 70 percent of children born with birth defects show signs of brain damage.[15]

I learned something from my dismissal of the villagers' observations. Researchers, medical or historical, pass through a settlement such as Sludorudnik quickly, gather data, and return home. Why did I not assume that people also study their own environments and communities up close on a daily basis? And that perhaps their insights, born of a long, painful examination, might be legitimate? After that, I started to listen more carefully.

28

The Zone of Immunity

Why was there no urgency to evacuate the irradiated villages along the Techa River? It's not that the generals who ran the plant didn't know how to uproot and resettle. For two decades Soviet security agents deported millions of people in mass actions.[1] Almost every big military leader of the atomic project dirtied his hands with some mass expulsion. Most of these operations required just a few days, but resettling ten villages along the Techa, a mere ten thousand people, took a full ten years. Why did construction bosses so proud of having built a plutonium plant in two and a half years need a decade to build a score of villages?

The order to evacuate came in 1953, the year Stalin died. In Ozersk, the post-Stalin power struggle caused a power vacuum for a critical five years. During that time, enterprise leaders edged out security leaders running the city, and these busy executives felt no urgency to obey Moscow's orders to evacuate the villages of the Techa River. In the United States in the fifties, pundits characterized Stalinism as an overinflated state, with a bloated central bureaucracy planning and controlling every aspect of life. But in Ozersk, management of the plant and city was decentralized and largely left to local bosses. Residents decried the "lawlessness" of their leaders and especially the corruption of enterprise chiefs. Even before Stalin died, they demanded more policing, more planning, and more oversight of municipal services, housing, and construction. They wanted, in short, more government, not less.

Ozersk did not have the usual government or city administration. In most towns and cities of the Soviet Union, party members elected local party committees. The committees met regularly to discuss community issues, solve local problems, supervise public morality, and allocate public funds. The party committee acted, in a constrained fashion, like local governance in a democracy. There was, however, no such "socialist democracy" in Ozersk, as the city wasn't incorporated.[2] No body or set of laws guided this strange company town. Under General Ivan Tkachenko's leadership, a Political Department of nine military officers oversaw the day-to-day management of the city, while the bosses of individual "establishments" (factories and workshops) wielded fiscal and judicial control over their employees' working and living conditions, much the way corporate managers ran Richland. Residents petitioned for funds from "the factory leadership" and appealed to the Political Department to correct abuses and mismanagement. A handful of overtaxed industrial managers and security officials oversaw the administration of a growing city of fifty thousand.[3] As a result, the city was highly undergoverned in an arbitrary, personal way that left the door open for corruption.

In 1952, one man, speaking of the recent arrest of eleven shop managers for embezzling an astonishing one million rubles, pointed to this lack of self-government as the reason for the crime: "It seems to me that around this issue there needs to be more social control. Especially in our city where there is no city council. Like we need oxygen, we need social control."[4]

City prosecutor Kuz'menko agreed, criticizing the "anarchic power" plant bosses wielded over their workers. He complained of bosses who punished their employees without legal grounding. "We found that in one very important establishment, 53 people were sentenced for violation of labor discipline—for drinking, for confrontations with neighbors, for uncultured behavior. 'Uncultured behavior,'" Kuz'menko continued, exasperated, "is a very malleable notion. How can you press charges for that?"[5]

Kuz'menko also complained that the security zone was responsible for the absence of free trade. "In the stores," Kuz'menko declared, "you can always buy vodka—that is how store managers meet their sales quotas—but it's hard to buy meat and potatoes. It's different in the big world. In a normal city, if a factory shop runs out of supplies, workers can go to the peasant markets and find whatever they like. Here, we can't do that."[6]

In 1952, Ozersk residents petitioned Moscow to grant them a party committee so they could govern themselves. Beria at the First Main Department denied the request, but to address these complaints, Tkachenko formed advisory commissions to carry out the various functions of a typical Soviet city council. He founded a commission for education, culture, trade, and public dining, and talked about a commission for public health.[7] In this way, he created a surrogate city government, which, like ersatz coffee, was bitter and less satisfying than the real thing.

In 1952, Tkachenko, visibly aging, his belly bulging under his military tunic, could not command order and obedience as he used to.[8] That year prisoners in the zone's large Kuznetskii labor camp lit up in a "massive disturbance" lasting three days. The convicts disarmed and killed guards, took over the camp, and refused to work.[9] Tkachenko's forces, unable to suppress the rebellion, finally had to call in the army. The rebels were arrested, but insurrection continued. Witnesses described violent clashes between prisoners and guards, and between rival prison gangs and ethnic groups. One melee in 1953 kept doctors working for three days on a virtual conveyor belt of surgeries, stitching knife wounds, extracting bullets, and reconstructing a chest caved in from a hammer blow. In the post-Stalin years, Gulag rebellions mounted. The Ministry of Internal Affairs (MVD) had trouble keeping order in garrisons and prisons across the Soviet Union, but in 1952 few establishments rivaled the plutonium plant's Construction Division No. 247 for disorder.[10]

In the spring, Tkachenko stood before members of the city's Communist Party and denounced a woman who had set herself up in Ozersk as a mystical healer. She sold amulets and crosses and christened babies, all for a price, he said, and she had a lot of takers. A former POW, a Baptist from Ukraine, Tkachenko continued, held secret prayer circles and preached a coming third world war, in which America, "the White Horse," would win. Tkachenko suspected these religious fanatics were plotting to plant enemies in production and sabotage the factory.

Tkachenko had ruled Ozersk by careful cultivation of fear and suspicion, but in 1952 something strange happened: people began to reject Tkachenko's vision of capitalist encirclement. During Tkachenko's speech, a man in the audience shouted that he wasn't so worried about "an old aunty selling crosses." A second heckler agreed, declaring Tkachenko should instead do something about the shortage of books in the bookstore.[11] Asking for more government and less surveillance, these voices undermined Tkachenko's power.

In early 1953 Tkachenko tried again to assert power through fear in what would be his final months of service in Ozersk. Closely following the developing scandal in Moscow over Jewish doctors allegedly poisoning and killing top Soviet leaders, Tkachenko announced in a closed meeting of party activists that "Jewish nationalists" led by American intelligence had infiltrated the closed city. "Foreign spies desire the secret data of our establishment," Tkachenko intoned. He pointed to the trials of "Zionists" in Czechoslovakia and to the arrest of "medic-murderers" in nearby Cheliabinsk. "They all were Jewish nationalists," Tkachenko fumed, "and they obeyed orders from a hidden Zionist organization in Moscow. They probably are plotting other conspiracies, all organized by covert American operatives." Tkachenko listed a number of the plant's remaining Jewish employees, some decorated scientists in failing health from radiation poisoning, and denounced them as saboteurs. Yet, Tkachenko complained, when he had tried to arrest these Jewish agents, they had been saved by factory bosses who argued the scientists were too important to incarcerate. "What insolence, what scoundrels!" Tkachenko raged.[12]

Tkachenko had never sounded crazier. His raving presaged the distant rumble of the collapsing Stalinist edifice, as did the crowd's heckling of Tkachenko, the rebellions among the prisoners, and the refusal of plant bosses to hand over employees he denounced. Even before Stalin expired, the order he had established, shored up by the politics of fear, hatred, and social exclusion, but most fundamentally by a sense of sacred, collective mission, had already begun to disintegrate.[13]

Tkachenko's speech on the Doctors' Plot was his swan song. Before the next party meeting, on March 5, Joseph Stalin died. Party members and loyal citizens throughout the country mourned his death and feared for the future. People wept because Stalin's image spelled order in a country where in many places it was in short supply. Without Stalin, the state became a bit more rudderless. Gulag prisoners rose up, workers slowed down, collective farmers feared their bosses a bit less, and exiles started to think of packing up and heading home.

Lavrentii Beria gave license to this growing sentiment by firing off a series of reforms. Long impatient with the inefficiency and shoddiness of Gulag labor, Beria called for a return to "socialist legality." He put an end to torture in police investigations, and he freed suspects accused in the Doctors' Plot. Beria ended wasteful Gulag construction projects, initiated negotiations to end the war in Korea, and took measures to liberalize the regimes in Eastern Europe.[14] Most important for Ozersk, Beria issued a mass amnesty of more than half of the Gulag population just three weeks after Stalin expired.[15] In April, thousands of prisoners in Ozersk labor camps left, thanks to Beria's amnesty.

Fewer prisoners also meant fewer construction workers at the same time that several major construction projects were under way. To produce a new round of thermonuclear weapons, Soviet arms builders ordered new reactors and processing plants at Maiak. At the same time, plant construction firms were also charged with building ten new villages for evacuees from the Techa River.[16] To make up for the loss of prisoners, Beria transferred nuclear construction projects from the MVD-run Gulag to a new civilian ministry earmarked for nuclear construction. To address the labor shortfall, the Ministry of Defense allocated one hundred thousand soldiers to the nuclear industry; seventeen thousand, accompanied by a thousand officers, went to Ozersk.[17]

Unruly, rebellious prisoners led by criminal warlords filed out, while disciplined soldiers commanded by Red Army officers marched in. This exchange of military for penal labor and civilian for prison administration should have looked like progress. But, within a month of Beria's reforms, the worried messages about unruly prisoners shifted to anxious communiqués about rioting soldiers, while several major construction projects at the plutonium plant fell seriously behind for lack of labor.

The arriving soldiers were housed in the just-vacated Gulag prison barracks along with lice, vermin, uncollected garbage, and other signature features of the camps. The barracks were some of the first built, and they had no plumbing or mess hall. The soldiers were not issued winter uniforms or a change of linen. The bathhouse was out of commission. The young men worked their shifts and then were sent back to their zone, a walled-off garrison of barracks and tattered tents. There were no clubs, books, films, games, newspapers, or radio to distract the young men.[18] Unhappy, bored, and with a shortage of officers, the soldiers went wild, drinking and carousing. In June, the provincial party boss reported to Nikita Khrushchev in Moscow that the arriving young conscripts had committed 891 crimes in one month.[19]

Someone looked into the matter and discovered that most of the soldiers came from the troubled Baltics, Ukraine, and Belorussia, territories that had been annexed by the Soviet government in the war years and where dissent was still sharp. More than a thousand soldiers in the group had criminal records. Among the arriving soldiers were national minorities, including Iranians, Greeks, and Bulgarians. A number of soldiers passed around religious pamphlets published in New York and sneaked off the base to attend church.[20] None of these men—the religious believers, the ex-con servicemen, foreign-born soldiers, or conscripts from annexed territories—were supposed to serve in the special regime zone. When security forces gathered up the worst troublemakers and brought them to the train depots in nearby Kyshtym and Kasli to ship back, the soldiers toasted their departure with vodka, started fights, and smashed whomever and whatever they could.[21]

Meanwhile, the amnesties of 1953 released prisoners convicted of minor crimes and left in the camps recidivists and violent convicts, inmates bitter that the amnesties had bypassed them. In Ozersk, Gulag officers worried about this concentration of hardened convicts and predicted the inmate population would become more volatile in the coming years. They counted a rise in violations, refusals to work, and escapes, and observed that criminal warlords were "terrorizing other prisoners and Gulag staff." Officers also feared that scarcely educated Gulag guards were "going native" after working too long in remote camps with only inmates for companionship. They suspected that some guards had become allies of prison gangsters. During a couple of escapes, suspiciously, the guard dogs didn't bark and the guards slept. When a camp rebelled in 1954, the guards on duty did nothing to warn about it.[22]

Beria's reforms backfired internationally as wildly as they did locally in Ozersk. In East Germany, workers went on strike over low wages and chronic shortages of consumer goods. The photos of young men facing Soviet tanks seemed to provide, as American president Dwight Eisenhower put it, "a good lesson on the meaning of communism."[23]

Nikita Khrushchev worked with impressive speed and political acumen to turn these events against his rival. A few weeks after the Berlin riots, on June 26, amidst troubling news about construction delays and rebellious soldiers in the Urals, Khrushchev hatched a putsch against Beria. Just a few hours before the ceremony to honor Beria and his Special Committee that had directed the Soviet atomic bomb project, Khrushchev had Beria arrested.[24]

Khrushchev had fixed up a case against the small, dark shadow governor of the Gulag. He disclosed that Beria—the man behind the wartime heist of atomic secrets, the great conductor orchestrating the Soviet bomb—was, in fact, an American agent, a traitor, "paid to restore capitalism to Russia."[25] Moreover, Politburo member Lazar Kaganovich charged that Beria had spent money on the atomic branch wantonly, building "not cities but spas."[26]

This news was hard to believe. As head of the Special Committee, Beria had a great deal of power and prestige, especially in nuclear circles. In Ozersk, where Beria had a lakeside villa and a main street named after him, he was seen as a wise, charitable patron. But few people outside the nuclear industry knew of Beria's role in the development of the atomic bomb. To undermine Beria's power base on the day of his arrest, Khrushchev dissolved Beria's Special Committee and set up a new civilian body cryptically titled the Ministry of Medium Machine Building (MSM) to direct nuclear weapons installations.[27] He placed Viacheslav Malyshev in charge, having him report directly to Khrushchev. Beria's longtime deputy, Avramii Zaveniagin, was promoted to direct the first division of the new ministry. Dozens of officers in Beria's circle, including Tkachenko, were fired.

In Stalinist fashion, Beria's former colleagues added "facts" to warrant his death sentence. At the celebration of the Special Committee, Malyshev slandered Beria as "an enemy of the people." Zaveniagin, Beria's faithful sidekick for fifteen years, testified that Beria was stupid and "thick." Igor Kurchatov alone refused to bear witness against Beria. When asked to do so, he replied, "If not for Beria, there would have been no bomb."[28]

To topple such a powerful minister, Khrushchev had to make the charges outrageous and unforgivable, and so he used Beria as a scapegoat for the excesses of the Stalin years. It was a risky maneuver, but it worked. Beria's mass amnesties of prisoners had triggered panic among Soviet citizens, concerned about crime and random violence as the prisoners returned from the Gulag. Soviet citizens cheered when Beria was arrested and executed, not because he had been the minister of the infamous Gulag or Stalin's "henchman," but because they suspected him of treason for going soft on crime, for releasing prisoners, and for freeing suspects arraigned in the Doctors' Plot.[29]

In Ozersk, however, peace did not ensue after Beria's execution. In 1954, 15 percent of the soldiers in Ozersk were caught committing crimes. The young conscripts slipped from their garrisons and went to neighboring towns and villages, where they drank, cursed, picked fights, brawled, and stole.[30] Villagers charged the boys did worse—raped, beat children, and murdered.[31] After raising hell, the boys slipped back into the regime zone, into which the pursuing village police could not follow because they had no pass to the closed zone.[32] Immune from prosecution, the soldiers would go out and raise hell again the next night.[33]

The inexperienced young men worked sluggishly, broke machinery, and crashed trucks. Some soldiers refused to work at all. It was standard in Soviet industries to falsely report overfulfillment of production quotas, but the construction divisions in Ozersk performed so poorly that supervisors had to admit to completion of only a third to half of the targets.[34] The soldiers were so insubordinate that in 1954 contractors refused to take soldiers as workers, requesting prisoners instead.[35] At least the hardened criminals, who had spent years toiling in the Gulag, had job training.

The men who had to answer for the moral failure of the conscripts were the leaders of the Political Department. Gathering information, the military men got busy writing reports, scheduling lectures for the soldiers, and issuing a questionnaire with blank spots to fill in the number of accidents, beatings, and rapes at each base and camp.[36] This bureaucratized the problem but did not solve it. After Beria and Tkachenko were gone, the Political Department had little power.

From a security point of view, this was all bad. The brawling soldiers in local towns were exposing the nature and location of their secret work in the plutonium zone. From the point of view of public health, the rioting soldiers greatly slowed progress on two important construction projects: a new processing plant, the Double B, which was needed to replace the highly contaminated original plant, and new villages for Techa River evacuees.[37]

The person immediately responsible for building the resettlement villages was Colonel P. T. Shtefan, director of Construction Division No. 247.[38] Shtefan was a reviled man in Ozersk because his crews lagged behind on every major construction project. Shtefan pointed to his labor force of soldiers and convicts as the reason for missed construction targets.[39] Complaints about conscripted labor had bounced around the zone for several years.[40] "They come, we train them, and they leave. We are nothing but a trade school," a contractor complained. "If you look at this from the point of view of regime secrecy," he added, "does it make sense that 100,000 various people have worked here and left, and now wander the globe?"[41] In 1954, Sergei Kruglov, the chief of the Ministry of the Interior, ordered that nuclear installations phase out forced labor in favor of hourly construction workers. He proposed that demobilized soldiers and freed prisoners be hired and given a subsidy to move their family members to the southern Urals to join them—in other words, to replace the unruly bachelor prisoners and soldiers with family men.[42] This plan made sense. Employees rooted in nuclear families had made for a relatively stable labor force producing plutonium in Ozersk.

The hitch was that hired construction workers lived in their own separate town, another staging ground, called the Constructors' Hamlet. In the walled-off hamlet, the dorms were filthy, the bathhouse and laundry were out of service, and bread, milk, and meat disappeared from the hamlet's one shop, often stolen by staff before they hit the shelves. The cafeteria had leaking walls, the wooden sidewalks were rotting, and there was no park, garbage collection, or radio service. With no proper maternity ward, women had to travel the pitted road to Kasli to give birth, some delivering their newborns right on the road.[43] Construction foremen were desperate to keep the skilled workers in the hamlet, but the living quarters were so miserable and the cultural services so grim that people fled or grew demoralized.[44] When asked why he failed to build decent housing for his employees, Shtefan bemoaned that he was short of money and supplies and bedeviled by volatile workers.[45] Indeed, in July 1955, soldiers looted another village, assaulting and terrorizing villagers in a clubhouse, and prisoners exploded in yet another camp mutiny.[46]

Yet something else was going on, something hinted at in party meetings. Shtefan and other construction bosses were including in their annual construction plans projects they never intended to build. They did this because the greater the budgets, the more Shtefan and his cronies could skim from state coffers.[47] Shtefan's creative accounting, one critic charged, made the "annual [construction] plan nothing but pure fiction."[48] Soldiers were idle not because of misbehavior or moral failings but because they sat on jobsites waiting for instructions and supplies that failed to materialize. The slower they worked, the longer a project lingered, and the more funds Shtefan could request in subsequent periods for cost overruns caused by "delays."[49]

Embezzling money and supplies left few resources to erect villages for evacuees, build housing for construction workers, or fix rotting barracks. When Shtefan did finally build them, he skimped. He signed off on new villages that had no bathhouses, medical clinics, kindergartens, barns, or dairies, all of which were called for in the blueprints. He put up high-priced prefab houses, built so poorly that they collapsed on their dwellers. "Those houses cost 43,000 rubles apiece," a collective farm chairman complained, "and they are worthless. They fall down. I don't understand how they steal money, on the left and right, and the Political Department doesn't even notice."[50]

In most Soviet cities, the secretary of the provincial party committee would oversee and punish the corruption of industrial bosses in his province, but because of the regime of secrecy, the secretary of the Cheliabinsk party committee could not enter Ozersk or go anywhere near the plutonium factory to investigate.[51] Once General Tkachenko was fired, the Political Department became largely bureaucratic window dressing for local governance. Officials in Moscow rarely followed up to see if their orders were carried out. They gave the plutonium bosses a great deal of independence, as long as plutonium deliveries rolled in on time.[52]

In short, the zone floated up a comfortable bubble of inscrutability for company bosses to enjoy cost-plus arrangements whereby the higher the plant's budget, the more riches the bosses personally commandeered. With the extra money, Shtefan built the plutonium bosses houses in a stylish lakeside neighborhood. He diverted soldiers from commissioned projects to put up garages and vacation houses.[53] Shtefan was so good to the plutonium bosses that, despite the many charges of embezzlement, he remained in his job for decades. The bosses also hired boats, threw parties, and appropriated public amenities as their own.[54] And no one could stop them.

Historians and memoirists often describe the nuclear regime zone as a prison, penning residents in, controlling and disciplining them, but the zone also gave people inside an unusual freedom to break laws and evade prosecution. Shtefan was simply doing on a large scale what a lot of people were doing in their own minor realms. Soldiers took advantage of the zone to loot neighboring villages. Shop managers used the zone to sell stolen food supplies outside the fence, while garrison commanders hawked purloined construction materials for impressive profits. Commanders' wives bought up scarce goods in the specially supplied socialist city to trade to friends in the "big world" at speculative rates.[55] Created to ensure strict adherence to security regulations, the nuclear regime zone enabled shocking corruption. At the same time, this zone of immunity left the Techa River dwellers and the young workers in the polluted chemical processing plant languishing for years.

29

The Socialist Consumers' Republic

I know of Ozersk's rampant corruption only because party members accused industrial bosses of embezzlement amidst the flaring tempers of Khrushchev's Thaw. In March 1956, Nikita Khrushchev gave an eight-hour speech exposing the beloved Stalin as the center of a cult of personality. Stalin was a dictator, Khrushchev said, who had laid waste to the egalitarian, democratic values of the 1917 socialist revolution. In subsequent months, Khrushchev eased censorship, pledged peaceful coexistence with the West, amnestied an additional four million prisoners, and argued that communism in Eastern Europe could take many national forms. In the spring and summer of 1956, first Polish and then Hungarian workers and intellectuals took Khrushchev at his word. They toppled statues of Stalin and rose up in the name of a national form of socialism and, more ominously, national sovereignty. In late August, Hungarians declared independence, and Western photojournalists gleefully snapped shots of Warsaw Pact tanks shoving civilians aside in the streets of Budapest.

In Ozersk, the Thaw was no less pivotal. On the wave of de-Stalinization, the people of plutopia won the right to leave the city—to go visit a sick mother, for example, or find work elsewhere. After nearly a decade of incarceration, this major reform radically changed the nature of life in Ozersk. Factory bosses could no longer hold workers in place by force. They had to work harder to persuade workers to stay at their dangerous jobs in the remote fortress. In 1956, Ozersk residents also won the right to have a city party committee, the main venue for local self-governance in the Soviet Union. After years lobbying for self-representation, the party members of Ozersk did not turn it down.[1] Communists met in groups across the city and happily christened their new "socialist democracy." At first residents found fault with the factory management over safety issues and their despotic, corrupt rule. Over the years, however citizens came to use their new voice to ensure safety not from radioactive isotopes but from criminals and drunks, and they sought security not from nuclear accidents but in terms of material well-being.

On the first day of self-governance in Ozersk, citizens spoke up boldly about what they saw wrong with their city. They criticized the "despotic rule" of the industrial bosses, denounced the former leaders of the city's Political Department as weak and inept, and demanded a voice in city planning and trade. To facilitate local democracy, they wanted a city newspaper, long banned for security reasons.[2]

Most daringly, party members questioned the safety of the plant. They listed the high number of accidents. "Managers automatically blame accidents on workers violating the rules," a union representative argued, "but when we asked them at Factory 25, 'Which rules?' they could not come up with any concrete violations. . . . Rather, management pressures workers to disregard safety regulations."[3] An engineer described how the radioactive waste division had long been underfunded and neglected. He said workers, who were not wearing the badges necessary to record their exposure, loaded waste into truck beds and pulled it by hand to storage containers. "This so-called 'technology' is shameful, but scientists walk by, see it, and pay no attention at all."[4]

The plant's chief engineer, G. V. Mishenkov, had to admit that poor safety practices were causing a labor crisis. The rules permitting workers to leave the closed zone opened a floodgate of resignations. In 1956, more than a hundred employees left each month, many of them engineers and technical people. "I asked a few why they were going," Mishenkov said. "They answered the work is stressful and dangerous." Trained workers were already in short supply because, after ten years of production, people were succumbing to a host of illnesses, which doctors associated with radioactive working conditions.[5] This labor crisis inspired a new consideration for plutonium workers. "We have to forget the earlier decrees that empowered us to keep workers on production by fiat," Mishenkov argued. "Now the situation is different and if we do not create better living and work conditions, the comrades will leave us."[6]

Party leaders agreed that the best way to keep employees was to tempt them with urban magnificence. In 1956, there were a few posh neighborhoods for the bosses and a nice downtown, but large sections of Ozersk consisted of shacks and shanties. The population was growing fast, by fifteen hundred babies a year. Families were living in communal apartments and barracks, five to a room. The housing committee determined the city needed more than four hundred thousand square feet of new living space.[7]

Party members demanded that the construction boss, Shtefan, account for his failure to complete any residential construction projects in Ozersk in several years. Working to stymie complaints and please his new constituents, who were charging corruption, Shtefan thought fast. His crews had not finished apartment buildings in Ozersk because he had stolen the money earmarked for them or spent it on luxuries for the city elite. Shtefan promised to make up for the shortfall by cutting the budget for construction of the Double B plant and new villages for Techa River evacuees. He redirected a full 20 percent from these projects to residential construction in Ozersk.[8] That move would in the coming years considerably ease housing problems in Ozersk, but at the expense of workers' health and the health of downstream villagers. In the new populist climate, housing and material well-being were popular, while safety concerns from invisible contaminants were less tangible.

Ozersk residents already lived far better than their neighbors outside the zone. One young woman, Taishina, pointed this out at a party meeting: "We were chased out to hear a lecture on Marx. But I was out in the big world and there people don't live so well. There is poverty. Why don't you give us a lecture on that?"[9] Publicly, party leaders told Taishina that the residents of the closed city deserved the state's largesse because their work and sacrifice defended the nation, and in turn, the nation needed to support them unconditionally. Privately, after her question Taishina became a potential security threat and security agents started a file on her. There was little room for socialist egalitarianism as it related to the "big world." What mattered was equality and democracy in the little world of Ozersk.

A team of architects arrived from Leningrad to redesign Ozersk into a "socialist city," with broad arteries to ease traffic congestion, generous tracts of green space, and broad squares for public processions. Soviet architects had been designing socialist cities on paper since the twenties.[10] Ozersk, with its expansive construction budgets and high-end government prioritization of consumer goods, offered a unique opportunity to see a socialist city emerge in three dimensions. Locals wanted a role in shaping it. "Why don't they consult us?" people asked. "It makes no sense to place the morgue next to the school and the auto shop so it fumigates the hospital."[11] Tired of renting and dealing with slow, inept maintenance divisions for repairs, Ozersk residents wanted to own their own homes. They asked for permission to lease lots and for financing to build private houses.[12]

Private property? Economic inequity? A liberal press? The Ozersk party committee, roused by Khrushchev's call to renew socialism, was leading the city away from the principles of Soviet socialism that plutopia had been created to defend. No one seemed to notice this trend until a studious thirty-year-old engineer, Anatolii Lanin, named it. Lanin, a ten-year plant veteran, speaking at a meeting in his lab, interpreted the city's housing problems, the bosses' corruption, and frequent plant accidents as a result of a "cult of personality" in Ozersk. He wanted to know why collective farms had failed year after year to feed the nation. He wanted more information about events in Poland and Hungary. In scientific journals, Lanin noted, scientists debate each other, but on political questions in the Soviet media "we have only consensus." Lanin asked why the Soviet press "lacquered" reality, especially about economic facts. "Why don't our lecturers tells us which country has a higher standard of living—the US or USSR?"[13]

Lanin's colleagues listened to him dumbfounded. He sounded like a conservative commentator on the Voice of America. Yet Lanin had read his Marx and Lenin. He quoted both in such an erudite fashion that his fellow party members were tongue-tied. Some even started to agree with him.[14]

Lanin's superiors responded with an emergency meeting to discuss the "Lanin affair." City leaders charged Lanin with agitation and for listening to jammed foreign broadcasts. He asked questions, they charged, not to fix practical problems but to seed dissent among his colleagues in an attempt to discredit the city's leadership. The province's party boss, Comrade Laptev, suggested that Lanin was plotting more: "You are a scientist, not a hayseed raised in the Taiga, and *you* can't figure out politics? Don't you see there is no difference between the questions you ask and those raised by the Hungarian coup plotters? Is that what you want—to haul out the factory director and the chair of the city party committee on a barrel because they are Stalinists?"[15]

That was the crux of it. If Lanin kept going, drawing on ideas and ideology from Khrushchev's 1956 speech, then the charges of gross overspending of public funds, corruption, embezzlement, and factory mismanagement might just stick, and the long-standing bosses, "the Stalinists," could lose their jobs.[16] No one in the entrenched industrial elite of Ozersk was willing to let that happen. After the invasion of Hungary, Soviet security officials carried out a wave of arrests of people who had taken Khrushchev too seriously. In Ozersk, the bosses followed suit. Lanin was not arrested, but his superior demoted him to the shop floor and expelled him from the party.[17]

After Lanin was demoted, only a few people publicly questioned the Soviet political order, but fault-finding remained a major pastime for Communist Party members in Ozersk, a process that residents called "inner party democracy."[18] Once Beria was ousted, Soviet citizens wanted more policing and social order, and Khrushchev gave it to them. He revived volunteer community policing programs called *druzhiny*. In Ozersk, several thousand people participated in the volunteer vice squads that patrolled the streets, looking to prevent crimes and correct unseemly behavior. In historical memory, these vigilantes are notorious for targeting young people dancing to "trashy" music and dressing in tight, pandemonium-colored clothing, but the vice squads also helped victims of theft, broke up domestic quarrels, sent drunks home, and retrieved lost children.[19]

Other Khrushchev reforms created workplace and community mediation programs in factories, garrisons, and even labor camps. Khrushchev encouraged party members to expand their work into the domestic realm to get at the root of rampant social problems.[20] In Ozersk, party officials got involved investigating and making public the private problems (alcohol abuse, philandering, domestic violence, and sex offenses) of fellow party members. Women's groups organized to work with troubled families, focusing especially on battered wives and neglected children. The Women's Committee made visits to homes, helped unemployed parents find jobs, and tried to talk sense into husbands who drank away a family's security.[21] Teachers, too, made home visits, investigating the domestic problems of troubled students. Local doctors and nurses taught courses on parenting.[22]

Several historians of the Khrushchev period have called this shift from Stalinist incarceration epidemics to community self-policing "invasive" and "intrusive behavioral engineering" as the state expanded into "the trivia of everyday life."[23] These historians follow a long tradition of Western journalists and scholars who have characterized the communist "regime" as "imposed" on the Soviet people.[24] And it is quite true: the modern state's usurpation of older forms of social control must have felt to many like unwanted meddling. This development, however, was not specifically Soviet, but a trend that occurred more universally as states stepped in to manage orphaned nuclear families in industrialized societies.

Social workers, teachers, doctors, parole officers, counselors, therapists, and volunteers took over from family patriarchs, village elders, and factory bosses in correcting and punishing social behavior. The new educated professional had a lot to say in the mid-twentieth century about how nuclear families should lead their private lives: how to raise kids, how to maintain a marriage, how to eat, drink, and have sex. Assuming a handcuffed populace, historians of the Soviet Union often view the modernizing Soviet state as exceptionally intent on hijacking individual freedoms. But to assume that Soviet community policing efforts were a way for the state to pry yet deeper into people's lives is to overlook the fact that Soviet citizens had been worried about crime and social disorder since the war ended, and that millions of volunteers readily joined the new programs because the reforms empowered them. Bullied prisoners had a chance to sit in judgment of their peers. Women in Ozersk used community organizations as a way to win, for the first time, a measure of local power in the male-run company town.[25] Women especially were interested in policing social behavior because they were often victims.[26] Husbands drank away their salaries, brought home girlfriends, injured others in car accidents, propositioned or assaulted women on the street, and landed in the drunk tank. Many men believed this behavior was their right.[27] As one man replied when neighbors brought up his poor treatment of his spouse: "She is my wife. I do what I want with her. If I want to beat her, or defile her, I will."[28]

In Ozersk, the new community programs reported successes. Professionals and volunteers handled cases that were too petty for the judicial system. They helped to decriminalize what Stalinist procurators used to call "traitorous" behavior and make it a function of social sanction, which meant far fewer people went to jail for misdeeds such as petty theft or truancy, yet more people who broke social norms and disturbed the peace were held accountable and sometimes even got help. Peer-led courts used public shaming, warnings, and oversight to punish and correct. Prisoner comrade courts managed to get more convicts released for good behavior. Volunteer policing brigades made the parks safer and more pleasant by pushing alcohol consumption into the pubs and beer kiosks. Women's groups reported advances working with broken families.[29] In a town run by distant, powerful, and unaccountable bosses, community policing and mediation were important ways for people to take control of their communities and feel safer.

In short, in the decade after the young engineer Lanin was demoted, party members used their new voice not to question politics but to question the city's order—particularly the social order in families and in teen clubs, but also the order of the checkout counters in the grocery store, the repertory for the children's theater, the assortment of bread and dairy products, and service in the cafeteria.[30] Over the years, Ozersk's city party committee dwelled most on issues related to families and consumption. After a decade in which Communist Party stalwarts had complained that Ozersk residents did not take an interest in their community, that they suffered from a "temporary" feeling, the reforms enabled residents to have a measure of self-rule, which in turn made the city, enclosed in barbed wire, feel a bit more like home.

In their concern over their community, the crusading citizens of Ozersk did not notice the villagers languishing in a contaminated zone three miles away, though they heard about it at party conferences.[31] The zones, dividing plutonium operators from farmers and construction workers who supported them, worked to create discrete spheres of knowledge, distinct realities—one of confused, dislocated communities cursed by illness and poverty, and another of select, outspoken plant employees, their growing affluence a signifier of their value to society.

In March 1957, researchers from the Ministry of Health reported that villagers remaining on the Techa were getting ill and that their exposures remained too high. In response, the Soviet of Ministers ordered the evacuation of eight more villages, plus 856 households and several orphanages and schools in three settlements not slated for complete evacuation.[32] Shtefan's construction firm responded even more slowly to those new orders, taking a full five years to accomplish them.[33] Only in 1960 did the construction bosses sign off on new buildings, though ones that were in "lawless disorder."[34]

Over the years, only one official looked out for the displaced, irradiated, voiceless Techa villagers. That was the chairman of Cheliabinsk province's Executive Committee, the weakest arm of the Soviet provincial governing apparatus. His staff wrote letter after letter complaining to Moscow and cajoling Shtefan and other construction bosses, who penned curt, dismissive responses that self-importantly used code phrases, referring to the "River T."[35] The executive committee staff recorded how, in the evacuations, crews missed some families, and others refused to go.[36] Inspections of the sixty families left behind on the river as late as 1961 found "all their personal possessions contaminated beyond permissible levels."[37] The chairman documented a rash of illnesses among villagers. Sickness became such a problem that villagers, at first resistant to move, began requesting their own deportation.[38]

By 1958, every man, woman, and child in Ozersk had on average a large room to call his or her own, a sign of astonishing opulence in the USSR at the time.[39] Villagers, however, on the radiated Techa were living through a crisis no one called by name. They had to wait up to a decade to leave their irradiated homes and crowd into drafty prefabs. The crews were especially slow in moving orphanages and schools from the contaminated river.[40] One boarding school, where radioactivity was five times higher than naturally occurring background radiation, was resettled only in the 1980s.[41]

As the years passed, the zone divided citizens into those who were protected and those who were not. By the early sixties, most other closed nuclear cities had phased out prison labor because it was explosive and inefficient, but Ozersk managers continued to deploy prisoners for canal- and dam-building projects to redirect plant waste from the radioactive Techa River. These projects required convicts and soldiers to dig, build, and haul in contaminated landscapes—work considered too dirty for wage laborers.[42] The prisoners created a network of dams and canals leading to a former swamp that became, with the addition of millions of gallons of radioactive waste, Lake Karachai. The lake was "hot" in two respects: large pipes sent in water that was both scalding and powerfully radioactive. Because the lake did not freeze, soldiers used it to bathe and launder. At the bank the soldiers received a year's dose of radiation in just an hour.[43]

In 1962, the chief of the regional KGB office complained that the Maiak plant's unfiltered radioactive gases were thoroughly contaminating two downwind communities of soldiers and workers who supported the plutonium city. Conditions were hazardous, the KGB officer warned, and threatened to get worse. He requested that the farmers and soldiers be moved out from under the radioactive drafts. Plant managers dismissed the investigation and denied any wrongdoing. They said their measurements showed levels "within the permissible limits." They did promise to install filters on the plants' smokestacks.[44] They were slow about it. Two years later, plant crews still had failed to install filters.[45]

If a "zone," a mere spatial arrangement, can have historical agency, then the regime zone played no small role in first abetting the illegal dumping into the Techa and then assisting the failure to protect villagers, prisoners, and soldiers from exposure. The zone, at first a stigma of incarceration, offered over the years a comfortable seclusion. Within the gated city, residents came to conceive of their dominion as ending at the posts walling off what they called the "little world." At the same time, the "big world" retreated from their field of vision, sphere of responsibility, and realm of action.

30

The Uses of an Open Society

"Progress is our most important product." That was the refrain of *GE Theater* host Ronald Reagan. Touring GE factories in the fifties, Reagan made an appearance in Richland. Black hair glistening, he looked handsome in a gray double-breasted suit, his fashion very much like that of the GE executives who employed him. It was easy for local farmers and city dwellers to place their trust in Reagan and the executives he represented because those executives had done so much for them. Looking around, the older ranchers could see how federal programs had remastered the barren, "desiccated" scabland of the Columbia Basin into a crop-rich, industry-driven landscape honed for production and profits. Seeing the green lawns surrounding orderly middle-class houses, Richland residents grasped readily how well GE had outfitted a comfortable suburban city from what had been a "jerkwater" ranch town.

The expanding industrial wealth of the West alongside the personally increasing prosperity of the American working class joined at a point where science, technology, and culture bolstered one another to send a message of competence, expertise, and trust. That is where the expensive cultural edifice of Richland and the highly subsidized, industrializing countryside paid off. Richland and its environs represented what was new, vital, and dynamic about the interior West. The city's prevailing culture, with its respect for education and expert knowledge, for hierarchies, regulations, and planning, had the power to mute doubts, override fear, and squelch rumor and plain fact alike. I have been pointing out how AEC and GE accountants allocated more money to Richland's school system than to waste management, public health monitoring, or scientific research at the plant. I mean to say that the cultural pillars bolstering Richland were well worth the investment. It helped people have faith in their compact with science and progress, no matter what those brought.

On his top-rated TV show, Reagan sold GE lamps, refrigerators, jet engines, turbochargers, and atomic safety devices.[1] During commercial breaks, Reagan showed viewers his home, exhibiting the "total electric kitchen," attainable at an employee discount in a Richland showroom. The appliances represented the accomplishments of industrial technology, which pushed scarcity, destitution, and disease to the margins and made human life so much more worry free while safeguarding time for family and leisure. Those appliances meant a lot especially in Richland. Well-running kitchen appliances helped residents accept the neighborhood monitoring stations, the antiaircraft rockets lining the road, the monthly air raid siren, and the lost time spent sweating in green buses during tedious evacuation drills. A 1964 alumnus of Richland remembered his hometown in 1999 this way:

> In Richland, the sun shined 300 days a year. The corner drugstore and grocery, little league field and service station were a few short blocks from every residence. Good doctors kept us healthy. Every lawn was mowed. Every house painted. Warm spudnuts [donuts], shopping at CC Anderson's and celebrations at the Bomber Bowl are memories prized by every kid from Richland. Why were those years so good? Protection. Parents raised children in an atomic city positioned far away from evil, urban blight and social degradation. It was serenity and goodness without conflict.[2]

Most people remember growing up in Richland as ideal. Courtesy of Department of Energy.

For this memoirist safety was spatial, encompassed in the capaciously zoned housing lots, carefully planned shopping matrix, and exclusive residency requirements. Since the Hanford Plant was off-limits to most observers, Richland became the public face of plutonium production. It was a soothing image. The neat green lawns, gently curving streets, stocked stores, and palatial schools filled with high-performing children instilled confidence and a sense of protection. The fact that the town was under surveillance, by the GE Patrol and by health physicists alike, was also comforting. The newspapers reported that Richland won national traffic safety awards and that scientists took 265 samples each month of milk and water to test for pasteurization and sanitation.[3] As one resident put it, "Living in Richland is ideal because we breathe only tested air."[4] GE plant doctors regularly assured the surrounding population that home appliances and medical X-rays were more hazardous than plutonium production. The volatile plutonium, they claimed, was safely contained behind barriers at the plant.[5] GE public relations staff pointed out that Richland had the highest birth rate and lowest death rate in the nation and lower than average rates of infant mortality and maternal death.[6] These statistics were not surprising for a community where employees passed a health exam to get hired and enjoyed universal health care, and where there were few elderly and no poor or indigent.[7]

What was surprising was when the health statistics began producing some ominous numbers. From 1952 to 1953, the number of fetal and infant deaths in Richland sprang to nearly twice the state average. Between 1952 and 1959, Richland, Pasco, and Kennewick had a higher number of congenital malformations than the state average. All three cities, but especially Richland, were off the charts for percentage of deaths among infants in the population. From 1951 to 1959, 20–25 percent of deaths in Richland were babies. The state average in the same period logged a steady 7 percent.[8] By 1958, there were four times more fetal deaths per capita in Richland than the rest of the state.[9] Among the children who survived, doctors in 1952 referred 25 percent of new kindergarteners for correction of "defects."[10] Although local reporters often published statistical health facts about Richland, they missed these statistics filed annually with the state board of health. General Electric doctors also failed to point them out. Bad news, in the atomic city, was often no news.

The Tri-Cities made for a small statistical sample, and Richland had an unusually high number of babies, skewing the statistics, but the infant death problem extended beyond the Tri-Cities. A Spokane mortician noticed years later a similar jump in the death of babies in 1953; 16 percent of the people he buried were newborns rather than the usual 5 percent. In Walla Walla and Spokane Counties infant mortality spiked in the early fifties and remained higher than average until the end of the decade. In 1993, a Spokane reporter discovered a hundred-year-old cemetery with a total of 680 graves, 261 of them occupied by babies who had died soon after birth between 1951 and 1959. Joan Hughes lost one of these babies in 1956. When the obstetrician told her that her infant had died of multiple congenital malformations, she asked to see the body, and he refused.[11] Just when eastern Washington emerged from decades of poverty and enjoyed a stable prosperity, infant death rates rose. What was going on?

There are a couple of possible causes behind these troubling statistics. A German measles epidemic swept Washington State in the early fifties. Babies exposed in the womb to the rubella virus can contract congenital rubella, which has a constellation of syndromes (cataracts, microcephaly, heart disease, and problems of the liver, spleen, or bone marrow) that might cause death. Also in the fifties, wartime chemical agents spread across newly industrialized, domestic American landscapes. Chemical plants produced ammonia during the war for explosives and DDT to combat mosquitoes for American troops in tropical climates. In the fifties, chemical companies used mass marketing campaigns to sell wartime chemical surplus to domestic markets for daily use.[12] Americans bought the new products, delighted with their miraculous properties.

In Richland, residents were handed bags of grass seed and told to plant lawns and water them to keep the (radioactive) dust down. City community organizations volunteered to plant thousands of trees to green their desert settlement.[13] Irrigation, sprinkling, and overwatering brought to the semiarid steppe pools of standing water warming in the sun, textbook breeding grounds for mosquitoes. In response, W. D. Norwood, the head of Hanford's medical services, embarked on a campaign of mosquito warfare. He sent out planes and Jeeps mounted with power sprayers that shot each year a ton of DDT at mosquitoes hovering over Richland's green lawns, onto which homeowners also poured nitrogen fertilizers.[14]

It is well known that DuPont and other wartime military contractors built Richland and Hanford, but less understood is the extent to which wartime contractors and their technologies created other features of Richland and American suburbs generally. To keep the steppe green and bug-free required an impressive volume of military-industrial machinery, including Jeeps, planes, and bulldozers, plus battle-ready petroleum and chemical agents—technologies so prevalent they became etched in childhood. Richland residents recall running after the fogging Jeeps to breathe in the sweet, rich smell of DDT. Farm children laugh about hitching water skis to pickup trucks to ski in irrigation ditches rippling with pesticides and fertilizers drained from their parents' fields. As the kids played, the wars on weeds and mosquitoes transformed into a new war, a war on cancer, which was fought with chemotherapies made from mustard gas and other chemical agents first devised to combat enemies.[15] The technologies for external wars migrated imperceptibly from foreign lands to the home front, from which they filtered into American bodies.

DDT, banned in 1972, is an endocrine-disrupting chemical that adversely affects reproduction by causing birth defects, increasing pregnancy complications, and decreasing fertility. In some studies DDT also induced chromosome mutations in humans and animals. DDT is also linked to lymphatic leukemia, liver cancer, and lymphoma.[16] High concentrations of nitrates in drinking water may also cause birth defects, cancer, nervous system disorders, and "blue baby" syndrome, in which infants suffer low oxygen content in their bloodstream.[17] In short, people of the Tri-Cities may have been poisoning themselves in their quest for agricultural fertility and suburban green on their arid landscape. If, however, German measles, DDT, or nitrates caused a rise in infant deaths, why did the spikes occur only in the southeastern corner of Washington State and end long before DDT and other chemical toxins ceased to be used?

The years 1951 to 1959 were the peak of plutonium production, when the Hanford plant produced the greatest volume of radioactive waste. In the period 1945–1952, parents-to-be were exposed to great clouds of iodine-131, xenon-135, and strontium-90 from the plants' unfiltered stacks and in intentional releases such as the Green Run. Medical studies have shown that parents exposed at low levels who have no obvious radiation-induced injuries can pass mutated genes to their offspring.[18] Other studies show how prenatal exposure to ionizing radiation produced significant increases in perinatal loss among offspring of survivors in Japan.[19] There is some evidence that radioactive isotopes work synergistically with chemical toxins such as DDT to exacerbate health problems.[20]

Epidemiology, the science of investigating illness in complex landscapes, is a blunt tool. It is very difficult to determine retrospectively the cause of the significant spike in infant deaths in southeastern Washington State. What aroused my curiosity was why there was no alarm about these statistics within the Hanford health community.

There are a number of reasons this crisis in plain sight might have been missed. The primary explanations are banal and institutional. First, Hanford plant doctors did not have a mandate or funding to engage in research on genetics or the effects of multiple radioactive environmental contaminants on human bodies. Postwar medical research prioritized lab work, where all factors could be controlled, over epidemiology, which looks at the interaction of variable risk factors in an environment.[21] Joseph Hamilton carried out lab research, for example, on "radiation and man" at the University of California until his premature radiation-related death, when his assistant Kenneth Scott took over. This arrangement made for a strange division of labor—Berkeley had few radioactive emissions, and Hanford was a world leader, increasingly exposing the surrounding public to long-term, low doses. Hanford's chief health physicist, Herbert Parker, had the mission of tracking radioactive waste spreading across the Columbia Basin, but he did not have a green light to determine what effects the increasing doses his researchers recorded had on the people exposed to them. That missing link is essential in understanding what didn't happen in eastern Washington during the peak decades of plutonium production.

Second, health physicists monopolized the job of monitoring the local environment. With their expensive instruments, the monitors alone had the power to make the invisible, silent radioactive isotopes stand up and be counted. They took over from individuals and communities the job of determining risk. This was part of a national trend. Increasingly the American public came to rely on officialdom—the weather service, the U.S. Public Health Service, and the news media—to inform them of dangers and hazards.[22] In eastern Washington, health physicists became the high priests of protection. Plant scientists checked the air, water, food, soils, plants, wild rabbits, and domesticated fowl. Plant doctors took regular urine samples and monitored employees' radiation exposure. Yet they had permission only to monitor, not to notify. Plant researchers penned statements such as "Drinking water in the Pasco and Kennewick systems contains more radioactive material now than has ever been measured at these locations in the past." They sent classified notes to water district employees telling them not to eat fish after certain spills, but they did not, could not, inform the public of higher than permissible levels of contamination as it surged from the plant in random, unpredictable jets.[23]

I do not mean to imply that the fifties was singularly a decade of silent conformity and deference to elites and experts. On the contrary, when nuclear power was less than a decade old, Americans, Japanese, and Europeans began to question the safety of nuclear weapons and the veracity of AEC statements. In 1954, a nuclear bomb test in the Bikini Islands produced so much fallout that the armed forces had to evacuate Marshallese from their island. Eighty miles from the blast, Japanese fishermen fell ill with radiation poisoning, and one died.[24] These events shifted public attention to the dangers of radiation from fallout and other sources. Soon after, thirty million Japanese signed petitions against nuclear weapons.[25] In the United States, women's groups questioned the safety of food sources contaminated by fallout, and a few veterans of the Manhattan Project accused AEC officials of making "reckless or unsubstantiated statements" about public health.[26] Former Manhattan Project physicist Joseph Rotblat calculated that the Bikini blasts discharged far more fission products than the AEC acknowledged. Groups of concerned citizens began to send letters and requests to AEC headquarters in Washington.[27] From 1954 on, AEC managers had a public relations problem on their hands that never really subsided. In responding to what they skeptically referred to as the "environmental crisis," spokesmen had to admit they knew little about the effects of chronic low doses on humans. Most AEC-sponsored research had been concerned with high blasts, the kind a person would get in a nuclear explosion or plant accident.[28]

Faced with public scrutiny, AEC officials tightened security on the commission's biological and medical research, while they set about trying to convince the public of the relative safety of nuclear weapons.[29] One way to do that was to sponsor research, the results of which would be guaranteed to soothe and reassure the public. For example, in 1955, the AEC took over the Atomic Bomb Casualty Study started by Japanese doctors after the war. In a closed meeting, Charles Dunham, AEC chair of biology and medicine, explained that AEC funding was necessary to ensure that "misleading and unsound reports of the effects of radiation on man emanating from Nagasaki and Hiroshima are kept to a minimum." Dunham continued, "If the United States were to pull out [of the study], the vacuum created would assuredly be filled by something possibly bad, even flavored with occasional tinges of red."[30] Dunham's plan succeeded. In subsequent decades, AEC-funded scientists in Japan determined there were no significant genetic effects from the atomic bombs. The Atomic Bomb Casualty Study set standards for health studies of American workers in nuclear plants for the rest of the twentieth century, and it is still frequently quoted.[31]

In Richland, with the public growing worried about nuclear fallout, Parker began to identify a new threat at the plant, that of public exposure. In 1954, the Redox Plant had another discharge of radioactive ruthenium particles, which coated the surrounding environment. Some of the tiny flakes produced an astonishing 40 rads an hour. Even at the millirad level, the specks reddened and destroyed skin tissue. Parker would have liked to have cordoned off the contaminated areas and test sheep outside the reservation, but he decided against it because of the "risk of exciting too much comment" and triggering "undue alarm."[32]

The particles attached themselves to dust and spread with the winds. They intensely contaminated areas inside the nuclear reservation, but monitors also found particles in Richland, Pasco, and farm communities in the newly irrigated former secondary zone, where, in 1954, farmers leveled ground for irrigation wheels, scraping and plowing the fine volcanic soils, raising great clouds of dust. The farmers remember that dust, how it plagued them. "It was tough," recalled Juanita Andrewjeski. "The men all worked the dirt."[33]

In 1948, during the first particle episode, Parker had worried about ingested flakes causing lung cancer. Six years later, in a professional climate that disavowed safety issues, he minimized the dangers of particles in the lungs. With no research to back him up, he guessed that the particles would "probably" be swept out of the lungs in a few hours. He focused instead on skin contact, a far less serious hazard. Since Parker felt he could not carry out a study of the local environs, he engaged instead in a thought problem: "One can picture the entire population of Richland lying unclothed on the ground for one day. There would be about 25 identifiable particles in contact with skin; not more than three would be in an activity type range that *could* produce a significant effect; not more than one would probably produce an effect."[34]

This passage is revealing. It shows how Parker had gone political after all, exchanging science for conjecture. He admitted that little research on radioactive ruthenium particles had been done since the first episode in 1948.[35] Peppering his prose with "probably," "possibly," and "reasonably expected," Parker nevertheless denied the dangers of the high-powered particles without any science to back his claims. At the end of the decade, Hanford managers would count the particle problem as the second most serious accident of the decade.[36] In 1954, the once-cautious Parker described the episode to his superiors as more "nuisance than hazard."[37]

Demoting the crisis to nuisance downgraded the response. Parker reckoned the man-hours entailed in sending out monitors to detect and safely dispose of every speck and dismissed this plan as "prohibitively expensive."[38] AEC officials decided to leave the particles in place, and instead drew up a program to rezone the nuclear reservation. Parker wanted to post eleven thousand "contamination zone" signs at the points where the flakes were located, but from Washington, Edward Bloch, AEC head of production, decided it would be "less exciting" to establish the entire reservation as a "radiation control zone." That way, Bloch reasoned, employees would not get alarmed when the contamination zone changed or expanded as the specks shifted around the reservation.[39] Parker and Bloch knew that in the strong, capricious basin winds, the particles would loft skyward and depart for areas outside the posted terrain. The men determined they could do nothing about the contaminated particles in Richland or Pasco, which were outside their jurisdiction.

By 1954, when Parker spoke about "danger" and "risks," he no longer had in mind just public health. Increasingly for Parker the "threat" came from what he called the "public relations situation."[40] In 1951, for instance, Hanford scientists began to work with U.S. Public Health Service officials to monitor the health of the Columbia River.[41] First conceived of as collaboration, over the years, Public Health Service officials started to contest the Hanford Lab's interpretation and control of data. Three years into the project, AEC chief Lewis Strauss asked Parker to prepare a report on contamination in the Columbia River. Parker laid out what he knew: that the river had hotspots of contamination, sources of drinking water and fish habitat were selectively contaminated, and counts of radioactive isotopes were on the rise as more reactors at the plant went online.

Aware of the value of the Columbia River as a center for sport fishing, Parker dedicated a large section of his report to public relations. He wrote that the Public Health Service had conducted an independent survey of the Columbia River downstream from Hanford. State health service officials testing the river were a real "threat," Parker wrote, because "adverse interpretation can be given by distinguished technical individuals, such as expert sanitary engineers, whose appreciation of the radiological hazards is perhaps limited to rather recent exposure to these complex problems."[42]

Parker described how he "defended" the plant. First, he wrote that the AEC had organized a Columbia River Advisory Group to act like an independent body overseeing the health of the river, when in fact AEC officials had chosen and controlled the group's membership. "The U.S.P.H.S. conducted an independent survey of the river from 1951 to 1953," Parker wrote. "The first draft of this report contained several statements that would have been highly detrimental to public relations. The combined efforts of the Atomic Energy Commission (HOO) [Hanford], the Columbia River Advisory Group and General Electric forces have led to a revision that should tend to preserve the present status. The final [published] report will be a valuable independent appraisal of the river condition."[43]

Animal testing at Hanford Experimental Farm. Courtesy of Department of Energy.

The initial Public Health Service report was indeed the first truly independent evaluation of the river since plutonium production began, but by applying pressure, the Hanford "forces" had the disquieting statements removed from the initial report. With the cleansed final draft, Parker won what he called a "valuable independent appraisal" assuring the American public that one of the nation's most beloved rivers was safe. Using the rituals and appearance of open debate, Parker shut out independent reviewers and won for Hanford a great deal of credibility.

Careful control of public information alongside a trigger-happy classification of data also proved useful in 1956 when Utah ranchers initiated the first lawsuits against the AEC charging that the Nevada atomic explosions had sickened and killed their sheep. Hanford had on staff a world specialist on radioactive sheep—Leo Bustad, a veterinarian in Parker's division. Since 1950, Bustad had fed plutonium-spiked pellets to sheep at Hanford's experimental farm. Bustad's classified experiments showed that the sheep on this diet grew fatigued, stupid, weak, and disoriented. They had difficulty moving, acquired ulcers, and gave birth to stillborn lambs. Even with low cumulative doses the animals developed tumors. Bustad also found that the sheep thyroids did not regenerate after exposure ended; the damage was permanent.[44]

In the Utah court, Bustad told a different story. He testified that the ranchers' sheep were exposed to levels of radioactivity too low to cause damage. He postulated that the animals died because they were malnourished—because, in other words, the ranchers did not know how to care for their livestock.[45] To bring this point home, Bustad published an article in Nature in 1957 asserting, in direct contradiction to his classified research, that daily doses of I-131 below very high levels (30,000 rads) did no harm at all.[46] Soon after, Bustad left Hanford to become a celebrated dean in veterinary sciences at Washington State University. He spent the second half of his life as a vocal spokesperson for animal rights, compassion for children, and, ironically, truth in the media.[47]

GE and AEC officials understood that silence on such a volatile topic as radioactive contamination would be suspicious. As early as 1947, AEC officials swore off the classification of all its scientific data and vowed "to provide [for] that free interchange of ideas and criticism which is essential to scientific progress."[48] In the United States, in contrast to Soviet society, there was plenty of talk about radiation, plutonium, and their possible dangers, and plenty of talk by authoritative experts minimizing, dismissing, and naturalizing the dangers. This "free exchange" of information led to trust, confidence, and faith in American nuclear experts, who staged a powerfully convincing performance of an open society.

31

The Kyshtym Belch, 1957

When I met Galina Petruva she was in her early eighties. She walked with a cane and had lost all but two front teeth. She smelled of age and illness. She said she feared talking to me lest the authorities learn of our conversation. "I've waited a lifetime to say what I am going to say," she half whispered. Petruva was sure she was being watched. She said that there were those in Ozersk who stalked and murdered old people.[1] There had been a few local news stories to that effect. She said all this in the first few minutes of our meeting. She was, I realized, what appeared to be the classic unreliable narrator.

Petruva described how she came to live in Ozersk in January 1957. She made a point of saying that she was from a settlement, not a village. That difference was important to her. After a two-year education as a medic, she was sent to a village to work. The village was a giant social step backward for Petruva. "After living in a settlement, I could not bear life in a village, the forest all around, in a cottage shared with two other couples." Her husband, in the military, felt the same way. They were both happy to get new job assignments to a secret destination.

Ozersk surpassed all their expectations. It wasn't a settlement but a city, and not just any city, but a pretty, orderly, well-stocked city. With hardly a wait, they were given their own apartment with electricity and running hot and cold water. Shopping in the elegant stores, standing in line with well-dressed couples at the movie theater, Petruva and her husband felt they were on their way toward the Soviet middle class.

Not that they were middle-class. Petruva's husband had a job supervising prisoners. Petruva worked as a lab tech at the Central Factory Lab, analyzing biopsies for cancerous cells. By 1957 the ten-to-twelve-year latency period for workers exposed to high doses during the start-up of the plant had passed, and many employees had lumps and nodes to be analyzed. "There was a lot of work," Petruva recalled. "We had to keep up a good pace." Petruva was issued a film badge to wear at all times. The other young female lab assistants, however, told her to take off the badge when she had "dirty" work. If she received more than the permissible dose, all the women on the shift would lose their monthly safety bonuses. Petruva did as she was told, but she was starting to suspect that something strange was going on.

A radiation monitor told Petruva that her boss was storing plutonium-239, plutonium-238, and americium in the wooden cabinet in her lab. After Petruva spilled a solution on her lab coat, a radiation monitor told her, "You need to tell your boss to remove all of those bottles from your lab." It was too late, though. One of Petruva's colleagues already had thyroid problems, which doctors at the lab were treating with an experimental medicine. "That stuff made her sicker and sicker. I told her to stop taking it," Petruva said. "I told her she didn't need to be their lab mouse." Petruva didn't veil her anger at the supervisors and scientists, at how they deployed security-sensitive hierarchies of knowledge to create hierarchies of risk that put low-level employees such as her in danger.

The plant's safety record had never been good. Nuclear power was a hopeful science in the USSR, a phoenix rising from the ashes of postwar ruin. Soviet propagandists dwelled on the benefits of nuclear power, such as medical isotopes and limitless energy. After starting up the world's first civilian nuclear reactor, they promoted the Soviet Union as the creator of the peaceful atom, rather than the deadly American-style atom.[2] As a result, there was little cultural impetus to focus on potential nuclear disaster. As the plant enlarged, however, it produced more plutonium, and the number of overexposed workers and fatalities swelled. In the first half of 1957, for example, the plant had twenty-three accidents. Each one spilled out radioactive materials that made the workplace even more hazardous.[3] Accidents were discussed in closed party meetings and top-secret safety reports, but most residents never knew of them because they occurred in the fenced-off production area among employees who had sworn an oath to keep secrets in order to preserve security. That changed in 1957 when a mishap occurred of such magnitude that it could not be contained within the plant confines. For several months, the city's well-heeled veneer gave way to expose the lethal and violent product beating at the heart of the city.

It was a pretty day, sunny and warm, September 29. A large crowd at the soccer stadium was watching a match when at 4:20 p.m. an explosion rocked the stadium. No one panicked. Only a few looked up. Spectators figured prisoners were blasting rock for some foundation work in the industrial zone. The crowd returned to the game, the players kept playing, the bartenders continued to pour mugs of beer, right through a megaton nuclear explosion.[4]

The source of the blast was an underground storage tank holding highly radioactive waste that overheated and blew, belching up a 160-ton cement cap buried twenty-four feet below the ground and tossing it seventy-five feet in the air. The blast smashed windows in nearby barracks and tore the metal gates off the perimeter fence. A column of radioactive dust and smoke rocketed skyward a half mile, blossoming into that distinct mushroom shape. Prisoners and soldiers, dazed, some bleeding, stumbled outside to watch the gray airborne mass waft away from the city over the plant grounds. In less than an hour, a strange, sooty precipitation began falling.[5] Someone muttered of sabotage; others speculated it was an American attack. Officers at the garrisons gave orders for battle alert. They beefed up the guard around the regime zone and locked soldiers and prisoners in their barracks. Few understood immediately that they were victims of friendly fire.[6]

No one knew what to do. There were no emergency plans prepared for a nuclear blast: no checklist, buses, ambulances, or sanitation points. The soldiers and prisoners did not know officially they were working with radioactive materials, and so they had no protocols, dosimeters, iodine pills, or respirators for a radioactive emergency.[7] The plant's managers were away on business in Moscow. After many hours of desperate searching, the factory director and his assistant were finally found at the Moscow circus.[8]

With no leadership or contingency plans, the emergency played out in slow motion. Several thousand people stood around, innocent of their ongoing exposure. After six hours, radiation monitors arrived to take readings of the territory and equipment, but they failed to check the soldiers.[9] After ten hours, the order came down to evacuate soldiers and workers located near the explosion. By that time, several inches of radioactive ash and debris covered all surfaces. The military units had no sanitation point to rinse off radioactive dust, so the soldiers were driven to a bathhouse, where they showered with ordinary soap, to no effect. A large dose of laundry soap and many more showers helped lower the needle on the Geiger counters.

Galina Petruva told me how she was called up on that day to ambulance service. As she raced to the garrison, she saw gray soot falling into Lake Irtiash, the source of drinking water for the city. The sight of the soldiers frightened Petruva. The young men were pale, vomiting, bleeding, and shaking, their hair already falling out. She had the job of calling a mother in Ukraine to tell her to come be with her son. "At least she made it in time," Petruva remembered, "so he didn't die alone."[10]

Prisoners, evidently overlooked, were not evacuated that first night. The canteen was under repair, and the inmates went out that night to their dinner served on rough wooden planks, covered with a couple of inches of ash, which the prisoners swept aside with their cuffs. The next day, the inmates sat around and watched soldiers withdraw guns and heavy artillery from a nearby garrison.

George Afanasiev was a young convict, arrested in 1947 for being twenty minutes late for work. He told his story to a *Moscow News* reporter in 1993 shortly before dying from cancer: "On the next day, sometime after 2 a.m. [October 1] we were awakened and told to prepare for evacuation because our camp was in a contaminated zone. We were given 15–20 minutes and ordered not to take anything with us (not even money or jewelry). That created a panic. We were loaded into open trucks and taken to a forest. On a large clearing, there were rows of tables piled up with new clothes and underwear." Radiation monitors took readings of the prisoners. Radiation had collected in Afanasiev's gold teeth and hair; the needle spiked at 800 microroentgens, which meant Afanasiev had become a dangerous source of contamination. A colonel addressed the prisoners through a bullhorn, telling the men that the radiation would work as a curative for their ailments. Then he told the men to strip and march to the bathhouse.[11] The pile of contaminated clothing, books, and harmonicas remained, moldering for several weeks, no one daring to go near it.[12]

When the plant director, Mikhail Dem'ianovich, returned from Moscow, his first instinct was to use the closed industrial zone to conceal knowledge of the accident from residents, while the plant kept running, churning out plutonium as if nothing had happened.[13] Shtefan, the construction boss, worried about delays on the Double B, the long-awaited, much-delayed radiochemical factory located closest to the explosion. He complained his crews could not work because the construction site was littered with radioactive debris.[14] Three days after the explosion, radioactivity in the area measured 4,000 to 6,000 micro-roentgens a second, hundreds of times the permitted dose.[15] Rooftops maxed out at 10,000 microroentgens; the edge of the crater recorded 100,000. The site was strewn with a liquid pulp amounting to 18 million curies of radioactivity, about half of it from strontium-90 and cesium-137, dangerous, bone-seeking isotopes with a half-life of thirty years.[16] The plant bosses discussed walking away from Double B and constructing a new plant on safer ground, but so much was already invested in the factory, so many years of delays and millions of rubles, and the plant was critically needed to replace the original contaminated plant; the bosses eventually decided to have workers and soldiers clean it up rather than abandon it.[17]

For several days nothing happened. Officers were afraid to order soldiers into the contaminated zone. Soldiers, when finally told to go in, at first refused.[18] Supervisors, from fear of contamination, stayed away from the site. A patrol was established to keep workers from leaving the job.[19] No one had experience cleaning up radioactive terrain. Russian officials call clearing away a nuclear disaster "liquidation," but that is a euphemism. There is no way to liquidate radioactive isotopes. You can only move them to places where they might do less damage.

When the "liquidation" started, it was performed at a run. Soldiers, plant operators, and construction workers were given a few minutes to dash in with a shovel. The soldiers first cleared the road of trash and sprayed it down. They used heavy wire brushes to scrub the roofs and walls of factory buildings, hundreds of feet long. They turned the earth, scraping up the topsoil, and buried it. The cleanup crews cleared away contaminated tools and machinery. Some of it they buried. They tossed into the swamp parts of the exploded waste storage container, fifteen million curies' worth.[20] In the first weeks, many workers had no special jumpsuits, and so they returned to the city after their shifts in their dirty work clothes.[21]

It took a year to somewhat contain the eighteen million curies of radioactivity spewed in the blast. The work went slowly because for health reasons shifts were short and followed newly adopted sanitation regulations. Most low-level plant workers and construction workers banned from living in Ozersk eventually had to pitch in.[22] Radiation monitors, often women, worked overtime. Students, recruited from local institutes, helped. Estimates of the number of soldiers used to clean up the accident range from seventy-five hundred to twenty-five thousand.[23] It is hard to say for sure because there was no real accounting of the "liquidators," no recording of radiation doses for unskilled workers at the bottom of the labor pyramid. Those most exposed were the least likely to be monitored.[24] The highly unstable conscripted labor force, which construction managers cursed for its high turnover, came in handy during the disaster. Using soldiers and prisoners as jumpers meant plant managers could maintain the official line that the accident killed "no one," meaning paid employees.[25] Eyewitnesses, however, describe every bed occupied in the hospitals and clinics, and sick and dying liquidators.[26] After treatment, soldiers were discharged, while prisoners received an early release.[27] The subsequent fate of 92 percent of the liquidators never made it into the medical record.[28]

For city leaders the most troublesome problem was just these soldier-liquidators, who "spread contamination and panic among the people."[29] In fact, city leaders expressed more fear of rumors and panic than of radioactive contamination. "Many people understand the accident not as they should," plant manager Dem'ianovich told a group of communists. "Some workers are in a panic. Many communists at the factory spread panic by exaggerating the size and effect of the accident many times over." Party leader N. P. Mardasov agreed: "Those who are spreading panic about the city are not communists."[30] Mardasov was planning the city's celebration of the fortieth anniversary of the Great October Revolution, featuring the Soviet Union's great technological accomplishments—among them the just-launched Sputnik.[31] News of an accidental explosion the size of Hiroshima did not fit into the celebration's narrative.

The city leaders' first instinct was to stifle all discussion of the accident using the sealed zones to contain information. Yet fission products did not recognize these boundaries. With no information about their exposure, employees unwittingly tracked radioactive isotopes home on their bodies, clothing, and shoes. Trucks and buses sullied city streets. People washed their contaminated cars in the lake, where others fished and swam.[32] The mysterious, slippery radioactive isotopes were irrepressible. Within a week of the explosion, contaminated waitresses in the city restaurant served contaminated food, for which diners paid in contaminated currency.[33]

Knowledge, too, could not be contained. With no official news, people exchanged information about the accident at crowded bus stops. As the rumors went around, valuable workers, fearing for their health and that of their families, quit their jobs and abandoned Ozersk. Nearly three thousand employees left in the months after the accident, about one in every ten workers, many of them engineers. The workers from several shifts at the critical and highly contaminated Factory No. 25 left en masse. Party members could not leave of their own free will, so they skipped work or violated a rule to get terminated. A few took the drastic step of handing in their party card.[34] This, indeed, was mass panic with very tangible side effects that could profoundly affect the city's and plant's fortunes.

Two months after the accident, party secretary Mardasov argued that hiding the facts only did more damage.[35] He insisted the city party committee make a statement about the accident. Deploying a budget of eleven million rubles for public relations, Mardasov sent out lecturers to assure residents of the city's safety. They admitted to the accident, explained that no one had been harmed, and asserted that rumors amounted to treason.[36] Meanwhile, technicians took measurements in the city and found that the most contaminated areas were Lenin and School Streets, where the plant management lived. After that, party leaders made it their mission to return the city to a preaccident state of cleanliness. Radiation monitors took readings in every apartment. They arranged for employees to transfer to clean buses at the city gates. People were told to take their shoes off before entering their apartments. Cars were to be washed weekly. Contaminated tools, clothing, and shoes were destroyed.

And this is when the city's gates—originally set up to seal in nuclear secrets—became useful for keeping out radioactive contamination. The double row of fences and guard towers barred many carriers of radioactive isotopes. The wind, of course, did not stop at the fence, but, fortunately for Ozersk residents, the prevailing breezes usually blew northeast, away from the city. At control points, vehicles, equipment, and transitory workers whose readings ran too high were turned away. The quarantine system worked naturally because it was built into the compartmentalized nuclear landscape. Soldiers, prisoners, and construction workers, the majority working on contaminated ground, had long been restricted to their own garrisons, camps, and hamlets, many miles removed from Ozersk. These distinct zones came to play an important hygienic role in keeping radioactive isotopes out of the closed city. As a result, the artificially created zones became very real and life-altering boundaries between the relatively clean socialist city and the increasingly sullied settlements for expendable, migratory workers.

In December, city leaders congratulated themselves on their successful battle with the radioactive contagion. At the big annual party meeting, G. V. Mishenkov, the new factory director who had replaced the disgraced Dem'ianovich, pronounced the city clean. But a plant scientist named Dolgii stood up and refuted this claim. He said his neighborhood was extremely contaminated. "It's all buried under the ice and snow now, but what will we do when it melts in the spring?" Mishenkov dismissed the scientist, saying the city had radiation levels lower than the permissible limit. "We could live here 150 years at this dose," Mishenkov insisted.[37]

And who could say otherwise? It was the easiest thing to deny—those impalpable, invisible radioactive isotopes, which, at lower doses, took years to undermine a body. It was simple for Mishenkov to say they weren't there, cast doubts on claims to the contrary, and deflect attention to other topics. To undermine Dolgii's assertions, Mishenkov did not question his science. Rather, he called into doubt the scientist's political maturity. "The city party committee has given over fifty lectures to plant workers explaining that the impulse to exaggerate the accident is treacherous. Dolgii clearly doesn't agree."[38] On this point, city leaders stayed on message: good communists didn't panic. Panic and exaggeration lent a hand to capitalist propaganda by undermining plutonium production and therefore the nation's defense.[39]

Instead, the party response was to use the accident opportunistically to get more materials and subsidies from Moscow. Mishenkov marveled: "The ministry is helping us a great deal. They have given us harvesters and fire trucks which we had not been able to acquire in ten years." In order to keep workers from leaving, party leaders proposed improved living conditions and city services, but also, for the first time, better, safer working conditions. The plant received respirators and safety gear. They made plans to automate dangerous work and renovate aging, contaminated workshops. They talked of training courses for workers. They also proposed to test the water supply and set up a radiation monitoring service for the city.[40]

With these measures, the emergency concluded in Ozersk. The party's public relations effort to assure residents of the cleanliness of their city worked. Employees gradually stopped leaving, and many families who had fled in fear requested permission to return after experiencing the relative poverty of the "big world" beyond the city gates. They wrote letters asking to come back to the closed city of stocked stores, excellent medical care, and spacious apartments. "We were stupid," they wrote. "Please take us back."[41] The petitioners preferred the unknown risk of radiation to the certain hazards of life in the Soviet provinces. That was probably wise. The city residents were, in the end, saved by their zone. The regime zone allowed city leaders to pull up the drawbridge and defend themselves from the siege of attacking gamma, beta, and alpha rays. The city, in short, took care of its own.

Petruva had agreed to violate her years of state-mandated silence about the 1957 disaster because she was stinking mad at her bosses, the state, and all the people who were responsible for depriving her of the peaceful retirement to which she felt entitled. "I live in an apartment with four other people, a two-room apartment. It is unbearable. They won't give me another room. I live on a meager pension." While we talked, Petruva repeatedly returned to this topic of compensation, pension, and housing. During her decades in the closed city, she had grown accustomed to thinking of herself as a person of privilege. To signal her status, Petruva listed all the countries she had visited as a tourist—nearly the entire socialist bloc. From the Soviet perspective, Petruva had come a long way, from villager to global tourist. She was a working-class success story. The collapse of the Soviet Union, however, and with it the nuclear weapons industry, had whisked away her entitlements. This loss was the source of Petruva's anger, the reason she talked to me. I wondered whether she would have been so critical had she not ended up on the wrong side of prosperity. I wondered if it mattered. Finishing our talk, Petruva leaned toward me. "There is just one last thing I want to say." Petruva's eyes grew wide. "And that is, I suffer from the fact that I saw it all clearly . . . and, still, I agreed to it."

32

Karabolka, Beyond the Zone

During the tense fall of 1957 city leaders made no reference to the fact that the plant's thick cloud of radioactive gas had headed away from Ozersk toward neighboring farmland. As it traveled, the cloud spread two million curies of radioactivity across a tongue four miles wide and thirty miles long.[1] Within this trace, the radioactive fallout landed in streams, fields, and forest, nearly an inch deep in soil, a complete blanketing in a territory where farmers in eighty-seven villages were busily harvesting the year's bumper crop. Ozersk city leaders did not mention the farmers in their meetings. Thanks to the zone system, the villages were not their problem.

Gulnara Ismagilova tells of working the harvest in 1957 in her village, Tatarskaia Karabolka, when she heard a blast so vast and encompassing that she and her classmates dove for the ground. She looked up to see a black cloud rise from the forest, spreading horizontally. Worried the explosion was the first salvo of a new war, men drove up in teams and told the kids to get into the wagons, and they rushed back to the village. No news followed, and that evening the villagers watched the opaque cloud linger above the trees and shift in the breeze. A light rain the next morning brought down a black snow, thick and flaky, the likes of which no one in the village had ever seen.

A few days later, men looking like cosmonauts, in jumpsuits and gas masks, dropped from a military helicopter onto the potato field. The men gave orders to the Soviet collective farm chiefdom, who then commanded the village women and children to continue digging potatoes and beets as they had before, barefoot and with no gloves, but to dump the harvest in pits dug by bulldozers that materialized in the remote Tatar village overnight. In November, the children and their parents harvested the wheat and rye and then watched as it was stacked in a great pile and burned into a toxic smoke. The children worked at these tasks all fall—becoming the first children to serve as liquidators of nuclear disaster.[2]

A week after the explosion, radiologists followed the cloud to the downwind villages, where they found people living normally, children playing barefoot. They measured the ground, farm tools, animals, and people. The levels of radioactivity were astonishingly high. S. F. Osotin, a monitor, remembered that a colleague went up to the children and held up his Geiger counter. He said, "I can tell with this instrument exactly how much porridge you had for breakfast." The children happily stuck out their bellies, which ticked at forty to fifty microroentgens a second. The technicians stepped back, shocked. The kids had become radioactive sources. The chickens had higher levels than the people; the cows, eating grass contaminated with fallout, surpassed all other creatures. The cows showed clear signs of radiation sickness, bleeding from mucous membranes, so soldiers shot them right away.[3] The scientists grew very anxious, for their own health and that of the kids swarming them. They estimated that one village, Berdianish, had spots with an alarming ninety thousand curies per square mile; the background radiation as high as 350–400 microroentgens a second—enough to get a life-threatening dose in a month in the village.[4] When the plant director, Dem'ianovich, heard about the readings, he said, "That can't be. Four hundred microroentgens a second! Impossible. Check it again!" They checked it again, and it was accurate.[5]

From Moscow, Efim Slavskii, the director of the Ministry of Medium Machines (the ministry in charge of nuclear weapons), ordered that residents of the three most radiated villages be evacuated within five days. Compared to the years during the Techa River disaster of wanton disregard for villagers' health, Slavskii's declaration of an emergency showed a heightened concern for the dangers to public health of radioactive landscapes. Speeding the villagers out of the irradiated tongue meant saving them, especially the children, from dangerous organ-seeking doses of iodine-131, strontium-89, and cesium-137.

The evacuation stretched out over two weeks, the delay evidently caused by a mundane holdup in financing reimbursements for the villagers' irradiated possessions. Historians explain this critical delay as due to a lack of experience in dealing with nuclear disaster.[6] But by 1957 plant managers had a great deal of experience in evacuating irradiated villages. They had spent the previous four years removing the villages along the Techa River. The 1957 evacuations followed, in fact, well-rutted patterns. Soldiers arrived in large, canvas-topped trucks. They ordered the mostly Bashkir and Tatar farmers, sometimes illiterate and usually poor, with many children, to pack up their possessions for resettlement.[7] Told they were being evacuated because of "industrial pollution," the farmers reasonably resisted. It was harvest time. The crops were bountiful. As far as they understood, there was no polluting factory for miles. Instead of evacuation, the action felt like occupation, as soldiers dropped their clothing, bedding, and household items into pits and buried them, while other soldiers led their livestock off to be shot at the edge of the forest. Many of the first evacuees were sent to the closed city's summer camp, Dal'naia Dacha. There they waited out the winter of 1957–58. A few wrote letters to Khrushchev: "In connection with some kind of accident in a closed city, Cheliabinsk-40, we underwent radiation poisoning and many of us are sick and we sit here without jobs and wait. What are we waiting for?"[8]

Several months later, the Moscow ministry issued an order to evacuate three more contaminated villages by May 1, 1958. One village was called Russkaia Karabolka; it was located next to Tatarskaia Karabolka, where Gulnara Ismagilova lived. The two communities, Russian and Tatar, divided by religion and ethnicity, were separated by less than a mile. The way Ismagilova tells it, she and her classmates were again called up to work one spring day in 1958. They were taken down the path to Russkaia Karabolka, where they saw that the village of 130 households had disappeared overnight, transformed into a leveled, emptied field. Only the remnants of a dynamited brick church remained. Police officers told the children to plant a row of trees between the road and the place where the village had been. The trees were to hide the disappeared village. Afterward, the children pulled bricks from the collapsed church, dug pits, and dumped them in.[9] Ismagilova is still bitter about it:

> We took those bricks in our bare hands. We didn't have boots or shoes. In those days if you wore your grandmother's galoshes you got a beating. Every day eight to ten people fell ill. Blood came from our mouths. That is how we worked and we ate there, too, cooked up potatoes, and drank that water which now they say had 6,000 microroentgens in it. All of us from our village worked on the radiated territory, on harvests, demolitions, guarding contaminated zones. No one from other places came to work. There were a lot of kids there. The police counted us and told us where to go and what to do.

By the end of 1958, soldiers had resettled seven of eighty-seven contaminated villages at an astonishing cost of two hundred million rubles. Because of the expense, ministry officials sought to keep evacuation to a minimum. In the remaining eighty villages they set up a brokerage system whereby state investigators purchased and destroyed produce that registered above the threshold.[10] Sanitation officers measured bulls and farm livestock, which in some villages "to a head showed visible signs of the effects of radiation."[11] The most contaminated animals, with open sores and molting fur, were confiscated. Ministry officials tacked up signs on village wells saying, "Dirty water, do not drink." Ismagilova is bitter about that, too: "What were we supposed to drink? How could we live here and not use the water? For appearances they put up those signs, but of course people continued to drink from the wells. Where else?"

Farmers also dug up the buried potatoes and sold contaminated beef at the market from cows they hid in the woods. After a year, sanitation officials realized that contaminated food was circulating through the province and that they could not stop farmers from using radioactive manure for fertilizer.[12]

Indeed, the consequences of leaving farmers on radiated territory and deploying them as liquidators became apparent within a year. By 1958, illness haunted many villages in the trace. In June, twelve-year-old Ismagilova grew ill with nausea, spitting up green phlegm. For weeks she slipped in and out of consciousness. The village had no clinic or medical personnel. Ismagilova's mother could only watch her daughter helplessly. Her mother had been pregnant while she worked as a liquidator. "When the baby was born," Ismagilova recalled, "the newborn was smeared, black, strange-looking. She lived only five days."[13] An estimated two thousand pregnant women worked on the cleanup.[14] Province officials penned reports about the high concentrations of radiation and the increase of illness among villagers. Villagers requested permission to move.[15]

Pressured by the Ministry of Health, the Soviet of Ministers ordered a third resettlement operation in 1959. They set a threshold—villages with twelve or more curies per square mile were to be removed. Twenty-three villages with ten thousand people met that threshold.[16] In 1960 these villages, too, slipped from the map of the southern Urals. Tatarskaia Karabolka, where Ismagilova lived, was to be one of them. Ismagilova pulled out a document from her files and showed it to me. There it was, plain as the weekend crowd of clamoring kids and men fixing their cars that had gathered in Tatarskaia Karabolka the August day I visited in 2009. The order stated that because radioactive contamination had made the territory uninhabitable, the twenty-seven hundred residents of Tatarskaia Karabolka were to be moved to a state farm on the other side of the district.[17]

Mysteriously, the evacuation of Tatarskaia Karabolka never occurred. The village disappeared from local maps, and its collective farm was closed because its food was "inedible," but the villagers remained. There are a number of theories about why. Some say it was the expense. Tatarskaia Karabolka was a large village; province officials estimated that the costs of resettlement would amount to 78.5 million rubles.[18] Others say it was an oversight, that officials confused Russkaia and Tatarskaia Karabolka, concluding that they were the same and had already been evacuated. The Russian Ministry of Health stated in 2000 that Tatarskaia Karabolka had less than 6 curies a square mile and so did not meet the threshold.[19] That can't be right. The village was clearly slated for removal in the 1959 order, and province officials who in subsequent years sent alarmed messages about illness in the village were well aware of Tatarskaia Karabolka's contamination.

Sitting in the log cottage where she was born, Ismagilova pulled out a map of measurements of radioactivity of the soil from the 1990s. "I got this map from a friend who works at the archive," Ismagilova said. Looking at me over the top of

Gulnara Ismagilova and Nadezhda Kutepova in Tatarskaia Karabolka, 2007. Photo by Kate Brown.

her glasses, she added, "I'm not supposed to have it." In the aftermath of the accident, Soviet scientists settled on a permissible threshold equivalent to about one-third of a curie per square mile. Ismagilova's map showed that thirty years after the accident her village recorded hot spots measuring up to sixty curies a square mile. Ismagilova pointed on the map to her house and those of her neighbors in the hot zones. "That's where we live, where we've lived all these years, farming, raising our children."

Ismagilova, a retired nurse, said the map explained the "whole bouquet" of medical complaints in her village: tumors, cancers, thyroid problems, diabetes, disorders of the circulatory and nervous systems, birth defects, strange and powerful allergies, intense fatigue, and fertility problems. A 1991 medical study estimated that people living on territory with from one to four curies had a 25 percent higher chance of dying from cancer than those resettled.[20] Ismagilova motioned around her to the village of log houses stretched along a road leading to a thick pine forest. Because of a tumor in her liver, Ismagilova said, she didn't expect to live much longer. "I'm the last one of my class. All the others who were child liquidators have died, mostly of cancer."

Ismagilova is one of many who believe her village was left in harm's way purposely to be medical subjects in an opportunistic experiment. Those of Tatar background point to the fact that the Russian village was moved, while the Tatars were left in place.[21] Soviet officials, however, were slow to monitor residents of Tatarskaia Karabolka, and never did so in any comprehensive way. They set up regular medical exams only in 1972, fifteen years after the accident.

The Karabolka case is puzzling. No document has been found to explain why Karabolka remained. Likely the plant's notoriously corrupt construction firm, tasked to build new houses and schools for Karabolka, ran out of money or time, or the money was embezzled or reinvested elsewhere; the construction bosses stalled, as they did with resettling the Techa villagers, and eventually let the resettlement of Karabolka quietly drop, an accepted oversight.[22]

It is true that nuclear officials were eager to capitalize scientifically on the accident. In the summer of 1958, Efim Slavskii proposed a major new research institute to be located somewhere in the nearly eight thousand square miles of the radiated trace. The institute would specialize in radioecology; its mission was to learn how to live on irradiated territory in order to survive a nuclear war. Moscow scientists, however, found the idea of working in the provinces and on contaminated ground frightful.[23] Instead, they founded the Institute of Radiation Medicine near Moscow to carry out lab work, and set up only an experimental research station on a former collective farm in the radiated trace.[24] Staffing the institute with local researchers, they set to work. Scientists found that pine trees, storing radioactive isotopes in their needles, were withered and jaundiced. By 1959, whole pine groves had died and remained crimson, like the later Chernobyl "red forest." Birch trees, hardier, survived, but turned a bluish shade, generated twisted or gigantic leaves, and produced fewer seeds. Grasses, meanwhile, grew intensely, producing three times more biomass. Scientists noticed that the most vulnerable animals were rodents who fed on the forest floor, where radioactivity concentrated. In the subsequent twenty years, mice's life span and fertility dropped markedly.[25]

But there was good news, too. The station's scientists figured out how to rehabilitate radiated soils. They learned which vegetables stored more radioactive strontium. They discovered that meat from animals fed contaminated fodder was safer for humans to eat than leafy vegetables, and the meat of pigs and fowl safer than that of cattle. In 1960, optimistic and intrepid plant managers dined on cucumbers, potatoes, and tomatoes grown in the contaminated trace. In 1967, scientists returned much of the original trace to the factory's management as an off-limits wildlife sanctuary.[26] The message was clear: even in the midst of a major nuclear disaster, life carried on.

Since the Kyshtym explosion became news in 1989, the Russian government has had to explain the presence of communities such as Karabolka within the confines of the original irradiated trace. The positive results from the experimental farm helped justify Karabolka's continued existence.[27] Medical investigators released classified studies claiming that none of the exposed villagers experienced radiation illnesses. The only epidemiological singularity they found was that children resettled from the trace were five to ten times more likely to develop thyroid cancer.[28] Western and independent Russian scientists have criticized the follow-up study for the small size of the population examined (1,059 people), the limited period of observation, and lack of an adequate control group.[29] Russian geneticist Valery Soyfer asserts that the Soviet government purposely under-funded and discouraged genetic studies of villagers in irradiated territories.[30] Other researchers have shown that the food villagers ate was contaminated, as are the bones of those who died.[31] In general, there are few health studies, and those that exist are government-issued, which assert the people are not sick, or if they are sick, then it must be from radiation phobia, or alcoholism and poor diets.

After the resettlement that never was, the Karabolka villagers remained with no collective farm, no income, no status or public existence in the Soviet economy and society. In a strange way, Karabolka mirrored the phantom city of Ozersk. The villagers lived within restricted grounds, outside of the tax struc-ture, off the map. But unlike the consumer-oriented residents of Ozersk, in Karabolka families engaged in subsistence farming. In Bashkir communities in the trace, villagers had a long tradition of gathering natural products from the forests, including berries, mushrooms, fish, and game—the foods researchers found to be most contaminated. They were banned from selling their produce at local farmers markets, although they did so anyway, in need of cash for clothes and other necessities. The farmers had to be sneaky, though. Radiation monitors ranged the local markets, checking for contaminated food. When they caught Ismagilova's grandmother selling radiated beef, they told her to take it home, boil it a long time, and then eat it.

Sanitation experts showed up in Karabolka over the years. They instructed the residents to whitewash their houses. They demonstrated how to clean and cook in order to protect against ingesting radioactive isotopes. They took mea-surements, and when they found objects that sent the meters ticking, they tossed them in pits outside the village—small local nuclear waste dumps.[32] The vil-lagers resented these medical investigators. "They'd knock on the door," Isma-gilova remembered, "in their masks and white coats. When you offered them tea, they'd decline, take out plastic to sit on, and drink their own water."

I glanced at the untouched tea and potatoes that Ismagilova had prepared for me, realizing my interview subject was sending me a message. I fell in the same category as the inspectors who invaded Ismagilova's privacy and stigmatized her and her home as contaminated. I, too, declined her food, questioned and gawked, and then, without fixing anything, left. Ismagilova told me she had dreaded talking to me, to tell her story yet again, as if she were an act in a circus freak show.

I asked Ismagilova if she would walk with me down to the village mosque, a small building of green clapboard surrounded by a cemetery. She paled at the thought. "No, I won't go there. Sick spirits roam in the graveyard after dark. I don't think you should go either."

On my departure, Ismagilova asked for money. I gave it to her. She asked for help getting drugs to treat symptoms of her cancer. I took down the name of the prescription. On leaving, I realized that, without intending to, I had become a disaster tourist with Ismagilova as my guide.

Disaster tourism was one of only a few career paths in Karabolka. With the local economy hollowed out, some people had found ways to make a living as purveyors of their misfortune. Neighbors made accusations about others profiting from their common calamity, and communities divided, often bitterly, which made it more difficult to find solutions other than a continued existence on radiated ground.[33]

I never saw Ismagilova again. I couldn't find her drug in the United States or Canada, and I sent word to let her know. Two years later, a mutual friend sent me a photo of her. I didn't recognize Ismagilova. The picture showed a woman emaciated and much aged. The photo haunted me. I realized how easy it is to get on with one's life and forget the little irradiated villages of the southern Urals and the sick people in them.

33

Private Parts

On December 1, 1962, in the Hanford Labs, subjects E4 and E5 walked into a room lined with 120 tons of steel. The subjects were seated in a heavy dentist chair. An attendant placed camera-eye monitors up to the subjects' chest and neck, turned on a TV, and left the room, closing the heavy door as he went.

E4, male, age thirteen, and E5, female, age nine, lived in Ringold, Washington. Ringold was a small farming community on the Columbia River, bounded on the north by the nuclear reservation and to the east and south by four-hundred-foot bluffs that isolated the community along a fertile riparian flat. The community had long been under surveillance. Police boats patrolled the river, and security guards manned the nearby gates into the reservation. Ringold used Columbia River water drawn downstream from the reactors for irrigation, and the hamlet was located thirteen miles leeward from the chemical separation plant, reactors, and fuel fabrication factories. For those reasons, in 1962, the farm families became objects of yet another form of surveillance.

Twelve of twenty Ringold residents agreed to sit in a whole-body counter at the Hanford Labs so that plant scientists could tally gamma rays inside their bodies. The parents of E4 and E5 were especially cooperative. They owned forty-five acres planted with peaches, apples, and pears. The fruit, fed with cool river water, grew well against the sun-baked bluff. The family ate from a kitchen garden, kept milk and beef cows, and dined on game—deer, quail, pheasants, and geese—bagged by the boy. Of the half dozen families in Ringold, the E-series family alone subsisted nearly exclusively off the land.[1]

This dietary choice was critical. Of the twelve subjects, the boy and girl had the highest counts of radioactive iodine-131 in their thyroids. Nine-year-old E4 had 120 picocuries (120 trillionths of a curie); thirteen-year-old E5 had 300 picocuries, "the largest amount observed in a child's thyroid to date."[2] The counter measured only gamma rays, not the beta rays also emitted by radioactive iodine, which concentrates in the thyroid. Both sources of energy can damage internal tissue. Children, because they are growing, efficiently absorb minerals and elements in their steadily multiplying cells. For similar reasons, in another family of an avid hunter and fisherman, a nineteen-year-old female, four months pregnant, had a high radioactive iodine count.

The Hanford scientists discussed the iodine measurements with the children's parents. They assured them that though the counts were elevated, they were well within permissible doses. They did not say that the standards they used were for adult workers in professions where they would be exposed to radiation.[3] The scientists suggested that the children might drink powered milk instead of fresh milk. Then, surprisingly, researchers concluded their published paper stating that the detection of only a few radionuclides among the Ringold residents was a "gratifying" conclusion.[4]

The unflagging optimism of the Hanford researchers is admirable. The researchers took the twelve-person Ringold sample as evidence that they were both vigilant in their care for fellow citizens and secure in the knowledge that the plant was safe.[5] These were important conclusions to draw because of a media storm that exploded in the late fifties when journalists reported radioactive Minnesota wheat, Iowa milk laced with strontium-90 from radioactive fallout, and oysters at the mouth of the Columbia River tainted with zinc-65 from Hanford effluence. Fear of radioactive food greatly shook public confidence in the AEC.[6] These fears were compounded by a rising American cancer rate and a dawning awakening, inspired by Rachel Carson's *Silent Spring*, about the possible long-term dangers to the nation's health caused by widespread environmental contamination.

For a decade, the U.S. Public Health Service had pressed Hanford scientists to disclose information on the contamination of the Columbia River, which fed a swelling network of irrigation canals. The canals brought water to farms east toward Ringold, and northeast, up and over the bluffs to the farms of Mesa, Connell, and Eltopia. Classified maps tracked prevailing winds from plant stacks through these communities; the breezes sometimes shifting due north toward Othello and the Wahluke Slope, or southeast toward Pasco and Walla Walla.[7] With an American public becoming more attuned to environmental pollution, Hanford health physicists used the whole-body counter to take measurements, for the first time, of radioactive isotopes in the bodies of people living near the plant. They fixed up a portable whole-body counter in a school bus.[8] The researchers took measurements in Richland, in neighboring farm communities, and in tiny Ringold. However, they only published the results of the twelve Ringold subjects.

In 2011, subject E-5 was middle-aged, living in Richland, and working for one of the contractors engaged in the Superfund cleanup of Hanford. E-5 told me she is healthy and has adult offspring and grandchildren. Her mother lived to be in her nineties. One of her brothers, subject E-4, died young, in Vietnam. Another brother died in his sixties. The rest of E-5's seven siblings are alive. In short, E-5 and her family are the thriving exemplars of the "gratifying" results of the Ringold study.

She and other participants told me they did not feel like human subjects of a medical experiment. Rather, they were comforted by the fact that plant scientists visited regularly, tested them in the whole-body counter, and gathered up samples of their farm produce. The scientists' attention led them to feel safe and protected. They remember being told that their test results showed they had no radioactive elements in their bodies at all.[9] One woman said that no one in Ringold got sick. She disparaged charges that Hanford emissions caused illness in the region and dismissed the lawsuits against Hanford contractors by groups known locally as "downwinders." She had known the downwinder activists since her school days, she said, and they were just the kind of people to bellyache.[10] The Ringold examinees I talked to were, in this way, the perfect subjects of the Ringold study. They took away the meaning the study was designed to convey, in order to soothe and reassure an anxious public.

But there are some puzzling qualities to the Ringold study. The authors do not explain why this study occurred so late, two decades after start-up, ten years after the worst dumping of radioactive waste had ended, at a time when plutonium production at the plant was winding down. The Ringold sample is tiny, yet they found a significant pair of outlier thyroid measurements.[11] Nor do the authors attempt to justify the scientifically anecdotal size of the study, even though they had also been tracking many farm families across the region. In the published version, the authors state, improbably, that the majority of the atmospheric radioactive iodine came from "global fallout," largely from Soviet testing.[12] In the published study, the scientists did not mention a large unintentional spill of 440 curies' worth of iodine-131 in April, nor an intentional test release of an additional 8 curies in September 1962. This test release triggered "very high" concentrations of iodine-131, not in Ringold, but up and over the hills in Connell and as far away as Ephrata and Moses Lake.[13] The Ringold study plays off the assumption of a map of radiating concentric circles, in which the closer people are to the source, the higher the dose. This has been a lasting association, but it is erroneous.[14]

The buffer zone around Hanford derived from the calculation that the most toxic gases, such as radioactive xenon and plutonium, would need a cube of air two miles wide and two miles high to safely disperse.[15] The radiating circles imposed over maps of Hanford predicted exposures fifteen, thirty, and forty miles away based on the notion that dispersal of gases would occur in a uniform manner, decreasing outward from the source. But that is a fiction. Over the years, Hanford meteorologists were frustrated by their inability to predict where radioactive effluence was heading and where it would touch down.[16] Tests showed that plumes took off from plant stacks zooming in one direction or lingering, switching course, touching earth only to bounce skyward again and land scores of miles away. Radioactive plumes often took the form of grotesque tongues reaching out from the source, twisting and thickening.[17] These variable conditions coated the earth in a mottled pattern of random hot spots against areas of normal background levels. The one thing the releases never did was form an evenly diffused, radiating circle as sketched in nuclear target maps.

In the early 1960s, Hanford researchers suddenly had plenty of money for research. Under pressure because of political exposure from the accelerating antinuclear movement, the AEC granted Hanford Labs a great deal more resources for medical research, including the first funds for studies on the biological

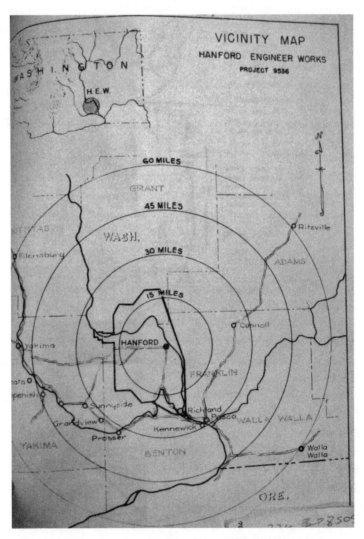

Hanford target map. Courtesy of National Archives, Atlanta.

effects on human subjects of chronic low doses.[18] At the time, AEC officials actively encouraged research on human subjects.[19] With the help of new technologies to count extremely small amounts of radioactivity, researchers felt they could conduct these experiments safely.[20] The effect of low doses on ambient populations was an extremely important research topic, one long overdue. Unfortunately, however, Hanford's transition to a national laboratory meant it acquired many of the features of long-standing AEC, Central Intelligence Agency (CIA), and army research programs on the effects of radiation on man that had blossomed in the decade following the Nuremberg trials.

This record was less than admirable. Before World War II, human radiation experiments numbered in the single digits. The most extensive experiments occurred in the state asylum in Elgin, Illinois, a Tuskegee moment in the history of radiobiology, when doctors fed radium-226 to thirty-three unwitting patients, all of whom died of cancer in subsequent decades.[21] In the postwar years, energized by new funding opportunities and the escalating arms race, human experimentation quickened. In the fifties, American doctors on federal research grants gave radioactive vitamin drinks to more than eight hundred pregnant mothers at Vanderbilt University and the University of Iowa. At university hospitals in Nebraska, Tennessee, and Michigan, doctors administered radioactive iodine to newborns. Radioactive iodine was also fed to more than a hundred native Alaskans. Researchers nourished disabled children in Waltham, Massachusetts, on oatmeal fortified with radioactive calcium. At the Medical College in Virginia, doctors injected radioactive phosphorus-32 into burn patients. At the National Reactor Testing Station in Idaho, scientists had volunteers drink radioactive milk, inhale radioactive gases, and swallow plastic capsules laced with fission products. Department of Defense investigators gave whole-body radiation exposures to indigent black cancer patients from 1960 to 1971 at the University of Cincinnati Medical Center, where doctors forged consent forms and exposed uninformed patients to more than 100 rads, which left them vomiting and writhing in pain.[22]

The list of experiments goes on.[23] By the 1960s, the AEC had dispatched more than half a million shipments of fission products to doctors engaged in 430 studies.[24] Adding the accidentally exposed Marshall Islanders, whom AEC researchers studied afterward, and the purposely exposed American soldiers in field tests in Nevada, the toll of human subjects of radiation experiments mounts to the tens of thousands. In most cases the subjects lived in remote areas, on the margins of mainstream American culture; they were poor, sick, conscripted, incarcerated, incapacitated, and underage.

Before the sixties, Hanford researchers had carried out only a few experiments on human subjects because GE executives were not interested in biology research.[25] With more funding and an AEC mandate to test on humans, Hanford researchers started their own experiments. They had GE employees ingest, inhale, or inject radioactive promethium, iron, technetium, and phosphorus to use in calibrating the new whole-body counter. They injected five patients in hospitals in Portland and Seattle with radioactive phosphorus-32, apparently without consent. Employee-volunteers drank milk from cows that grazed on radioactive grass. They also contracted medical student volunteers to ingest small amounts of plutonium and eat a weekly half pound of Columbia River fish.[26]

The largest study at Hanford involved prisoners as subjects. In 1965, workers renovated a basement storage room of the Walla Walla State Prison into a medical research lab.[27] In the basement, convict-volunteers entered a cell fortified by a massive wall of concrete and sand. They lay facedown on a trapezoid-shaped bed. As the men placed their legs in stirrups attached to the bed, their testes dropped into a small plastic box filled with scrotal-temperature water. Attendants then pulled a switch that shot X-rays at the testes from two sides.[28] The prisoners were paid $5 a month for their service, $25 for each biopsy, and $100 for a mandatory vasectomy at the end of the study so they would not spawn children with genetic mutations.

Radiation at the lowest dose level, ten rads, caused damage. At twenty rads, the prisoners became azoospermic (in lay terms, sterile). Despite the findings that even low doses killed all sperm, the researchers continued to experiment at higher doses of 25 rads, 40 rads, and 60 rads, up to a terrific 600 rads.[29] No matter how high the doses climbed, the University of Washington's Dr. Alvin Paulsen and the University of Oregon's Carl Heller got the same azoospermic results, testing on 131 prisoners, zap after zap, for a decade of contract renewals.[30]

The missing sperm were no medical novelty. Researchers discovered from human plutonium studies in 1945 that after microcurie injections, patients' sperm tails broke off and disappeared.[31] In 1962, however, the question of sperm took on new immediacy when three operators in the long-troubled Hanford 234-5 Facility saw the penetrating blue light of a criticality accident, an unintentional chain reaction followed by a surge of neutron radiation. With the alarms piercing the air, the men ran for the exit, already flooded with fleeing employees. Marge DeGooyer shot out of the building and took a seat in a car pulling up to whisk workers away. Harold Aardal, trembling and sweating after being the closest body to the shooting blue rays, sat down next to her. DeGooyer, wanting to be nowhere near the radioactive Aardal, leaped out of the car to find another ride. Left alone, Aardal was terrified. He knew that no one had ever lived after seeing the blue flash of a criticality accident.[32]

At the Kadlec Hospital, Aardal and his two colleagues were checked into the special, thick-walled ward for radiation injuries. The doctors, wearing protective suits, monitored the men hourly, took blood, urine, and semen samples, and ran them through a whole-body counter. They trimmed their fingernails, shaved radioactive hair from chest, pubic area, and posterior, and removed the men's gold fillings, neutron-rich after the accident. Aardal got the highest dose: 123 rems to his whole body and 218 rems to his gonads.[33] The accident sterilized him, left him seriously anemic for two years, and rendered him morbidly fearful.

Hanford doctors, however, displayed no fright. As the first acute criticality accident at the plant, they relished the chance to examine the exposed operators and were delighted to find that the men experienced only transient symptoms. "Study in the hospital of the three employees for a period of 8 days revealed no symptoms of injury attributable to radiation, except fear."[34] Fear was not acceptable, and for years afterward Aardal was scorned by his medical handlers as the "nervous guy."[35] For the doctors the only troubling symptom, which cropped up later, was their patients' missing sperm—a worrisome indication for a lab of male researchers. The scientists devised the study of prisoners' testes to find out more about how neutrons zapped male virility.[36] They naturally turned to the Walla Walla prison because they already had a relationship with that institution. A Richland psychiatrist had used prisoners in the fifties to test "truth serum" drugs for the CIA. Since that time, the prison warden had socialized on the Richland cocktail party circuit.[37]

Concerned about liability and publicity, AEC officials instructed the doctors to proceed "cautiously."[38] The Battelle Memorial Institute took over management of the Hanford Labs in 1965, renaming it the Battelle Northwest Labs. The new lab inherited the study from GE. The initial deal called for Dr. Paulsen to select the prisoners and manage their medical surveillance, while employees of GE Hanford were to set up the irradiation equipment and operate it.[39] Looking into it, however, Battelle technicians determined they could not fully shield from exposure prisoners' prostates and other organs near the testes.[40] Battelle lawyers also determined they could not fully insulate the company from legal exposure.[41] And so, for liability reasons, Battelle lawyers banned company employees from direct involvement with the prison volunteers. To overcome this obstacle, Battelle executives insisted that Paulsen would "push the button" on the X-ray machines.[42] But Dr. Paulsen was also not willing to push the button. Instead, he paid "inmate technicians" to move the controls that sent X-rays shooting into fellow prisoners' testes.[43] This solution had problems, too. The prisoners had no training in radiobiology. Sometimes they held the button too long on fellow convicts, out of ignorance or malice.[44]

The whole program smelled bad. In 1967, Parker and his colleagues concluded that Battelle Northwest should work toward separation from the study because of "technical findings" (the inability to protect prisoners from unintended irradiation) and "administrative findings" (the inability to defend Battelle from liability).[45] AEC officials concurred, stating privately out of Paulsen's hearing, "Let's finish this up and get out!"[46] Yet despite AEC and Battelle's qualms, Drs. Paulsen and Heller kept renewing their contracts for another six years, spending $1.6 million in federal funds.[47] Finally, Audrey Holliday, a Walla Walla prison psychologist, questioned whether in a prison atmosphere human subjects could be "volunteers." Under pressure from Holliday, prison officials ended the study in 1971.[48]

The prisoner testes irradiation study was a low point in Hanford medical research, a moral hiccup. There were not many high points, however. For all the expense, AEC funded research provided no real answers to the critical question of the health effects for humans exposed for long periods to low doses of radiation. In part, the lack of answers was due to the fact that good science takes a long time, but there wasn't much evidence of a culture of good science. Hanford scientists complained of the lab's "narrow research horizons" and their isolation from mainstream scientists.[49] The pronuclear climate of the lab meant that supervisors often dismissed studies as inaccurate or flawed that gave damaging results.

Hanford scientists also failed to ask vital questions that might refute the AEC position that fission products were safe within permissible levels. As I will show in more detail in a subsequent chapter, AEC scientists focused their research agendas on combating independent studies that questioned the safety of the Columbia Basin. The Ringold study served as an illusion, a palliative to support the assertion of safety, and the prisoner study served as a source of income for Hanford's labs. Neither project, however, was designed to address the vast epidemiological questions created when Hanford waste came in contact with local bodies in contaminated landscapes. On the contrary, the prisoner study put more people in harm's way and created yet more contested terrain.

In 1976, five prisoners sued the federal government for damages.[50] The plaintiffs charged they had received large doses of radiation administered to their testicles by unlicensed prison inmates. They said they had endured health problems, including back and gonad pain, ulcers, tumors, blood poisoning, and radiation burns, and that when they had written letters to Dr. Paulsen requesting promised medical aid, Paulsen refused to reply. The court ruled that the AEC and its contractors were not "sueable."[51] Eventually, the five plaintiffs settled out of court for a collective $2,000.

Surprisingly, when the study ended, a number of prisoner-volunteers objected to the halting of the experiment.[52] Over the years, some prisoners had come to rely on the study as a source of income. Recidivists often signed back up for the study after returning to jail for another sentence.[53] The prisoners, begging to have their study back, might well serve as an apt metaphor for plutopia in the waning years of the arms race.

34

"From Crabs to Caviar, We Had Everything"

In the summer of 1959, Soviet premier Nikita Khrushchev was up to his characteristic pranks. In July, he issued an ultimatum to Western occupying forces in Berlin, telling them to get out. "Berlin is the testicles of the West," he chuckled. "Each time I give them a yank, they holler." Soon after, Khrushchev accompanied Vice President Richard Nixon, visiting Moscow, on a boat ride down the Volga. As they motored, they passed bathers wildly waving. Eight times Khrushchev shouted to them, "Are you captives?" Eight times the swimmers smiled and shouted back, "Nyet!" Nixon could manage as a comeback only a grimace.[1]

Nixon was in Moscow for the American National Exhibition, where in his opening speech he championed American prosperity. "American homes have 50 million television sets and 143 million radios. . . . These statistics," Nixon asserted, "dramatically demonstrate that the United States, the world's largest capitalist country, has from the standpoint of distribution of wealth come closest to the ideal of prosperity for all in a classless society."[2] In 1959, Soviet headlines featured litigation over integration in Little Rock and the strike of a half million American steelworkers. To discerning critics, the United States didn't much look like a classless, egalitarian society. Aware of these charges, Nixon worked to turn the debate from discussions of political freedoms and civil rights to those of "free enterprise" and the freedom to consume.[3]

Sociologist David Riesman first imagined replacing the arms race with a consumption race, which he called Operation Abundance. If Soviet citizens were allowed to taste the riches of the American middle class, Riesman reasoned, the Soviet people would soon toss off their masters.[4] The American exhibit in Moscow featured a full-scale model of a suburban ranch house. The American planners hoped the model home would trigger a "minor feminist revolution" in Russia, as Soviet women, depicted in the American press as wearing scratchy woolens and stooped over shopping bags, mobbed communist leaders, demanding the freedom to buy American vacuum cleaners and dishwashers.[5] In the model kitchen, Nixon and Khrushchev famously debated which nation better provided for its housewives. To viewers, it appeared that the scowling world leaders, fingers jabbing, were in an argument. On the contrary, the emissary of the free world and the leader of the communist world were in complete agreement. Their debate sent the message that the winning ideological system was the one that served up the most goods, or, as they put it, the highest standard of living.

The concept of "standard of living" was an American-promoted set of criteria that prioritized per capita consumption and purchasing power over other qualities of life such as health, safety, and environmental and economic equity and security.[6] Khrushchev agreed to this standard when he announced that the USSR would overtake the United States in the production of consumer goods, a credible claim in the fifties, when the Soviet economy was experiencing an astonishing economic miracle, surpassing all but West Germany in economic growth.[7] By agreeing to the appliance race, however, Khrushchev lowered Soviet aspirations to those of the capitalist philistines, who boiled human life down to statistical averages. In its conception, the Bolshevik revolution was to deliver not just consumables but radical new cultural values and humanist goals.

When Khrushchev signed on for Operation Abundance, he agreed to forsake that revolution, yet the new message that freedom amounted to the freedom to consume was initially lost on many Soviet citizens. The communists in Ozersk, for example, had trouble accepting Khrushchev's appliance race. Since the late fifties, they had aspired to something better. Only later, after Khrushchev had been removed from power, did Ozersk citizens succumb to his remastered revolution. Over the years, as their socialist consciousness flagged, so did their interest in safety. In those years, residents of plutopia exchanged their biological rights for consumer rights.

In the early sixties, managers rushing to fill orders for plutonium had trouble attracting and keeping workers. The press for labor was exacerbated because a decade after start-up, many plant veterans had grown too sick to work, even at desk jobs. In 1958, several thousand people in their thirties and forties were retired on invalid pensions.[8] These ghostly people, leaning on canes, further undermined statements about plant safety. People asked about the mysterious illness, called cryptically "the fungus": where it came from, how to avoid it, why the plant doctors had no cure.[9]

To replace the invalids, managers brought in new recruits who went through security briefings and job training, but many left as soon as they started work. In some groups, all trainees departed, taking with them state secrets and valuable skills. They fled because they heard rumors of health problems and disliked the walled city with its troublesome pass regulations.[10] Experienced workers abandoned the plant, too. The more dangerous the shops, the higher the turnover rates.[11] In ten months during 1960, five thousand of about sixty thousand residents deserted the closed city.[12]

Plant bosses went to great lengths to keep their employees. When judges attempted to sentence lawbreakers, supervisors intervened, demanding their staff report to work instead of jail.[13] Citizens who committed misdemeanors could lose their pass to leave the city, but residents and even the KGB chief strongly resented this form of social control. They railed against the two bureaucrats who controlled passes, calling them "little tsars," and demanded the practice be restricted.[14] When party bosses sought to exile troublesome residents from the city, factory managers stepped in again, arguing they needed all hands on deck. As a consequence of the failure to prosecute, the city's crime rate rose each year from 1950 to 1962.[15] "Why," a city party leader asked, "in our city where we have special conditions, where we are safe from criminal reinforcements from the outside world, is the crime rate so high?"[16] Police responded with a war on "hooliganism," banishing teenagers to juvenile labor colonies outside the gates.[17] Too young to work, teenagers could be spared. Managers also freed up housing by sending away veteran workers classified as invalids.[18] Healthy workers, however, even criminal ones, were rarely sent away.[19]

Yet the most important way for managers to encourage employees to stay in Ozersk was to fulfill and overfulfill their employees' consumer desires. By the end of the fifties, the four closed nuclear cities of the Urals (with populations ranging from twenty thousand to fifty thousand) devoured 39 percent of the provincial budgets. While the majority of Soviet citizens lived in communal residences and waited a decade for an apartment, 70 percent of atomic city residents dwelled in private apartments, the largest contingent in unbelievably luxurious three-room apartments. In the "big world" young couples lived with in-laws. In Ozersk, a young couple had to wait no more than a year for an apartment.[20] In the sixties, while factory leaders quietly shifted budgets allocated for plutonium production into city services, housing, schools, and paychecks, city leaders loved to reflect on Ozersk's prosperity by listing, like Nixon, what residents owned: 1,500 television sets, 5,000 radios, 1,400 cars, 2,500 refrigerators . . .[21] They told audiences that Ozersk "was getting better every year, with more goods, more services, more staff to serve."[22]

In the forties and fifties, city bosses told residents they were in the right place because they were defending the nation. In the sixties, the city's unique affluence became its major selling point: "Our city isn't on the map, but many can envy the conditions in which we live." As one woman remembered, "It seemed that we were already living under Communism. In the stores there was everything, from crabs to caviar."[23]

If Communism, as Khrushchev asserted, was a confluence of abundant consumer goods and workers armed with wallets to purchase them, then Ozersk had indeed arrived. Soviet theorists believed that once society achieved Communism, humans would behave differently. Aided by wild leaps in technology, communists would use their newfound prosperity to advance civilization and improve the species. Ordinary people living under Communism would become well-rounded individuals, learning, as one resident put it, "to dress inexpensively and with taste, prepare healthy food, and raise good children."[24] In postwar Soviet culture, scientists, especially physicists, epitomized the new socialist man, dedicated, selfless, and armed with technology to improve human life.[25] Soviet science testified to the superiority of the Soviet experiment: the world's first nuclear power plant (1954); the world's first satellite, Sputnik (1957); the world's first civilian nuclear-powered ships (1959); and the first man vaulting through space to look back on a small, marbled earth (1961).

In the 1961 film *Nine Days of One Year*, the nuclear scientist Dmitri Gusev was this new kind of scientist-hero, a platonic warrior sacrificing his youth for humanity. In the film, Gusev, working in a well-heeled, closed nuclear city, selflessly toils night and day, exposing himself to radiation in order to continue his experiments, "any one of which," the film ads proclaimed, "might be his last." Gusev, a facsimile of the physicist Igor Kurchatov, dashes into reactor chambers and pores over calculations as his health fails and his lonely wife waits for him in their childless home. Gusev is indifferent not only to his health and fertility but also to his beautiful wife, fine car, stylishly furnished apartment, and lavish meals in elegant restaurants. He wants only to invent the world's first thermonuclear-powered reactor, an innovation that would produce "practically unlimited energy" and an unimaginable prosperity.

Ideally Ozersk, an exceptional city of young technocrats, should have been full of ardent and altruistic communists. With an eye to Ozersk as a model for a socialist city, party leaders in the early sixties founded a sociological service to carry out classified opinion polls among residents. The polls invited residents, aware of their select status, to annually calibrate their achievements.[26] As they read the results, however, city leaders stumbled across nagging issues. They discovered there were not many young Gusevs in their community. One communist expressed it this way: "We have everything: schools, excellent teachers, fine facilities, cultural organizations, and a solid material base," but, he added vaguely, "these possibilities are not being fully realized."[27]

That was the rub. Though they lived in a socialist paradise, it did not *feel* like Communism was anywhere on the horizon. Theft, alcohol abuse, violence, unruly workers, and delinquent children continued to trouble the community. In the early fifties, communists said that problems with wayward youth and crime were due to poor conditions: crowded schools, overworked teachers, and busy parents leaving children unattended. By 1965, these problems had been solved by generous budgets flowing from Moscow. "In other cities it is hard to buy tickets to the movie or theater," a city administrator pointed out, "while here we have trouble filling the hall. In other cities it is difficult to get your child into a good school or music program. Here we have more places than children."[28]

A group of young men became the poster boys for the failed new man in Ozersk. The group called themselves the "ailurophiles," or cat-lovers. They had graduated from good Soviet institutes and arrived in the city just after the 1957 accident. According to their supervisors, they had experienced no hardships in their lives, no revolution or world war. They had gone to school on the "backs of their parents" and been given everything, including jobs at the factory, excellent salaries, and promotions, yet they showed no gratitude. In their free time, the men did not attend concerts or lectures, and they openly scorned the Young Communist League. Instead, they went to restaurants and threw parties with women of poor reputation. The men paid too much attention to style, earning the derisive epithet *stiliagi* (hipsters). At the House of Culture, they drank and danced the boogie-woogie shamelessly, and then showed up for work the next day hung over. The educated twentysomethings did not read Soviet publications but would listen instead to the Voice of America, from which they picked up subversive ideas they repeated so brazenly a sympathetic supervisor warned them, "Quit saying those things openly. You'll get arrested."[29]

City leaders were at a loss to explain the cat-lovers' behavior. Soviet pedagogues did not believe in teenage rebellion or generational conflicts. They argued these were capitalist notions promoted to try "to undermine the solidarity of Soviet people around the Communist Party."[30] Mystified, the older communists asked, "Why do they, and others, turn their backs on the honor and worth of the Soviet people?"[31]

The dozen cat-lovers were not alone. City fathers polled youth and identified many others with "limited interests—drinking, music, dance."[32] Adults linked the passion for "stupid dances" and "dressing like a parrot" with "hooliganism," a term that encompassed breaches in loyalty, security, and discipline.[33] In 1960, a youth broke into the city's radio station and sent the community into a panic by giving an Orson Welles–style broadcast announcing the start of war—in Ozersk surrounded by Zenit missiles. Several young plant operators got acquainted with a young American guide from the American National Exhibition while in Moscow. Plutonium workers were banned from having contact with foreigners, especially Americans likely to be CIA informants.[34]

Party members asked each other: "How do you raise children in the spirit of revolutionary consciousness when they did not suffer the deprivations of the tsar, the Revolution, or the Great Fatherland War?"[35] "Why do these young people sing those vacuous love songs?"[36] "How can you be a successful communist, yet raise bums and parasites?"[37]

Gradually it dawned on party members that the conditions of advanced so-cialism were creating citizens who were entirely unsocialist. In fact, the accumu-lation of material wealth was not a remedy but an addiction that never seemed to be sated. The more Ozersk residents acquired, the more demanding and dis-cerning they became, in a way that would have made the creators of the Ameri-can National Exhibition proud. They wanted better apartments, with higher ceilings and bigger rooms. City dwellers owned two to twenty times as many cars and appliances as the rest of the country, but they had yet more desires. They wanted only clothing from Moscow or foreign manufacturers. They demanded an airline counter, jewelry and fur stores, better TV picture quality, and home delivery of milk, bread, linens, and flowers. Though paid twice as much as average citizens in the "big world," they wanted even higher wages and refused jobs that didn't pay enough. They wanted more dignity in the form of self-service grocery stores and bus service without ticket controllers "to show we are trusted." They wanted a marble gazebo on the lake, more sailboats in the yacht club, and more resort choices for vacations.[38] Overwhelmed, the director of the city's supply complained, "The population makes so many demands, we can't possibly satisfy them."[39]

Ozersk elite took to discussing their "opportunist" and "materialist" neigh-bors. A party activist in 1960 grumbled: "Some people start to see the purpose of their life in the acquisition of a car, a suite, television, rugs, while work and social responsibilities become secondary affairs. We have to cure people of that."[40] The growing embourgeoisement of plutopia generated anxieties in part because mass consumption carried with it a dangerous leveling quality. When blue-collar workers with more generous paychecks and cheaper mass-produced goods began to afford to outfit themselves like the Soviet profes-sional classes, the elite found it deeply disquieting. They believed they knew how to consume with restraint and taste in a way that maintained peace and order. Lower classes indulged wantonly, riotously, and would drown society in avarice and acquisitiveness.[41]

And so communists worried about the emergence of a petit bourgeois men-tality in their socialist utopia: "These people with fashionable jackets and boots, furniture and cars . . . Some workers stop seeing life itself, so lost are they in a philistine haze." Yet, with Khrushchev announcing the USSR would bury the United States in terms of consumer goods, these criticisms made little sense. "Why do they sell all this stuff in the stores," one man said, defending himself, "if we aren't supposed to buy it?"[42]

Worse, some people took advantage of their city's unique affluence and the zoned-off territory to acquire yet more. There was no farmer's market in Ozersk, as farmers were banned from entering, so people with garden plots had their kids selling berries and vegetables on the street for high prices—"the blood-suckers."[43] Researchers at the experimental farm on the radiated trace were caught selling produce rejected as radioactive.[44] Residents bought up scarce goods in Ozersk's shops and drove them outside the gates, where they sold for prices two to three times higher. Reselling cars, which normal Soviets had to wait years to buy, was especially profitable.[45]

Party leaders charged their comrades had become "bourgeois" and "materialistic."[46] For them, this was not irony, but the fruits of a deliberate American plot to turn Soviet citizens toward capitalist materialism. Did not the American government spend millions on exhibits, radio broadcasts, and films targeted especially at weak-willed Soviet women and youth?[47]

Meanwhile, in distant capitals, the globe-trotting Khrushchev blustered, binged on alcohol, banged his shoe, and folded his friends and adversaries alike in bear hugs. He proclaimed unilateral moratoriums on bomb testing and promised eternal peace while walling up West Berlin and leading his nation to the brink of nuclear Armageddon over Cuba. In 1964, Khrushchev's Politburo colleagues, embarrassed and anxious about his contradictions and volatility, asked him to retire.

When Stalin died there were tears and fear in the closed city. When Khrushchev was deposed, no one mourned, though his attention to economic security, self-governance, and community policing had greatly improved daily life in Ozersk. After a decade on the rise, crime fell by a third in 1963 and was cut in half in 1964; in the following three years, no violent crimes occurred at all.[48] By 1965, most crime had been reduced to "hooliganism," which involved misbehaving teens and drunks.[49] Prisoners no longer visibly worked in penned-off construction sites in Ozersk, but were safely locked in workshops, out of sight. Ex-con residents, hated and feared for years, had assimilated into the community. By the late sixties, the city at last was safe. Residents remember leaving keys under the front mat and allowing their children to roam freely.[50]

As the Brezhnev years passed, life grew even better. The workday shortened to six hours for factory workers, and everyone but prisoners got five-day weeks. The childcare centers stayed open on Saturdays to free up mothers for shopping. After years of having their pay linked to monthly quotas, which meant that workers lost wages during slowdowns or accidents, plant operators finally won guaranteed salaries, plus generous bonuses and pensions.[51] Employees were also awarded paid vacations of four to eight weeks a year, which they spent at plush resorts reserved exclusively for nuclear workers. Imagine it: people who had worked since childhood, who had experienced in their lives only rest, not leisure, were getting paid not to work. It was an amazing transformation for people who had grown up underfed, cold, and poorly shod to live in abundance, with choices and a discretionary income, one they could count on. After years of turmoil, they finally enjoyed the prosperity they had been promised.

Group walk of well-paid workers from the reactor factory. Courtesy of OGAChO.

With leisure time, people who had long been defined as "workers" acquired new identities and pastimes. They fished, gardened, tinkered with the car, sang in a choir, played sports, and sat in front of their new television sets. At school gatherings, parents proudly watched their scrubbed children recite poems and accept awards. With the excellent education the city offered, parents became aware that their children had options. They could do something with their lives other than take the first job they could find.

The growing prosperity and peace of Ozersk translated into a new trust in the plant leadership and the closed status of the city. Architects won prizes for designs of apartment buildings, theaters, and hotels in Ozersk.[52] The city's architecture, in fact, was so successful that it is now unexceptional. The same designs were repeated in hundreds of Soviet cities, and so even though I have never entered Ozersk, photos give me the impression that I have already lived there. At meetings, people no longer spoke up against the gates and the guards as creating a kind of gilded prison. In polls, they said they understood why those measures were important.[53] As the city's affluence grew to outpace the standard of living in the surrounding countryside, the gates increasingly were seen as necessary to keep the riffraff out, rather than the residents in.[54]

The plutonium plant was still plagued by accidents, but ordinary ones, at least for a plutonium plant.[55] After two decades of the tumult of the Gulag order, construction, start-up, and furious deadlines, no catastrophic mishaps occurred in

May 1 demonstration, late 1950s. Courtesy of OGAChO.

the sixties and seventies. And that was a good thing. People took comfort in the stable, unchanging rituals, the parades for May 1 and the October Revolution, as the congregated community passed by the same unflinching rostrum of city leaders, the same banners with slogans no one bothered to read, the same speeches with only a few changes over the years, switching out American imperialism in Korea for American imperialism in Latin America, Vietnam, and the Middle East. The essence of the leaders' speeches, the cadences and complex jumble of manufactured compound nouns, became as lovably familiar as the twilight summer croaking of bullfrogs and the lazy slap of fish on the lake, the gentle chug of the truck delivering fresh bread, the fine tap of rain on autumn leaves, and the crisp slicing of skates on ice, the sound traveling long distances in the dry, frigid air of a Urals winter. The seasons cycling round in ordinary beauty alongside the recurring rituals and rhetoric of public holidays made Soviet society seem timeless and eternal, and so proper, just and right.[56] It was as if it were indeed as Karl Marx had promised, the end of history.

Of course there were complaints in the Brezhnev years. According to the professional-class elite, not everyone spent their leisure time well, in a way that advanced civilization. On paydays, the drunk tank continued to fill.[57] The city's actors and musicians continued to play to half-empty halls. There was no church in Ozersk, but informers counted about sixty Ozersk residents, "mostly old women," each Sunday in the crumbling gold-domed cathedral in Kyshtym. Adolescents, some of them, tediously refused to care about the important things. They didn't read Lenin, and they only feebly understood Soviet history. They stayed at home watching TV rather than join the parades on national holidays.[58] When asked why he skipped meetings of the Young Communist League, a young man responded: "We go and they collect dues. Better to spend that money on a bottle." The party boss, Viktor Podol'skii, charged that youth had an atrophied understanding of class hatred. And how could it be otherwise in the opulence of Ozersk? "Without the landowners' knout, without life's difficulties, it was impossible," he complained, "to properly orient youth."[59]

The majority of teenagers did, however, agree to live within society's boundaries. City officials succeeded in channeling young people's energies into sport and competitions. The city had dozens of gyms, pools, and playing fields. Recreational gymnastics, soccer, basketball, hockey, and soccer drew large crowds of participants and spectators. The city's sports champions, who regularly and soundly defeated youth in surrounding towns, became local celebrities and affirmed the superiority of Ozersk over its provincial neighbors.[60]

Gradually the *druzhiny*, volunteer vice squads, disbanded—not because Brezhnev-era turpitude meant that no one cared, but because they weren't really needed.[61] By the late sixties there were more effective, informal mediating institutions to set young people straight long before they had to be banished. Vladimir Novoselov described how he used to love the Rolling Stones, and he asked his mother to sew his pants so tight he had to grease his legs to squeeze them on. Strutting in front of the mirror, he liked how he looked, until he walked outside and fingers pointed at him from several directions. The fingers belonged to watchful adults who told the hipster to get back inside and change. That semester in school Novoselov got a D in behavior for those few minutes of fashion.

Novoselov knew if he had many more grades like that he could be expelled from school and exiled from the city. That wasn't a risk he was willing to take, not over a pair of pants.[62]

Sure, the young people still traded bootleg tapes of Western bands, smuggled in by their parents following trips to socialist-bloc countries. Young Communist League activists played the music at the House of Culture on Saturday nights. The teenagers' boogie-woogie bothered only party stalwarts such as Podol'skii. For him, rock music, nightclubs, and fashion were part of a Western imperialist plot to turn Soviet youth away from politics and toward idle, brainless diversions.[63] Podol'skii's views corresponded with those of American Sovietologists, who also displayed an astoundingly naive faith in the subversive, anti-Soviet power of American music and fashion.[64]

Only party officials still believed that rock and roll was America's Trojan horse. Teenagers just liked the beat and the way the songs inspired a feeling of belonging to a larger world beyond their small closed zone. The lyrics about justice and peace validated young people's feelings about the rightness of their socialist society.[65] At any rate, Podol'skii need not have worried. Pollsters found that the younger generation blamed the arms race on the United States and expressed overwhelming support for nuclear defense and its outsized budgets. The younger generation was even more hawkish than their defense-minded parents.[66]

Youth and adults alike came to identify closely with the closed city and its affluence, which was a source of status and self-worth. They saw themselves as "chosen" people. As one person put it, "We take pride in the fact that the state trusts us enough to live and work in Ozersk."[67] This message hit home when residents left the city and saw the bare shelves of provincial shops. They viewed their neighbors' poverty with scorn and condescension.[68] One resident told me that while outside the city, he could always recognize residents of other nuclear cities by their air of confidence and autonomy.[69] One woman, giving an example of how Ozersk was superior, declared that she had never in her life worn Soviet-made footwear.[70]

Ozersk residents' superior purchasing power translated into a sense of ascendancy over their shabby fellow citizens, but also into feelings of loyalty, belonging, and physical safety. Paid and living like Soviet middle-class professionals, working-class residents began to identify with their bosses and city leaders. Affluence supported the conviction that plant managers and city leaders were competent, trustworthy and caring. This consumer-generated confidence led at last to a more stable labor supply in the sixties, when employees generally overlooked the fact that they made a dangerous product in an accident-prone factory. The plant was no safer; nonetheless, people sought out positions in production for the higher pay.[71]

It was easy to forget about the plant's hazards, as city leaders tabulated Ozersk's excellent health record. Life expectancy, they pointed out, was higher than Soviet averages, mortality rates were lower, fertility rates were double the national average.[72] These statistics are not surprising. Employees passed rigorous health exams before being hired. Sick people too feeble to work were sent out of the city. The population's average age was twenty-seven. Ozersk had no poverty. Ozersk should have topped the national charts for good health.

Consuming superiority also translated into better health. Unlike their downwind and downstream neighbors, Ozersk residents did not live off a contaminated landscape. They purchased food in shops after monitors inspected the goods. They drank tested water. Children played in parks that had been swept by monitors. Residents had regular health exams and checked into sanatoria for health problems. Meanwhile, healthy residents could go to prophylactic care centers, where they rested, exercised, and ate good diets. Special schools and summer programs served children with disabilities. Pregnant women received excellent prenatal care. Gynecologists were quick to catch signs of a weak or irregular fetus in time for prophylactic abortions, which were more common in Ozersk than elsewhere in the nation.[73] These terminated pregnancies lowered the number of children with genetic abnormalities, which contributed to the picture of the city's rosy health.

Even so, there was a steep spike in infant mortality in Ozersk from 1950 to 1959, and female workers exposed to radioactivity had children who were twice as likely to die before they reached twenty.[74] Plant workers did get sick and some died unusually young, but on the whole the population thrived. This statistical creation enabled people to feel safe. In fact, residents had so much confidence in the safety of their town that when a third major accident occurred in 1967, spreading contaminated dust for miles from the dried-up banks of radioactive Lake Karachai, no panic ensued in Ozersk, as it had in 1957. Nor was there much concern for the several accidents that occurred each year at the plant.[75] There was no general alarm either in 1968 when a criticality accident killed three workers and wounded one man, Yuri Tatar, who took in 860 rads, a dose that should have killed him, too, but he lived on after sixteen operations and the amputation of his legs and one hand. In subsequent years, Tatar's long life was touted as evidence not of the dangers of the plant but of the survivability of radiation.[76]

In the late sixties, as the concern for plant hazards largely faded from the city transcripts, managers and engineers at the troubled Maiak plant gained a reputation as knowledgeable experts in nuclear technology.[77] When in the mid-sixties Soviet leaders planned a major new nuclear power plant near the sleepy Ukrainian town of Chernobyl, they chose the Maiak engineers to furnish designs.[78]

During the Brezhnev period, Western journalists searched for acts of resistance to Soviet violations of human rights and civic freedoms. But Soviet dissidents were a tiny minority in the USSR. Westerners instead missed a larger story—the contented years of late state capitalism when most people just lived, joking among themselves about the calcified leaders on TV but still marveling at the eradication of hunger in Russia, the steady paychecks, paid vacations, and retirement pensions. As a consequence, most Soviet citizens believed, as they were told, that they lived in the most advanced, fair, just, and egalitarian country in the world.[79]

Peace, contentment, and tranquility reigned in Ozersk, this major front of the Cold War, as if it had slipped the collective mind that the city existed to produce plutonium, not the other way round—that plutonium's existence was there to ensure the city's prosperity.

Part Four

DISMANTLING THE
PLUTONIUM CURTAIN

35

Plutonium into Portfolio Shares

In 1964, in the wake of the Cuban missile crisis, with a glut of stockpiled pluto-nium, President Lyndon Johnson announced that the U.S. government would gradually shutter the aging Hanford plutonium plant.[1] A few weeks later, GE ex-ecutives declared they were pulling out of Hanford. The statements sent shock waves through the Tri-Cities, especially Richland. The one-industry city relied almost wholly on the hazardous plant for jobs. In the previous few years, Rich-land residents had been forced by congressional fiat to incorporate their city and buy their rented houses, albeit at highly discounted rates.[2] It was a "dirty trick," *Tri-City Herald* publisher Glenn Lee seethed. "They sold the town, and then can-celled the jobs."[3] Without the plant, locals feared the regional economy would go bust and real estate values would plummet.

Soon after the announcement, Lee sent AEC commissioner Glenn Seaborg a telegram:

> Regarding story in Washington press this morning concerning 45 per-cent cut back on nuclear weapons. Will greatly appreciate it if you will wire me today as to what bearing, if any, this will have on future of Hanford situation. Believe it most imperative that the Tri City area public have some information as to the possible effect on Hanford, or some assurance that it does not affect Hanford. So we need a statement from you.[4]

I was struck by the imperious tenor of Lee's telegram. Lee ran a regional newspaper and he was writing to the chairman of the powerful Atomic Energy Commission, a man knighted with a Nobel Prize, usually referred to as "Honor-able." And Lee was making demands. How could he do that? What did Lee have over Seaborg?

Lee was a founding member of the Tri-City Nuclear Industrial Council. The council members, local bankers and businessmen, had sought for several years to "diversify" the regional economy in order to relinquish the company-town grip GE and plutonium had on Richland. Lee normally wrote solicitous letters, making requests, deferentially suggesting funding for this or that project. Council members, in this way, prevailed in the late fifties on the AEC to spend $200 million on a new dual-purpose N reactor to produce both plutonium and electricity. They, with Senator Henry Jackson leading the charge, asserted improbably that the United States needed more plutonium and the Northwest more energy. They also argued that the Russians had several dual-purpose reactors and plans to build more, so the United States had to keep up.[5] The Tri-City boosters won that battle. In September 1963, President John Kennedy appeared for an hour in Richland, the first U.S. president to grace the city. Kennedy waved a uranium wand, which fed an electric charge that automated a clamshell bucket that broke ground for the new N Reactor. Then he got back into his helicopter and flew away.[6]

The reactor's energy was to feed the regional electrical grid, which was already engorged with currents from the overbuilt Columbia Basin dams, also backed by Lee and his fellow boosters as a way to keep ahead, they said, in the peaceful competition with the Russians for hydropower.[7] Council members such as Lee and Richland banker Sam Volpentest had a lot invested in the regional economy, and they had unusual access to Senators Warren Magnuson and Henry Jackson.[8] The boosters raised money for the lawmakers and the governor. They were regional kingmakers. But Seaborg's scope, in contrast, was national and international.

Nevertheless, Seaborg made the long trip to eastern Washington, met with council members, and promised that the AEC would not abandon the Tri-Cities after the plant was mothballed.[9] He did so because Seaborg needed Lee as much as Lee needed Seaborg. The AEC had a credibility problem, an accelerating one, complicated by the fact that AEC officials were promoting the spread of civilian power reactors in resistant American communities that were nervous about nuclear spills. At the same time, the Federal Radiation Council was decreasing permissible levels of civilian radiation exposure, which left newly formed environmental protection agencies sniffing around the edges of AEC installations. Seaborg required positive testimonies from satisfied, long-term neighbors of nuclear power. Richland qualified as one of the largest, oldest, and most loyal atomic cities. Lee and his newspaper had stood behind the Hanford Plant and its eight reactors for years in a stalwart, unswerving devotion to all things nuclear, but in 1964, when the news of the closing broke, Lee made some ominous speeches. He complained bitterly in Seattle that the closures would leave Richland "boarded up," a "ghost town." Lee said the Tri-Cities had been the AEC's "slave and captive." He coined the phrase "plutonium curtain" and asserted the AEC had a "hammerlock on the plant, its secrets, its people."[10] And that was just the beginning. Lee had his hand on stories that could drown the

AEC in controversy. Seaborg deferred to Lee because the AEC had a public image problem, one so dire it flipped the long-standing power dynamic, reversing the roles of hostage and keeper. Seaborg and the AEC eventually became captives of Lee and the Tri-Cities.

In 1961, the U.S. Public Health Service placed Hanford on notice that the average radiation exposure of nearby downstream populations was too high, with some people, especially local farmers and Indians, eating fish from the river and local produce irrigated with river water, taking in up to 115 percent of the permissible limits. "The capacity of the Columbia River," the report stated, "to receive further radioactive pollution appears saturated." In the past, the advisory group expressed its concerns privately, but in 1961 they were threatening to go public.[11] A year later, Hanford scientists published the Ringold study to answer these concerns, but the small study had little power to convince.

In 1963, Ernest Sternglass, a nuclear instrumentation specialist, published an article in *Science* claiming to find a significant increase in infant mortality due to radioactive fallout. The Federation of American Scientists supported Sternglass.[12] That year, the city of Richland started to draw drinking water from the Columbia River. In 1964, Public Health Service officials, concerned by monthly reports showing higher levels of contamination in the water in Richland than in Pasco, demanded that Hanford reduce radioactive dumping into the Columbia by 50 percent within twelve months. Worse, they announced this recommendation to the public. AEC officials balked, replying that a year was not enough time, nor did they have enough money. Public Health Service officials responded that Hanford spent $1.9 million a year just to monitor the river and had just piled $200 million into the N reactor; surely they could spare a half million for a cleaner river.[13]

In 1965, as AEC officials were opening the Columbia River above Richland to swimming and boating, Robert Fadeley, director of the Colorado Center for Environmental Research, published an article stating that people living in counties that bordered the Columbia River in Oregon suffered from higher rates of cancers than elsewhere in the state.[14] Seaborg quickly requested a National Cancer Institute review to refute Fadeley point by point, but the damage was done.[15] In 1966, the Subcommittee on Air and Water Pollution reported that radioactivity in the soft tissues and bones of people living near Hanford had risen by 50 percent, and the Federal Water Pollution Control Agency dubbed the Columbia "the most radioactive river in the world."[16]

In 1964 and 1965, the plant was plagued by expensive and hazardous accidents. High-level radioactive waste leaked into "swamps," essentially open waste reservoirs, and gave off 20 rads at the banks. There were fires, explosions, slug failures, and extensive contamination, as in the past, but with stricter reporting requirements, the AEC could no longer command silence on these routine calamities, and news of them rippled through science conferences and press accounts.[17] Working to counter reports of negative health effects, AEC officials commissioned Thomas Mancuso, a renowned epidemiologist, to carry out a study of cancer rates among Hanford and Oak Ridge employees.

In the midst of this credibility problem, Glenn Lee's demands for continued federal support to the Tri-Cities had an attentive audience in Washington, D.C. No one in the AEC administration wanted Lee, with his infamous temper and his litigious impulses, to get upset. So Lee and his colleagues got what they wanted, which was for the AEC to "diversify," choosing new contractors to run what was left of the plutonium plant. They lobbied for an arrangement in which private businesses would win lucrative AEC contracts based on pledges to dedicate funds to bring new industry to the Tri-Cities.

Unfortunately, retrofitting the aging Hanford plant for new business was easier said than done. AEC consultants found that Hanford's facilities were too obsolete, too specialized for plutonium production, and too contaminated for plowshare projects. The Columbia Basin itself was also none too attractive for business. The area lacked transportation, natural resources, industries, and markets. Meanwhile, premium defense-contract pay scales meant that local wages were inflated, a situation Lee blamed on unions (the region "yoked" to a "labor collar"), but local unions charged in court that GE paid less than prevailing wages on the government project. The bloated wages more broadly reflected white-collar and management salaries.[18] The consultants predicted that about two thousand people would lose their jobs with the reactor closings, which spelled "serious repercussions" for the local economy. Given the gloomy prospects, AEC reviewers determined it was the AEC's duty "to think of ways to establish a large number of new jobs to alleviate the resultant economic displacement of the Tri-City economy."[19] Following the directives of Lee and the Tri-City Nuclear Industrial Council, AEC officials decided to award contracts to corporations promising to spend the most money in the Tri-Cities.

Hanford boosters expected to cash in on glamorous, profitable businesses related to nuclear power, missiles, and aerospace. The Isochem Company, which undertook to build a $9 million plant to repackage Hanford waste into marketable isotopes, was just the kind of enterprise they wanted. But the project stalled. A year after signing on, Isochem decided against building the plant, and an angry Lee persuaded AEC managers to cancel Isochem's contract and award it instead to Atlantic Richfield, which pledged $3 million, not for nuclear work, but to refurbish the city's Desert Inn. Instead of high-tech industries, diversification funds went for a feedlot and meatpacking facility, a conference center, an auditorium, and vague community grants.[20]

In terms of stable employment and housing prices, however, diversification worked. A year after Johnson's announcement of the Hanford's closing, Richland housing starts, which had dropped in 1964, were back to their usual figure, while Kennewick developers confidently broke ground on a large new indoor shopping mall.[21] Before shutdown, GE had on the payroll 8,277 people. In 1967, 8,140 people passed through the security gates each day.[22] After the closure of three reactors and one massive processing plant, nearly the same number of employees was getting paid at the plant. I found that baffling. If they were no longer making plutonium, what were all those employees doing?

It took me a while to figure it out. Apparently some money went to make-work projects. One former worker told me they used to be called out for overtime at the end of fiscal reporting periods. The workers would clock in, stand around, be dismissed, and get paid for the whole shift. "It was a use-it-or-lose-it situation," Ed Bricker recalled, "and the contractors didn't want to have to give the money back."[23] A growing number of people were employed on contracts for medical and environmental studies that ballooned in the mid-sixties. Some of this research was of questionable scientific value, but, with an annual budget that was over $20 million and growing, science at the new Battelle Northwest Labs kept people employed.[24] A lot of money and jobs also went to an unexpected recipient—waste management, long the stepchild of the nuclear family. In 1959, the annual waste management budget for Hanford was all of $200,000. From 1965 to 1972, Hanford contractors had spent or planned to spend two thousand times more—$436 million—on waste management, much of this for research and development.[25]

The new direction of work changed the composition of the labor force. As a fully operational plutonium plant, Hanford had employed a majority of blue-collar workers, but after 1964, union members got laid off and well-paid supervisors, scientists, and managers were hired.[26] Richland was finally becoming the middle-class city of engineers and scientists it had long claimed to be.

In 1967, just months after Senator Jackson had promised there would be no more closings, the AEC announced plans to shut down another reactor. Lee lobbied hard to keep the aging, troubled D reactor. "Stall," Lee directed Senator Jackson, "any change at Hanford." With a $73 billion defense budget, Lee said to Jackson, "What's a few more million dollars to keep our reactors running?"[27]

As they cast about for development ideas, Tri-City boosters requested projects that placed more people in danger. Locals asked the AEC to open more farmland in the closed buffer zone of the Wahluke Slope.[28] They wanted a highway through the plant and a bridge on its border to speed transportation networks. Boosters urgently requested more reactors. The Tri-City Nuclear Industry Council promoted Hanford as a "total energy center" or "Nuplex," with an envisioned fifteen to twenty nuclear power plants. Boosters made a value of the region's undesirability. What better place, they asked, to locate potentially dangerous reactors than in the desert far from the populated cities and fragile ecosystems of the coast?[29] AEC officials signed off on the highways and subsidies for agribusiness and irrigation, but they were leery of handing over the Hanford site to civilian contractors while the last PUREX plant was still churning out plutonium buttons.

As Richland's economic prospects dimmed, Hanford's public image problems magnified when attention shifted from the Columbia River to Hanford's waste. In 1968, the Government Accounting Office skewered the AEC for its parsimonious management of long-lived radioactive effluent. In terms of radiation's effect on genes, a National Academy of Science (NAS) panel asserted, no dose was safe, and it charged the AEC with devising a permanent solution to safely store radioactive waste.[30] The NAS report, made public in 1970, relayed that the AEC had functioned for decades without an office of waste management, a long-term waste plan, or central oversight of radioactive effluent. AEC officials were not sure exactly how much had been dumped, where, or when. They had left it up to their contractors to take care of waste independently, and the contractors' responsibility ended after the radioactive materials had been "safely" released into the environment.[31]

In the early seventies, as Vietnam and the Watergate scandal were shattering public trust in government, scientists leaked documents to journalists about the "staggering" scope of Hanford's waste management problem.[32] In the articles that followed, the public learned that when engineers discovered that corroding toxins had bored holes in waste tanks holding highly radioactive waste, AEC accountants, cutting corners, had failed to commission new storage tanks after GE managers requested them. Instead, in 1964, when all the tanks filled, operators topped off self-boiling, highly radioactive liquids in leaking tanks jammed beyond capacity. The oldest containers, journalists reported, had discharged a half million gallons of high-level radioactive waste into the soil, while other tanks boiled and "burped," so the earth above quivered like Jell-O. American readers discovered that Hanford engineers were trying to recover two hundred pounds of plutonium intentionally deposited in open trenches over the course of three decades, because engineers feared the accumulated plutonium might go critical and blow a volcano of radioactive mud for miles. Journalists reported that "plumes" of radioactive effluent were heading for the Columbia River and toward underground aquifers shared by regional farmers; that teenagers swimming in the Columbia accrued 53 millirems and fishermen standing on the shore received a gonad exposure of 8.5 rems; and, finally, that the problem of runaway Hanford effluence reaching drinking water was so obvious, Russian scientists had warned of it at an international conference in the fifties.[33]

AEC officials were not inclined to acknowledge these problems. NAS scientists grew irritated with Seaborg's "vague, evasive," and "over-confident" responses peppered with rosy, unsubstantiated statements such as "radioactive waste management practices have not resulted in any harmful effects on the public, its environment, or its resources."[34] At press conferences, AEC officials blandly repeated that radiation in the Columbia River and ground water was well within permissible levels, but the statements were increasingly less reassuring.

The shock of news of leaking fission products was compounded by new estimates of health effects. When Ernest Sternglass in 1969 projected that the United States had suffered four hundred thousand excess deaths due to fallout, he was championed by a growing number of antinuclear activists. AEC officials swiftly tasked two trusted insiders at the Lawrence Livermore Labs with refuting Sternglass. After a quick review, John Gofman and Arthur Tamplin found that Sternglass' numbers were indeed exaggerated, but they added their own estimate, stating that fallout had caused thirty-two thousand excess deaths to date, and perhaps thousands more would die with the spread of nuclear power plants.[35]

This statement was disastrous. The official AEC line was that radiation below permissible doses caused no health effects. AEC officials tried to silence Gofman and Tamplin, but the two former AEC contractors broke ranks and made the controversy public. The scientists spoke on TV, on radio, and in magazines ranging from *Science* to the *National Enquirer*. They told the nation that AEC officials tried to suppress their research, cut their funding, squelch publication of their work, and demote them. AEC officials denied the charges. The exchanges grew shrill.

"The primary problem faced by the AEC in the matter of the environmental 'crisis,'" Robert English, an assistant AEC general manager, stated in a 1970 memo, "is one of credibility." For AEC officials, the "crisis" was not real but a matter of perception that derived from a need to educate the public that "radiation hazards are controllable." According to English, the people making charges were "ignorant of the program—in some cases, willfully ignorant."[36]

Well, perhaps not willfully. AEC officials vigilantly controlled federal regulators' access to the Hanford site, especially the highly polluted processing and waste storage areas. Scientists connected to the AEC were not free to talk to the press, and if they did so, they often withheld their names because AEC supervisors had a habit of dismissing and discrediting dissenters.[37] They were at it again in 1974 when Sam Milham, a Washington State Department of Health epidemiologist, noticed a 25 percent increase in cancer among Hanford workers. As in the past, AEC executives sought to counter this bad news with a rival study. They pressed Thomas Mancuso, working on the AEC-funded study of Hanford workers, to endorse a press release that contradicted Milham's findings. Mancuso replied that it was too soon to draw conclusions because his study had been delayed in getting health records from Hanford contractors, who he suspected were purposely withholding or doctoring the records. In response, AEC executives cancelled Mancuso's contract. Mancuso then teamed up with epidemiologist Alice Stewart, and they published a study showing that Hanford workers suffered from an abnormally high number of cancers. AEC officials replaced Mancuso with Hanford scientists Sidney Marks and Ethel Gilbert, who insisted in a competing study that Hanford workers in fact died from cancer at lower rates than the general population.[38] Mancuso continued to work on his study using his pension to fund it, becoming a hero of the antinuclear movement along with Karen Silkwood, killed in a mysterious car accident on her way to meet a *New York Times* reporter to hand over documents about health hazards at the plutonium processing plant she worked at in Oklahoma.[39] As these scenarios played out in the national media, AEC executives and their contractors appeared to be the ones who fostered ignorance of environmental and health issues at nuclear installations.[40] For many Americans, already suspicious because of the Watergate scandal, the charges of a government cover-up resonated deeply.

This was the house that plutonium built. AEC, the agency that licensed and managed the growing number of nuclear reactors and their lethal waste, had a major credibility gap. Even David Lilienthal, the first AEC chairman, no longer believed in the cause. The man who once had promised Americans limitless nuclear energy spoke of the "rash" of proliferating power plants as "the ugliest cloud hanging over America."[41] An incredulous reporter asked, "How can they sell the public on the safety of new reactors, when a nuclear citadel like Hanford can't even handle its own garbage?"[42]

In Richland, confidence and hopes faltered, too, but for different reasons. As AEC executives announced another reactor closing, the press conferences became a long, slow funeral dirge to the once proud plant that had saved the free world from fascism and Communism. In 1971, AEC officials revealed they would mothball the newest, unprofitable N reactor, which cost more to operate than it produced in electricity.[43] The reactor closing would cost the community fifteen hundred jobs; another forty-five hundred positions would be lost with the end of plutonium processing. Locals lobbied to save the N reactor by turning it over to a local power company, but, bizarrely, the AEC-built reactor did not meet AEC licensing requirements for commercial reactors.[44] It turns out Hanford designers had saved several million dollars by skipping the containment structure, which meant the N reactor could not meet the safety code.

Despite the news about the closings and the AEC's environmental problems, Richland residents remained exceptionally patriotic. In the sixties, civil rights protestors marched to city hall in Pasco, but people in Richland largely sat that battle out, secure in the knowledge that their city, home to but a handful of minorities, did not discriminate.[45] During Vietnam, hippies gathered in Pasco to protest the draft and the war. They had some skirmishes with the police. In

Richland, where the patrol gave kids tickets for riding double on their bikes, there was not much room for dissent. Richland teens did not grow their hair or wear bell-bottoms. No one staged antiwar protests. People understood that joining a political protest would mean that the family breadwinner might lose his or her Q clearance and job.[46] The environmental and antinuclear movements that took the country by storm in the early seventies also got no traction in Richland. On the contrary, hundreds turned out to heckle Ralph Nader.[47] Many young men enlisted in the armed forces. In short, loyalty and conservatism reigned in Richland, until it came to the local economy.

Plans to board up the last operating N reactor volatilized Richland residents. Forming a group called the Committee of the Silent Majority, protestors were anything but silent. They organized a thirty-five-thousand-strong letter-writing campaign. Teachers had their students write form letters: "Dear President Nixon, Please don't shut the reactors down. My daddy will not have a job." Even Gene Murphy, chief of the local Sierra Club, championed the N reactor. Murphy admitted that his paying job as a communications specialist with Douglas-United Nuclear, a Hanford contractor, was a determining factor. "Sometimes," Murphy said, "you wonder if you're letting the pocket book dictate your conscience."[48]

In 1971, while even *Penthouse* ran articles about nuclear power with titles such as "Our National Death Wish," Richland residents raised money, wrote letters, and organized meetings to save their reactor. They needed to, or felt they did. In 1972, the Hanford payroll was down to sixty-three hundred workers running one reactor and one processing plant, the lowest employment rate since the late forties. As unemployment climbed, the housing market slowed. Facing yet another bust, Richland's future looked grim.

But Richland was saved yet again, this time by the 1973 oil crisis and the 1979 war in Afghanistan. Predicting a future energy shortage, the Kennewick-based Washington Public Power Supply System (WPPSS) sponsored a plan to build five civilian reactors in Washington State, three on the Hanford reservation. The approved budget called for $660 million in construction. In 1974, thousands of workers streamed into the Tri-Cities for construction jobs. By 1978, twelve thousand were employed on the reservation, $4.5 billion in construction was under way, and the *Washington Post* raved about Hanford's nuclear frontier and golden investment status. In the mid-seventies, while the U.S. economy was suffering from a severe economic recession, the Tri-Cities boomed.[49]

The WPPSS project, bankrolled by state bondholders, was plagued by cost overruns and delays. In 1978, costs soared, while construction was delayed. The Three Mile Island accident dampened already sagging public enthusiasm for nuclear energy, and after several lawsuits, WPPSS in 1982 became labeled "WHOOPS" in the press when it defaulted on its bonds and abandoned half-built reactors. Only one of three reactors at Hanford was completed.[50] Reeling from this disappointment, Tri-Cities boosters next pressed for the area to become a national nuclear waste storage site. They lobbied successfully to spend millions of federal dollars to study basalt caverns under Hanford as a possible permanent repository, although from the start geologists predicted the plan would fail.[51]

In the end, it was the Soviet Union that saved Richland from going bust. When Soviet forces invaded Afghanistan in 1979, President Jimmy Carter ordered a stockpile increase that upgraded N reactor production to weapons-grade plutonium. Construction workers piled in to renovate the shuttered PUREX plant to process spent fuel into plutonium.[52] Thanks to the end of détente, Hanford was back in the plutonium business.

In 1964, when President Johnson had announced the first Hanford closures, he declared the reactors would not be a "WPA nuclear project, just to provide employment when our needs have been met."[53] But that is exactly what happened. Every time Hanford was about to fail, a new pulse of money came in to save it. A decade after the first reactors were shuttered, more people were working at Hanford than in its heyday, when nine reactors and two processing plants kept eight thousand people on the job. Pay-to-play and make-work contracts artificially propped up the regional economy until the end of the century.[54]

Novelist Joan Didion argues the West was won in a spirit of optimism and careless self-interest, the kind that kept people in the Columbia Basin attached to their dangerous nuclear installations.[55] But there is more to the story than that. The easy deniability of radioactive contamination combined with AEC officials' curatorial vigilance over science plus their stubborn statements that not one person had been harmed by the Hanford Plant made it easy to believe that the Hanford Plant—and, by extension, nuclear power and radioactive garbage—was perfectly safe. Conflicting accounts of health effects led many to dismiss the whole issue in confusion.

Increasingly in the seventies, the people of the Tri-Cities were in the minority regarding opinions on nuclear safety, but they were a united, geographically defined minority, and the derision they felt from outsiders in Seattle and Spokane drove them into an ever more defiant posture. Richland's godfather, Senator Jackson, fumed that environmentalists were ruining the country, bringing on a "New McCarthyism."[56] Locals resented the jokes about "glowing in the dark." They felt they had sacrificed without complaint to protect the nation, yet the feds, liberals, and ecologists were trying to break their part of the nuclear contract and walk away.[57] Like shifting tectonic plates, Richland rafted away from mainstream American politics in lonely isolation. When a presidential candidate finally emerged who looked like a man of the fifties, who harked back warmly to the days of surefooted allegiances to just causes, the people of the Tri-Cities rallied to support Ronald Reagan, the man who had visited their city, knew their plant, and seemed to understand their alienation. Reagan brought back the clear polarities of the Cold War, resacralized weapons production, and promised a bright future for nuclear defense economies with Star Wars.[58] In so doing, he ordered up more plutonium and gave Richland back the status and pride taken from it in the seventies.

It was infuriating and humiliating all those years to be misjudged and maligned, to be both national scourge and regional laughing stock.[59] A Richland high school alumni listserv, Alumni Sandstorm, pointed to these feelings. Normally alumni on the listserv shared their happy memories of growing up in Richland, but in the late nineties a few people brought up health problems and environmental contamination. Unhappy with these submissions, the listserv editors gradually ceased to post them. Several listserv members complained of "censorship," while other alumni defended the need to cleanse the listserv of "political" comments so it could remain "a forum to share with others 'the good times.'"[60]

I had never realized how much sorrow is embedded in nostalgia. To read the memories posted on Alumni Sandstorm as statements of "good times," a reader had to overlook the posts about fertility problems, children undergoing heart surgery, and families flush with cancer; these accounts were often dismissed with a cheerful "But I just flow with the 'glow'!! LOL!"[61] The mournful call to censor the alumni listserv preserved the memory of happy childhood, if one that played out on a lethal landscape. As a woman lamented, "I just want to have one place [the listserv] where it's safe to love my hometown."[62]

36

Chernobyl Redux

Natalia Manzurova was born in Ozersk, and for as long as she can remember she wanted to leave it. She didn't like living behind a fence. When she received a passport at age eighteen, which gave her the right to live elsewhere, she swore she would depart and never return. She went to Cheliabinsk to study agricultural engineering. After graduating, she took an assignment to work in a small town in the Baikal region, where she was amazed to find nothing on shop shelves. How did people feed themselves? she wondered. Manzurova married young, as did most of her peers, and soon had a child. But it was hard to keep food on the table given the meager provisioning of provincial Siberia. Finally, in the seventies, Manzurova asked her husband to move to Ozersk, where supplies, health care, and wages were bountiful.

Her husband never got used to Ozersk's gates and guards. After a few years, he left her for another woman outside the zone. Manzurova and her daughter stayed on in Ozersk alone. Manzurova found work at the experimental research station on the trace left behind in the 1957 explosion. Manzurova liked the job. The researchers raised animals and planted crops and then measured them for radioactive isotopes. The small team experimented with ways to farm so that fewer radioactive isotopes reached the food chain. They also tracked the genetic evolution of animals and plants living on the radioactive terrain, a territory that at the time was unique in the world. As one scientist from the station described it; "This wasn't a lab where a fruit fly is isolated in a petri dish. We had biological subjects in nature interacting with fission products and that made it an experimental station like nowhere else in the world."[1] Manzurova carried out research, working toward a dissertation.

She was scheduled to defend her dissertation in the spring of 1986, but the defense never occurred. In April 1986, Reactor No. 4 of the Chernobyl power plant overheated and blew up, hurling the reactor's twelve-hundred-ton core, packed with radioactive isotopes, into the forest, swamps, and lakes of northern Ukraine. Manzurova and the group of researchers at the experimental farm were the world's leading specialists in radioactive landscapes. After three decades working in the Urals' trace, they knew how to contain the spread of contamination and how to protect people and animals from exposure. In 1986, their knowledge was suddenly much in demand. Scientists from Ozersk took the first measurements in the contaminated Chernobyl area and drew the map that outlined what became the Chernobyl Zone of Alienation, from which 350,000 residents were evacuated.[2] In 1987, Manzurova dutifully went to Ukraine to help. She was part of a mass mobilization of over a half million citizens doing their part in the midst of the national disaster.[3]

Disasters have a way of stripping society down to naked truths.[4] The Chernobyl catastrophe exposed Manzurova to the harmful, authoritarian qualities of the secretive nuclear security state located at the core of Soviet society, and she was never the same again.

The modern city of Pripiat, built as a reserve for the Chernobyl nuclear power plant's operators, was too contaminated for habitation. Manzurova's team instead reported to the old shtetl of Chernobyl. The team set up a makeshift lab in a kindergarten, pushing aside the cots and toys to make room for their equipment. Manzurova lived in a barracks, where liquidators slept in bunks in shifts. The Soviet economy was limping in the mid-eighties, and consumer goods were scarce. The Chernobyl liquidators were fed well, but Manzurova had trouble getting basic materials. She cut up old blankets from the preschool to make safety gloves and scarves. She and her colleagues wrapped boots with removable plastic tape because they could not buy new ones if the old ones became contaminated. She scavenged among the possessions in the abandoned houses in Chernobyl for a TV and radio.

The team's first job was to take measurements of radiation levels in the evacuated zone. They started in the abandoned city of Pripiat, where Manzurova came across a disturbing postnuclear landscape. Soldiers pushed furniture and personal articles from the windows of high-rise apartment buildings into dump trucks below. The men cried as they drove coveted family cars, the symbol of Soviet masculinity, to a dump and buried them. Hungry family pets, with bleeding sores, stalked the city. In a maternity ward, Manzurova found containers of fetuses aborted after the explosion.[5]

Manzurova's team took measurements and located graves of heavy machinery, tombs of bulldozed peasant cottages, and mass burial grounds of farm animals shot in the weeks following the accident. The animals, at the top of the food chain, had concentrated long-living radioactive isotopes in their bodies. Buried underground, the decomposing carcasses were leaching fission products into the soil and groundwater. Working with soldiers, the team dug up the carcasses and buried them in cement containers. They eventually built a mass nuclear grave complex in the zone. As the number of bodies mounted, so did the radiation, to dangerous levels. To avoid getting harmful doses, workers placed the animal corpses in leather caskets in the back of dump trucks. The driver backed up to the landfill, quickly released the lever to dump the bodies, and raced away. Then other workers used remote control bulldozers, designed for the army, to push the bodies into the mass grave. When full, the tomb was capped.

In Ukraine, liquidators had no knowledge of the Maiak plant, the research station in the Urals, and the work of its top-secret lab. As they tried to figure out how to manage the disaster, they were not privy to the Maiak group's classified research.[6] Manzurova could not even obtain a copy of her own dissertation to consult in Ukraine. She grew frustrated at how knowledge about radiation was classified in the Chernobyl zone. Manzurova met prisoners offered reduced sentences to perform extremely dangerous work tunneling under the blown Reactor No. 4. Officers told her she could not inform the prisoners about their exposure. Manzurova did so anyway.

Knowledge of risk was a closely guarded secret. Manzurova was appalled that she and her colleagues were issued personal radiation detectors they could not read themselves. Without legible detectors, the liquidators blindly stumbled into hot spots and radiation fields as they worked. In the makeshift lab in the kindergarten, Manzurova placed her hand on a table and felt a shock run up her arm. Her finger swelled and turned blue, with the skin peeling off, where she had touched a tiny radioactive particle. Manzurova made a map of radiation levels in their labs, coloring in red the hot spots to avoid. After she gave it to the Department of Radiation Security to be checked, the map disappeared. She redrew her map, making two copies this time, and repeated her request for personal dosimeters. Her request was denied, and security officials told her if she persisted, "there would be consequences."[7] With this official gag order in place, Manzurova was not surprised to find that officers directing the cleanup and the people working for them knew little about radiation safety. She started to teach informal classes on how to live on irradiated terrain: which foods to prepare and how; how to eat, bathe, and store food and clothing; how, even, to urinate (washing hands *before* as well as after).[8]

The Chernobyl liquidators found the specialists from the Urals uncannily knowledgeable.[9] They recommended burying cottages in pits, shaving away the top layers of soil, spreading fertilizer to stave off radioactive mimics of essential minerals, cutting down forests turned red from exposure, and rinsing streets with special chemicals devised for radioactive spills. How did they know so much? Most liquidators in Ukraine had no idea that Chernobyl was not the nation's first nuclear disaster or that, from a scientific perspective, there was little that was new in the Chernobyl cleanup. The emergency actions in Ukraine had all played out before in 1951, 1953, 1955, 1957, and 1967 in the Urals.

In other ways, too, Chernobyl repeated the experience of its progenitors in the military nuclear industry. The Chernobyl RBMK reactors were designed to produce both electric power and plutonium. The Chernobyl plant exploded because of problems that had plagued the Maiak plant for decades: irresponsible management, poorly trained workers, rushed and faulty designs, and procedures that emphasized economy over safety.[10] As in Ozersk, the sole whistle-blowers at the Chernobyl plant were KGB agents, who had wide powers to access classified records and identify problems but were largely ignored when before the disaster they had complained of alarming accident records, faulty repairs, work areas so dangerous employees refused to enter them, and irradiated fish from power plant's cooling ponds sold to the public.[11] The compartmentalization of information, the secrecy, the failure to inform the public of radiation dangers, the evacuations that occurred with critical delays, the deployment of expendable prisoners and soldiers on the most dangerous jobs, the failure to inform these "jumpers" and other employees of ways to protect themselves, the unpredictability of radioactive fallout in concentrated hot spots outside the neat zones of concentric circles—all were eerie repetitions of the plutonium disasters of the previous four decades. The only new feature in 1986 was that the catastrophe occurred while the cameras were running.

For four and a half years Manzurova worked in the zone, twenty days on, ten days off. The research group collected soil and plant samples, more than they could use, and stored them in a house that grew so radioactive it had to be razed.[12] They cut and buried pines that had died from accumulation of radioactive isotopes. They tracked the migration of radioactive water toward reserves of drinking water. The nature of the research team's work called for them to dwell in the zone's more contaminated areas. Often, upon returning home, Manzurova and her colleagues felt chilled, weak, and faint. Their heads ached. They vomited. Over time, the scientists grew forgetful and had trouble concentrating. Their immune systems weakened and they suffered frequent colds and infections. Manzurova's speech slurred and she had trouble keeping her balance. Radiation sped up the liquidators' metabolism and heightened their sex drive, and workers coupled up, forming "zone marriages." Many liquidators developed nervous system disorders, depression, or both. Self-medicating, some took to drinking heavily.

Photographs of Manzurova in the seventies show a lithe blonde with a tomboy's casual beauty. After Chernobyl, Manzurova's hair had lost its luster; her glazed eyes were circled with dark moons over sunken cheeks. She developed thyroid disease. Exhausted and ill, in 1992 Manzurova left her job in the Chernobyl zone and returned to the Maiak Plant to work as an engineer, but because of persistent illness, she didn't last long. Manzurova did not defend her dissertation because her advisor, also working in Chernobyl, died of radiation-related illness. Other colleagues followed. When asked about fatalities, A. M. Petrosiants, head of the Soviet Committee for Atomic Energy, replied, "Science requires victims." In 2010, Manzurova was the only one of her group of twenty from the Chernobyl zone still alive. As the nuclear catastrophe laid waste to the assurances that Soviet leaders and Soviet science would protect and defend its citizens, Petrosiants failed to see that Chernobyl's greatest victim would be the Soviet state.

37

1984

Ed Bricker drives a bus in Olympia, Washington. He has an Orwellian story to tell about lives caught in the aging American arms industry just as a handful of courageous people struggled to shut it down. In 1984, Bricker was working in Hanford's Z plant, newly reopened with Reagan's revival of the arms race. The plant was designed to turn plutonium nitrate solutions into hockey-puck-sized "buttons" for bomb cores to add to an already glutted U.S. stockpile. Bricker likened starting up the long-shuttered plant to trying to run a threshing machine that had lain rusting in the field for a decade. "The process hoods were so dirty you couldn't see in them," Bricker recalled. "The vacuum system didn't work. When it did, you couldn't be guaranteed where the plutonium solution would go. The place," Bricker added, "was one big mess. One engineer was so afraid he would stay in his trailer all day, rather than do his job."[1]

Bricker's father and grandfather and many of his six brothers worked at Hanford. Bricker started in the seventies at the plant in the "tank farms" or waste storage area. He moved to Z plant in 1983. "I could see the problems clearly. It was horrifying. Z plant should never have started up when it did." Bricker was just a low-level operator with a two-year associate's degree, but he took safety seriously and presented a list of hazards to Jim Albaugh, head of Safety and Quality Assurance for Rockwell, the plant's federal contractor.[2] Rather than addressing the problems, the next day Bricker's superior handed him an unscheduled, negative job appraisal. Undeterred, Bricker continued to complain, citing state and federal safety regulations.

Soon after, Bricker put on a moon suit to work in the plutonium finishing plant's highly radioactive processing canyon. The suit, checked and laid out the night before, had Bricker's name on it. When Bricker entered the canyon, his oxygen hose suddenly detached from the tank. Bricker grabbed for his backup air canister, but the handle was taped shut. Suffocating, Bricker ran for the exit and fell in a heap just over the canyon's threshold. An investigation followed. Off the record, a Department of Energy (DOE) investigator told Bricker that his suit had been tampered with, but no culprit had been turned up.

Most people would have quit at these first acts of intimidation, but Bricker had the fervor of a believer. He had grown up in Richland, was proud of the plant, sure of its quality. "When asked about safety," Bricker told me, "I'd devoutly repeat the party line: 'We have the best and brightest scientists in world. We have things calculated down to a gnat's eyebrow' . . . that kind of stuff." But in the years when a battery of contractors managed Hanford, Bricker witnessed disturbing lapses in safety. He saw a friend get killed, crushed under the counterweight of a swinging crane. He watched workers open waste tanks and breathe in noxious vapors without face masks. He had been sprayed with highly radioactive solutions that burned his leg. Bricker viewed the safety problems as partly due to the parceling out of the plant to competing contractors. "The various contractors all had a 'production first, safety second' attitude. They were there to do a portion of the job, and no one had any perspective on the whole project."

I asked Bricker why, after the attempt on his life, he still persisted. "I may be one person," he replied, "but I didn't like the way they were treating me, the public, or taxpayers' money. To me, it was clear they didn't know what they were doing or how to fix the problems, and they weren't interested in what the workers had to say."

Bricker, a Mormon and father of six young children, didn't give up. He wrote letters to the top brass at Rockwell and badgered Mike Lawrence, Hanford's DOE manager. "They said," Bricker recalled, "that I was crazy." To prove it, Bricker's superiors sent him to the plant psychologist, and his manager made clear he could have Bricker's security clearance pulled for mental health problems if he kept "nitpicking." Bricker's refusal to shut up was taken as "tattling" and as a betrayal of his fellow workers. In retaliation, Bricker's foreman and coworkers harassed and harried him. While Bricker worked the swing shift, callers phoned his house and whispered death threats to his wife. They called him "whiner," "cocksucker," and "spy." To management, Bricker constituted a major threat to the corporation and to their personal wealth. Reports of safety violations could get the plant shut down temporarily, which would cause workers to lose wages and the top brass to forfeit extremely generous bonuses for meeting planned production targets.[3] If the matter was pushed too far, Bricker's reports could cause Rockwell to lose its lucrative federal contract. In short, the stakes were high and a lot of people had it out for Bricker. "I'm surprised," a Rockwell manager later told a Labor Department investigator, "that one of the guys hasn't killed him by now."[4]

Bricker's whistle-blowing also caused problems at home. Cindy Bricker, who worked as a secretary at Rockwell's engineering office, wondered how her husband, who had no special training in nuclear physics, could be right and all the competent engineers she respected could be so wrong. Bricker and his wife had a lot of arguments about it. Cindy's father, Harvey Earl Palmer, had worked as a

Hanford senior scientist, and earned national recognition by treating Harold McCluskey, the "atomic man," who had radioactive americium embedded in his body from a 1976 Z plant explosion. Palmer argued with his son-in-law, telling him the plant was safe. Finally, worn down, Bricker in 1985 asked for a transfer back to the tank farms. The personnel manager granted Bricker the transfer, as long as he promised to stop complaining. Boxed in, Bricker acquiesced.

One of Bricker's first tasks at the tank farms was to rope off and post an area contaminated by a major leak, but a couple of days later Bricker was ordered to take down the familiar yellow nuclear hazard signs in preparation for a visit by Washington governor Booth Gardner. Bricker watched, astounded, while the governor and his large entourage were led over freshly contaminated ground.[5] Watching this scene, Bricker realized that the lawless disregard for safety was not limited to Z plant. He also saw how the secretive nuclear reservation gave the federal contractors who ran it the power to flaunt federal and state laws and expose not just wage laborers but the governor himself to deadly contaminants.

Bricker wasn't the only one troubled by the reactivation of the decrepit plutonium plant. In 1984, Karen Dorn Steele, a reporter for the *Spokesman Review* in Spokane, got a call in the newsroom from a woman in Pasco who said she was a nuclear scientist at Hanford and that while she shouldn't be talking to any reporters, she was worried that her bosses would blow up Pasco. Steele thought the woman was crazy, but, just to be sure, she made the long drive to Pasco to check it out.[6]

The woman, who is still an anonymous source, turned out to be wholly sane. She was an engineer at the PUREX chemical processing plant that had been restarted on the cheap after the Soviet invasion of Afghanistan in 1979. Steele's source claimed the PUREX plant was a mess—poorly run, with fifties-era technology and unsafe equipment. In addition, she said, PUREX had a lot of MUFs, the Hanford acronym for "missing unaccounted for"—shorthand for misplaced stocks of plutonium.[7]

Steele started to make calls and filed Freedom of Information Act (FOIA) requests for documents about the MUFs and Hanford environmental monitoring. Steele was the first *Spokesman Review* reporter to investigate Hanford. Though a major regional industry, Hanford had long been off the paper's radar, left to the coverage of the *Tri-City Herald*, where reporters soft-pedaled Hanford news.[8] Steele's editor, a cautious corporate climber, was displeased that Steele was working on the Hanford story. Steele and her editor clashed a great deal, and eventually Steele did her Hanford work on her own time.

Not long after her FOIA requests, an FBI agent showed up asking Steele a lot of questions, especially why she wanted to know about plutonium. Steele asked the agent to leave the newsroom and interpreted his visit as an attempt to scare her off the story. But Steele, a petite redhead, did not scare easily. The fact that the FBI wanted to frighten her made her even more convinced she was on the scent of an important investigation.

Steele confirmed her source's reports of missing plutonium. Meanwhile, a Rockwell safety inspector, Casey Ruud, was also disturbed about safety problems at Hanford, and Rockwell management's refusal to correct them. He leaked an audit report to a *Seattle Times* journalist, whose subsequent articles supported Steele's.[9] Together the reporters investigated further and filed articles about mismanagement and contamination at the PUREX and Z plants.[10] Those stories suggested that the aging plants were a danger to workers and the communities around them. Picking up on the coverage, a peace group met in a Spokane church to discuss their unease over the renewed Reagan-era Cold War rhetoric and the 39 percent increase in defense budgets. The group members decided to target Hanford as a local manifestation of the arms race.[11] Bill Hough, a Unitarian minister with a PhD in chemistry, suggested that instead of protesting at Nike missile sites, the group should channel their resources into finding out about the sealed-off Hanford plant. They strategized that with knowledge of the secret site, they could acquire a voice about nuclear weapons production in their region. Forming the Hanford Education Action League (HEAL), members joined up with Robert Alvarez of the Environmental Policy Institute in Washington, D.C. Together, they filed a large FOIA request asking for information about radioactive emissions from the plant.[12]

By that time, Steele's editor saw the value of her Hanford reporting and gave her time to work the story. Steele and other reporters were aided by people in the Tri-Cities who, after decades of silence, began to talk, or at times whisper. Steele, for example, interviewed Herbert Cahn, head of the health authority in Franklin and Benton Counties. Cahn had long wanted to distribute iodine tablets to the local population in case of a reactor accident, but he was told he would get fired if he did.[13] As her stories appeared, Steele started to receive death threats. She made sure she drove the long distance between Spokane and Richland during the day, and sometimes took a colleague along. She had two small girls at home, and Karen Steele didn't want to become the next Karen Silkwood.

After the January 1986 *Challenger* explosion and subsequent revelations that problems with the shuttle had been covered up, Ed Bricker decided he had to break his vow of silence. Cindy Bricker learned about the Whistleblower Protection Act, in which an employee had the right to disclose information about employers' lawbreaking or actions endangering public health or safety. Figuring they had the law on their side, Cindy agreed to collaborate with her husband.[14] In the evenings, after the children went to bed, they typed up reports of safety violations at the plant, which they sent as undercover informants to a nonprofit watchdog group, the Government Accountability Project, and to various congressional committees.

HEAL's pressure, the media exposure, and the damaging whistle-blower reports to congressional investigators created a gathering storm for the Department of Energy (DOE), which had taken over nuclear sites after the AEC was disbanded in 1977. The pressure finally forced DOE officials to play their hand. At a press conference in February 1986, Michael Lawrence, manager of the Richland DOE, dropped nineteen thousand pages of declassified FOIA documents before expectant activists and journalists. Declaring that Hanford had "nothing to hide," Lawrence stated that the documents would prove that no one had been harmed by Hanford. Privately, Lawrence hoped to scare his critics off by the sheer volume and technical complexity of the five-foot-tall stack of papers.[15]

That tactic blew back phenomenally. The HEAL group and local reporters were loyal Americans who believed that the essential undergirding of an open society was information. Undaunted, they dug into the mountain of dense, technical data. Working quickly, Steele and HEAL activists discovered documents about the poisonous Green Run, which dwarfed the Three Mile Island accident.[16] Worse, they discovered that the Green Run was no accident but intentional, and that the plant had, as part of operating procedure, dumped a colossal volume of effluent on a daily basis in the forties and fifties. They concluded that total contamination from spilled radioactive isotopes mounted into the millions of curies, more than any other place known at the time.[17]

Just a few weeks after these stories broke, Chernobyl blew. Americans watched televised images of liquidators heaving radiated graphite blocks. Viewers saw helicopters fail from the intense gamma rays and dive toward the smoking reactor. They witnessed a whole community of middle-class Soviets filing breathlessly into buses and fleeing for their lives, while Soviet officials safely in Moscow minimized the catastrophe with bland platitudes. The Chernobyl accident sent shock waves through the American nuclear establishment because Chernobyl's graphite-moderated, water-cooled reactors were based on designs stolen from the United States in the forties. The dual-purpose plutonium-producing Chernobyl Reactor No. 4 was a close replica of Hanford's N reactor, and DOE officials quickly closed it a few months later. Bob Alvarez, a congressional investigator at the time, said lawmakers were anxious because of a number of other disturbing similarities between the Soviet and American nuclear complexes: reactors without containment, dirty processing plants, plants run outside the boundaries of regulatory commissions, wanton expenditures of public funds, and the handling of nuclear weapons sites as national sacrifice zones.[18]

Alarmed, congressional investigators stepped up probes and conducted the first outside safety reviews of U.S. nuclear weapons installations.[19] The DOE carried out its own safety reviews, which found problems at Hanford so serious that lawmakers ordered a shutdown in the fall of 1986 to fix them.[20] Press stories described the lawless quality of the nuclear zone, with workers using and selling marijuana and heroin on the job and carrying concealed weapons onto the Hanford reservation, where local police could not investigate.[21] Besieged with FOIA requests and plagued by leaks and whistleblowers, the DOE released more declassified documents. In 1987, Rockwell was sacked, replaced with Westinghouse. The revelations continued, however, about gross mismanagement and pollution at all DOE weapons installations nationwide, the story making headlines day after day in the *New York Times* from late September 1988 to early March 1989. The headlines had the effect of a carpet bombing, each shot puncturing the accepted truths that had held communities like Richland together for decades.

In the Tri-Cities, a small peace group protested the renewal of weapons production and encouraged whistle-blowing and discussion over silence.[22] But many more locals seethed over the negative national press coverage. If whistle-blowers kept complaining and journalists continued to pander to them, the plant could be shut down permanently, and then what would happen to their jobs, real estate values, and communities? The news coverage also got personal. Some commentators likened the failing weapons plants to the generally declining auto and manufacturing industries, implying that Americans no longer could master the science and technologies they had invented.[23] With tensions in the community on the rise, at work Bricker became "that fucking Bricker." He was given demeaning tasks and belittled. A coworker hit him in the face. In January 1987, assistant general manager Clegg Crawford met with Whitney Walker, assistant director of DOE's Safeguards and Security Office, to discuss the "Bricker problem." Whitney, who directed a sizable army of four hundred armed private security officers, drew up an action plan for the "timely termination" of Bricker in a memo called "Special Item—Mole."[24] Firing Bricker, however, wasn't so easy. A ten-year veteran, Bricker had seniority and the backing of the union. Whitney sought compromising information on Bricker to justify his termination or, via blackmail, to induce him to quit.[25]

Pursuing this mission, security agents asked Bricker's coworkers to wear a wire to record conversations with Bricker. They bullied Bricker's friends into informing on him. The Brickers heard clicks on their phone line and noticed an RV parked in front of their house. They wondered if they were being photographed and recorded. In what the security officials called the "Bricker War Room," security officers analyzed the data they collected, but they could pin nothing on Bricker, a church-going family man, that would justify his termination. When Westinghouse took over the Rockwell contract in 1987, security agents continued to gather what amounted to an astonishing eleven volumes of information on Bricker. Westinghouse managers demanded that Bricker undergo two more psychological exams to keep his job. "Pronouncing dissidents insane is the kind of behavior seen in Russia," Cindy Bricker remembered. "I never dreamed it would happen in America."[26]

Meanwhile, in Richland, residents organized to defend their plant. They formed a group called the Hanford Family. They printed brochures promoting the excellent health of the community's residents and the plant's safety. They staged rallies and joined hands across the Columbia in solidarity for their plant and national defense. But the fact that East Europeans were taking apart their Communist Parties and Mikhail Gorbachev was cheerfully dismantling the Soviet Cold War arsenal made the appeals for secrecy, defense, and nuclear vigilance sound antiquated at best. As the Berlin Wall came down in 1989, the aging American nuclear weapons complex collapsed from the sheer weight of its obsolescence.

In 1989, DOE officials shuttered the plutonium plant and admitted they had a serious environmental catastrophe in need of cleanup.[27] DOE officials also acknowledged an oversight and leadership problem that required fixing with, in part, a new vocabulary. They vowed in the future to work with state and local "stakeholders" and to practice "transparency" and "environmental stewardship." President Bill Clinton appointed Hazel O'Leary as secretary of the DOE, and O'Leary promised to protect whistle-blowers. Casey Ruud got a job as a DOE safety inspector. The Washington State Department of Health hired Ed Bricker to inspect and regulate the Hanford site. DOE auditors estimated the cleanup would cost $100 billion and take fifty years.

While the price tag stunned lawmakers, it brought an audible sigh of relief from people in the Tri-Cities. The longtime Hanford booster Sam Volpentest bubbled tactlessly to the *Wall Street Journal* that the reservation's radioactive contamination was a "gold mine." "The green stuff," he enthused, "is just raining down from heaven." Indeed, by the early nineties, more than eighteen thousand people, a forty-year high, were working at Hanford on the cleanup, earning an average salary of $43,000. School enrollments were up 30 percent. The community was building a new golf course.[28]

It would be nice to think that after the Cold War ended, after the secrets Hanford had kept for decades had been disclosed, after lawbreaking contractors had been fired and new contractors hired, that a whole new culture of openness, community participation, and environmental stewardship commenced at Hanford.

But it didn't work out that way.

Instead, it was as if the international Cold War, neutralized abroad, came home to roost in America's heartland. The harassment of Ed Bricker continued, and his brother, Bill, who also had complained about safety problems, was laid off.[29] After Casey Ruud testified before Congress, Westinghouse managers fired and blacklisted him.[30] When engineer Sonja Anderson warned a waste tank could explode, she was silenced, and the tank indeed blew.[31] A crew of pipefitters was fired for refusing to fit undersized valves on pipes running radioactive effluent.[32] Anderson, Bricker, and other whistle-blowers, including Inez Austin, Paula Nathaniel, and Gary Lekvold, complained their phones were bugged, they were tailed, their houses were broken into, family members were intimidated, informants surrounded them, and they were demoted or fired for raising questions about safety. In pursuing dissidents, security agents applied tactics of psychological warfare that had been honed for Cold War enemies. Inez Austin returned to her house in Richland to find someone had broken in, taking nothing but leaving every door and window open and every light on. Security agents questioned Lekvold's girlfriend about his sexual performance and asked clerks at the 7-Eleven about his purchases of beer and lottery tickets. Lekvold likened the company's surveillance of his private life to the tactics of the KGB.[33]

"We do not conduct surveillance on any of our employees," Westinghouse Hanford president T. M. Anderson told reporters in 1991, "nor do we possess such equipment to do so." A Department of Labor investigation found in Westinghouse's possession a spymaster's arsenal: helicopter gunships, bionic ears, pinhole video cameras, time-lapse VCRs, listening devices for a network of two hundred phones, and an RV modified as a spy center.[34]

Inertia is proportional to an object's mass, and the mass of the nuclear weapons complex was, and is, enormous. Although contractors and managing directors came and went, most of the staff remained from the pre-Chernobyl days or cycled through from other Cold War defense institutions such as the nuclear navy.[35] These employees, trained to produce plutonium, not to make it disappear, had trouble readjusting to the plant's new cleanup mission.[36] They also retained other old habits: the urge to keep secrets, cover up accidents, minimize dangers, and discipline dissenters. New cleanup contractors, as in the past, were also plagued by fraud, cost overruns, massive charges for overhead and bonuses, pay-to-play arrangements, and pressure to conform, all bolstered by a calm faith in authority, science, technology, and enduring federal payments.[37]

Even without these congenital problems, the cleanup job was not small and the path was extremely dangerous. The mission to safely contain seventeen hundred pounds of plutonium-239 scattered among fifty-three million gallons of other poisons and fission products had never before been attempted. The uncharted nature of the assignment compounded the technological problems. Auditors realized that the site held three times more plutonium than originally reported.[38] Workers came across surprising discoveries such as a buried train car loaded with the contaminated carcasses of lab animals, a storage room of dirty baby diapers, and soil so radioactive it could kill on contact.[39] In the nineties, Hanford contractors ran through scores of billions of dollars only to admit that they needed twice as much money and time. In 1993, Westinghouse was sacked, replaced by Fluor.

Fluor made little progress and became mired in delays, cost overruns, and safety fines. At the end of the decade, Fluor officials acknowledged that they had made little progress in fixing dozens of leaking underground waste tanks, cleaning up reservoirs of toxic waste just above the Columbia River, sealing forty-five miles of open trenches of dumped radioactive effluent, or devising a method to safely contain in glass blocks the millions of gallons of radioactive waste gurgling in temporary storage in barrels and tanks.[40] After 2000, Battelle took over designing a new waste treatment plant. But as Battelle bogged down in design problems, the job was reassigned to Bechtel Jacobs. Bechtel enjoyed a bountiful stream of new funding with President Barack Obama's stimulus program, promoted by Senator Patty Murray of Washington.[41] But in 2011 Bechtel, too, was facing delays, cost overruns, and charges that the corporation skimped on design and safety to earn multimillion-dollar bonuses. Bechtel also faces lawsuits from repressed whistle-blowers.[42] One of these whistle-blowers, Walt Tamosaitis, a former chief designer, is among those who claim that the $12 billion vitrification plant, designed by Battelle and fast-tracked by Bechtel so that the company could collect bonuses, will not work and has a high chance of exploding after start-up.[43]

In 2011, Ed Bricker was driving a bus in Olympia because as part of his Washington State Department of Health litigation settlement he can no longer work as a state watchdog of the nuclear industry. He suffers from a number of work-related health problems: on his skin, melanoma and cysts from dripping effluent, and inside his body, chronic obstructive pulmonary disease from breathing toxic waste-tank vapors. Karen Dorn Steele took early retirement after the *Spokesman Review*, troubled by financial problems, downsized its investigative division.[44] Casey Ruud worked as a DOE safety inspector in the nineties but was sacked soon after his high-level protector, Hazel O'Leary, left office. A number of other whistle-blowers followed him.[45]

As grim as this picture looks, there are reasons to take heart, for once the plutonium curtain was torn, it became harder to hide fraud and safety violations. As long as the journalists, watchdog groups, and congressional committees continue to monitor and investigate, then the stewards of the country's largest Superfund site can be held accountable. Historians tend to celebrate singular moments of triumph, when walls crumble, dictators fall, and long-sequestered truths finally shine through. Reform and revolutions, however, are drawn-out, tiresome, messy affairs best suited for those with dogged patience, steely courage, and unflinching determination. And those people, fortunately, exist.

Twenty-five years after breaking the story, Karen Dorn Steele continues to file articles on Hanford. Tom Carpenter, of the Government Accountability

Project, first defended Bricker and Ruud in 1987. He is still tirelessly watching and reporting on the Hanford cleanup and violations of safety and labor, as is Bob Alvarez. Those are just a few of the people who made it their mission to monitor the state and its contractors in order to see through the labyrinth of bureaucracy and security regulations and focus a light on the fission products that are so easy to hide, minimize, and deny. Unbidden, these individuals shoulder the burdens left by four decades of what historian Richard Rhodes calls an "arsenal of folly." In the 1980s, members of the Hanford Family took up American flags and claimed patriotism as theirs alone against their detractors. And many plutonium operators certainly gave their lives and health to defend their nation. But it also takes a stubborn faith in democracy to hammer away for decades at the plutonium curtain. These guardians of national health are also heroes, if largely unknown and generally unsung.

38

The Forsaken

I showed up in Muslumovo on a Saturday morning in August 2009. Muslumovo is a big village, sprawled inside a crooked elbow of the Techa River. The village center has a train station, a few apartment buildings, and a corner store. Marat Akhmadeev met me at the station in his dented maroon vintage Lada. We jolted up and down on the choppy seas of the unpaved streets. Muslumovo is a strange village—half there and half disappeared. To the left and right, many houses were abandoned, partway dismantled, exposing weathered wallpaper, discarded clothing, and overturned appliances.

Muslumovo is one of three large villages on the Techa River that were not evacuated after the 1949–51 flood of radioactive waste. The villages remained because of their size and because contractors maintained they would be expensive to rebuild. For six decades, residents have lived near the river, a reservoir of radioactive waste.

There's no work in Muslumovo. Either a person works in Cheliabinsk or farms a patch of land of the long defunct Muslumovo collective farm. My host, Marat, farms, living off the land—a term that takes on new meaning in Muslumovo, where radiation levels are alarmingly high. After we pulled up at Marat's house, his teenage son silently trailed us. Noticing a twitch in the boy's step, I turned to look at him. The boy's mouth drooped and his fingers twisted as he mouthed a stuttered greeting. "This is Kareem, *nash luchevik*" ("our radiant one"), Marat said in an offhand manner, as if every family has a *luchevik*.

Marat showed me to a table full of food—veal, goose, salads, beets, potatoes—then rushed out to heat up the sauna for his American guest. I figured the wood was cut from along the Techa, the only trees in sight. I protested that I didn't want to take a sauna. Marat insisted, lighting the fire. I eyed the smoke bending toward us in the yard, a little Chernobyl right there.

Murat wanted to eat first and talk later, "in Tatar tradition." I didn't want to eat. It was nine in the morning and I wasn't hungry, but mostly I wasn't brave enough to ingest his home-grown food. This was the elephant in the room, about which neither he nor I nor his quietly attentive wife spoke; their food, which they must eat every day, must live on, was too irradiated for me to eat for just one meal. Increasingly agitated, Murat pulled out a bottle of vodka, offering me some. I turned that down, too. I didn't know it at the time, but Murat was doing all he could to secure my health on entering his contaminated home. In the villages, as in Ozersk, vodka was seen as an important bodily cleansing agent, a natural elixir. Villagers viewed the sauna, too, as essential for purifying a body.[1] Soon a neighbor arrived, a refugee from the Chechen War who had settled in an abandoned house in Muslumovo in the late nineties. The two men started to drink. In a few hours, Murat was howling drunk.

There is a legal contest going on over the health of the people of Muslumovo, whether they are sick, and if so, whether they are ill from the radioactive isotopes dumped in the river or from poor diets and alcohol abuse. Medical evidence has been contradictory. In 1959, A. N. Marei wrote a dissertation in which he argued that the Techa villagers were in poor health because of their poor diets.[2] In 1960, in contrast, the Cheliabinsk provincial executive linked the river dwellers' illnesses to the contaminated river.[3] Clearly this debate between nature (radiation) and nurture (lifestyle) has been going on a long time.

In 1962, the Cheliabinsk branch of the Institute of Biophysics, FIB-4, started regular medical exams of the Muslumovo population.[4] FIB-4 doctors invited village children playing on the streets to a clinic room to give blood and tooth samples.[5] In Cheliabinsk, they set up a repository of irradiated body parts: hearts, lungs, livers, bones.[6] They started a collection of genetically malformed babies who died soon after birth, each infant preserved in a two-quart glass jar. A Dutch

Farmers in Muslumovo harvesting potatoes. Courtesy of Robert Knoth.

photographer, Robert Knoth, visited the repository, where he saw hundreds of babies in jars. He photographed one infant with skin like rough burlap. Another boy had eyes on top of his head like a frog.[7] The doctors did not inform the people they examined of their exposures or of diagnoses of radiation-related illness. Instead FIB-4 doctors told patients they had vegetovascular dystonia, a term that is a vague description of a premorbid state between sickness and health.[8]

The most immediate targets of the river's radioactivity were village men whom MVD officers hired in the early fifties to guard the river. These men, who stood for eight-hour shifts on the riverbank, died young. Ramila Kabirovaia's father was deputized to watch the river in 1952. After two years he fell ill, and he died seven years later. Kabirovaia's mother, left with seven children to support, got a job that seemed easy, collecting water samples from the river for the scientists who visited periodically. She stored the radioactive samples, which she had no idea were dangerous, in glass jars under the bed where her children slept. Five of her children developed radiation-related illnesses. Two died in their forties.[9]

In the 1990s, when villagers found out about their exposure and FIB-4's long-term medical research, they formed organizations called the White Mice and Atomic Hostages and charged that the Soviet government had left them on the river to use them as human subjects in secret medical experiments. But the story

Malformed infant preserved at the Ural Research Center for Radiation Medicine (formerly FIB-4). Courtesy of Robert Knoth.

is not so simple. The river dwellers presented a unique opportunity in the history of health physics—what scientists call a "natural experiment"—that promised to answer an important civil defense question posed by Soviet leaders concerned with how to survive a nuclear attack.[10] The experiments in Muslumovo were not premeditated. Rather, it was what police investigators would call "a crime of opportunity."

In fact, having a population with successive generations exposed to radioactive isotopes in a natural setting came to have great monetary value. Hoping to attract foreign money, a 2001 Russian Ministry of Health brochure promoted the "Muslumovo cohort" as a data set with "worldwide significance for the evaluation of the risk of the carcinogenic and genetic impact of chronic radiation in humans."[11] The U.S. Department of Energy invested heavily in this data set from the Urals. Japanese researchers, however, found the dosimetric records too unreliable to be of use.[12]

It reminds me of *The Truman Show*, a 1998 film about an insurance adjuster who discovers his entire life has been staged for a reality TV show. Imagine waking up one day and realizing you have always been watched as part of a medical experiment, and that is why you are in Muslumovo and not elsewhere, and why doctors at a remote clinic know you and your extended family by name, and perhaps why you don't feel so well. Dr. Glufarida Galimova had such an awakening in 1986 soon after the Chernobyl disaster when she was working as chief doctor at a pediatric clinic in Muslumovo, her native town. She was puzzled by the saturation of illness in her community. The illnesses were rare, strange, complex, and often genetic: children with cerebral palsy, hydrocephaly, missing kidneys, extra fingers, anemia, fatigue, or weak immune systems. Many kids were orphaned or had parents who were invalids.

Galimova asked other doctors about it. They said the villagers were sick because of their own doing, from poor diet and alcohol. Doubtful, Galimova investigated and learned that FIB-4 had a fifty-year-old registry with Muslumovo's health records. She requested the records be opened to the public. Her requests went unanswered. She went to the press and helped organize citizens' groups. The security services accused her of disclosing state secrets, and she was fired from her job. Undaunted, Galimova teamed up with the chief of genetics of the Siberian Academy of Medical Science, Nina Solovieva. The two doctors tracked neonatal and pediatric health in Muslumovo. When, in 1995, Solovieva died of breast cancer, Galimova continued alone. She found that more than half of the children born in Muslumovo in the 1990s suffered pathologies. In 1999, 95 percent of infants born that year had genetic disorders. Meanwhile, 90 percent of Muslumovo's children suffered from anemia, fatigue, and immune disorders. Galimova examined the records of the city's adults and found that all of 7 percent could be described as "healthy."[13]

In 1992, FIB-4 doctors finally declassified Muslumovo residents' health records. In their public statements, the researchers focused on chronic radiation syndrome and largely failed to address genetic health effects, which Russian researchers have long considered to be the most worrisome outcome of radiation exposure.[14] FIB-4 doctors had over the years diagnosed 935 people on the river with chronic radiation syndrome.[15] Of them, 674 people, located forty-five years later, were still alive. FIB-4 researchers elaborated that the CRS patients' mortality rate was the same as that for an unexposed control group. The only difference was that the CRS patients had higher rates of disorders of the circulatory system and cancer than the control group, and a statistically higher rate of leukemia. The message, however, was that CRS could end in recovery.[16] Dr. Angelina Gus'kova, in 1991 the chief official voice in evaluating Chernobyl-related health problems, disputed the FIB-4 conclusions. She argued that in fact there were only sixty-six cases of chronic radiation syndrome among the Techa River people. The rest, she claimed, suffered from more prosaic diseases such as brucellosis, tuberculosis, hepatitis, and rheumatism, caused by poor diet and sanitation.[17] Meanwhile, officials charged that many people started to dream up illnesses or connected their quotidian illnesses to a real or imagined radioactive past in order to sue for compensation.[18] These people, they said, had no chronic radiation disease but were chronic welfare cases looking for handouts.

Both positions—that the villagers are sick from radiation or that they are sick because of other sociocultural factors—can fit into one schema. Radiation illness is not a specific, stand-alone illness. Its indications relate to other illnesses. Radioactive isotopes are known to weaken immune systems. Radioactive isotopes also damage organ tissue and arteries, causing illnesses of the circulation and of the digestive tract. The river people in the postwar years indeed suffered from poor nutrition, occasional hunger, and stress, while they engaged in heavy physical labor from a young age—all of which made them more vulnerable, both to the harmful effects of radiation and to other diseases.[19]

As I considered this contradictory research, I kept thinking about two sisters I had met in Cheliabinsk—Rosa and Alexandra, born in Muslumovo. When I met her, Alexandra was forty-four. She clicked out her teeth and showed me how, except for four, they are all false. She lost her teeth at age thirty-two. She pulled out a form from FIB-4, which gave her a body count of radioactive cesium of 190. Cesium concentrates in bones. Alexandra said several times, "My father had a count of 195." I didn't understand the significance of this number until later when she and her sister described how their father had died, his bones crumbling from within: at first he started to limp, then walked heavily only with help, and eventually could no longer move. The doctors diagnosed him with Peugeot's disease, which is quite rare. Only later, when her father was near death, Alexandra said, did the doctors at FIB-4 call it bone disease and determine it was related to the radioactive Techa flowing peacefully past Muslumovo. Alexandra didn't say it, but I imagine she anticipates she will share her father's fate; as she pointed out, she has a similar cesium count and only four teeth.

I asked if they had swum in the Techa as children. Both women were incredulous. "We didn't swim in the river! We didn't go near it, nor did we drink milk from local cows. Our father drove a tractor for the collective farm and he did liquidation work, plowing under irradiated soils after the accidents. He knew all about the Techa catastrophe and he made sure we didn't go near that river."

"We used to call the river *atomnaia*. Everyone knew it was dangerous."

"So," I asked, "you knew about the atomic plant nearby?"

"Sure, of course we knew."

Now I was incredulous. The people their age who grew up in Ozersk told me they had no idea that the factory where their parents worked made plutonium. One woman said she believed her father made candy wrappers.[20]

I asked the sisters if they kept a garden. "Yes, sure, everyone did, and there were houses that they bulldozed from surrounding evacuated villages, and my father would go and get the wood and we burned it in our stove." I thought of Joseph Hamilton's 1943 research on offensive radioactive warfare, how enthusiastic he had been about smoke laced with strontium, which he estimated would be "a million times more lethal than the most deadly war gases."[21]

Alexandra said as a child she would get horrible headaches and lie in bed for days. She and Rosa told me they had thyroid disease and autoimmune disorders, but the biggest problems had occurred in their children.

Alexandra gave birth to four children. One died the day after he was born, and one daughter had severe diabetes.

Rosa bore two children. Her daughter, Ksenya, had medical problems as an infant. Her leg twisted as it grew. She had circulatory and digestive tract disorders. Weak and crippled, the girl could not walk. Rosa carried Ksenya everywhere until she was thirteen. She would lift the girl onto the commuter train in Muslumovo, then carry her from the station in Cheliabinsk to the bus stop, onto the city bus, and finally into the FIB-4 clinic. Rosa and Ksenya made that tiring expedition many, many times, but the doctors gave them no answers. They just told Rosa to give her daughter vitamins and massages. Rosa had relatives in Tatarstan and finally she flew there, where doctors told her that Ksenya needed operations for her legs. Rosa borrowed money for the surgeries. After the first operation Ksenya could walk a bit. After the second, the fifteen-year-old could move about with a walker.

Then one day Rosa's boy, age nine, who had never had medical problems, went to sleep at bedtime and never woke up. He had not been ill, had no fever. The doctors said he died from a virus complicated by an immune disorder.

Rosa told me that when she was in college, before she married and had children, she dated a medical student. After a year of courtship, he broke up with her because he heard a classified lecture at his Cheliabinsk institute about genetic problems among the Muslumovo cohort. He said, after hearing the lecture, he did not want to have children with her. He advised her never to have children. "At the time, I was very hurt. I thought he was being cruel," Rosa recalled, "but now I see he was trying to help me."

Nadezhda Kutepova, a lawyer representing the claims of Muslumovo residents, said that growing up in Ozersk, she learned to have a certain disregard for the neighbors in the poor agricultural villages outside the nuclear city. "We were told that they were ignorant, that they drank too much and married each other and that is why they were always sick—because of inbreeding and fetal alcohol syndrome. But when I started to work with them, I realized they were like any other population—smart, informed, sober, and yes, some of them drink too much."[22] In the 1990s, when the radiated Techa became a national news story, Muslumovo was often in the spotlight. The TV cameras showed old women milking their irradiated cows, pulling up mushrooms from the forest as the dosimeter ticked excitedly. The next shot included kids blithely diving into the river and men hauling in hot fish. News stories of ignorant villagers unknowingly poisoned by a secret, high-technology nuclear complex had just the kind of polar juxtapositions of ignorance and knowledge, of rich and poor, on which the media thrive.

But the story is more complicated than that. Kids from village families told the journalists they would swim in the river for twenty dollars, and the journalists obliged. The men, too, agreed to be filmed fishing if they were paid. Few people in the village ate the fish or the mushrooms, because they knew those items were dangerous; instead, they took them to the Cheliabinsk-Yekaterinburg highway and sold them there. Few people in Muslumovo were ignorant, as many were in Ozersk, of the existence of the plant, or of the accidents contaminating the region. They did not underestimate the medical consequences of long-term exposure to low doses of radiation, for the people in Muslumovo lived with the consequences. They witnessed it daily in their close-knit community. The plant's invisible radioactive contamination had long embroiled the community in the pursuit of medical knowledge for a growing list of complaints.[23] Alexandra worked in Muslumovo's pharmacy and saw daily this search for solutions among people struggling to find relief.

The trope of ignorant, genetically deficient, and drunken villagers is a common one in Russia. In the southern Urals in the past few decades the cliché has been useful in glossing over the human suffering connected to uncontrolled dumping into the Techa River. In conferences debating the number of victims of the Chernobyl accident, I heard the same charges from officials who drew

paychecks from nuclear lobbies.[24] Indeed, trauma and fear have been major factors in the lives of the Muslumovo sisters Rosa and Alexandra. The stresses of
family illnesses exacerbated by low wages, uncertain access to transportation
and health care, and social stigmatization have taken their toll. Environmental
desecration often collides with poverty, afflicting communities that have fewer
resources to respond to traumas and make decisions that will ensure a better,
rather than a declining future.[25]

In Muslumovo, exposure and lack of resources exacerbated the plutonium
disaster. Murat cannot sell his irradiated cottage and farmland for any price, and
so he does not have the opportunity to move elsewhere, nor does he have skills,
at age sixty, to find another job. His handicapped son, Kareem, will live dependent on his parents as long as they live. The family's other son has recently married. The young couple live with Murat and his wife. As soon as they get enough
money, they plan to set up house in a cottage in Muslumovo, they said, because
they want to be near their friends and family.

And so it goes. In the 1990s, Boris Yeltsin appeared in Muslumovo and proclaimed the village had to be moved. Yeltsin left, and no action was taken. Finally
in 2008, some residents of Muslumovo were resettled, but to the other side of
the river, which government officials asserted was safe.[26] Local critics claim the
move was to keep the valuable Muslumovo cohort together for future medical
research.[27] Indeed, it is hard to believe that one side of the river would be much
less radioactive than the other. Alexander Akleev, research director of FIB-4, told
me the Muslumovo cohort needs to stay put in order to receive proper medical
surveillance. "These people already got large doses of radiation in the fifties.
They need good medical care now. They already have strontium in their bones,
and if we moved them to New York or Maryland, how would that help them?"
When I asked Akleev if they have good medical care now, Akleev replied, "No, of
course they don't, and they didn't have it before either."[28] Meanwhile, people will
continue to live in Muslumovo because of family ties and necessity.[29] As they do,
villagers are assaulted by a "multimedia dose" of radioisotopes from the air, surface, well water, dust, soil, food, and ambient gamma rays. In 2005, researchers
found even the hair of villagers to be radioactive beta emitters.[30]

After I failed to eat Murat's meal, Nadezhda Kutepova showed up with a bucket
of marinating pork in the trunk of a cab. Murat's older son grilled the meat, and
finally at midday we all sat down to share a meal. Afterward we took a walk down
to the river. We descended a high, grassy bank alive with yellow yarrow, purple
clover, and bees mining them. We came to the ruins of a nineteenth-century mill
on the bank of the river. I recognized the old mill from a 2005 report in which
monitors recorded radiation levels eighty-three times higher than background in
front of the mill, where we were standing.[31] It was midsummer, and the river was
low, a clear stream of sweet brown water gently bending back feathery grass.

The stream was no wider than a sidewalk, no deeper than my knee. "Is this it," I asked, "or just a branch?" Marat assured me that the creek in front of us was the infamous, feared, highly radioactive Techa.

Usually when you are looking at an environmental catastrophe, you know it. Disasters have the look and feel of the natural order disassembled. In my mind, disasters should smell, smoke, or produce ugly scars. Yet nothing was out of place along this inviting little stream. The air was fresh. Swallows darted back and forth over the current. The afternoon was turning hot and, as if a siren were calling me, I had a desire to slip down and run my feet over the smooth stones on the river bottom. There were no fences or warning signs to stop me. I had to remind myself that I stood before the world's most radiated river. I had never encountered a disaster more lovely and tempting, one less worthy of its name.

39

Sick People

In the late eighties, letters flooded Karen Dorn Steele's box at the *Spokesman Review*. Arthur Purser wrote from his dairy farm in Ringold that he had a tumor on his thyroid. Laverne Kautz counted up ten cousins, five aunts, nine friends, and a mother suffering with cancer. Betty Perkes, living on a farm in Mesa, said that in 1960 she had lost an infant and a doctor removed tumors from the throat of her kindergarten-age daughter. The rest of the family had thyroid disease. Melvin McAffee stood in his wheat field in mid-1986 and considered his prostate cancer, his wife's thyroid cancer, and the thyroid disease with which two of his four children suffered. "A lot of the old-timers have had cancer here," he told Steele. "They should have warned us." McAffee's son Allen drew the line more directly: "They are killing us."[1]

These scenes repeated in Oregon, Idaho, and elsewhere in Washington State. The late 1980s newspaper headlines about Hanford contamination led thousands of people in the interior West to worry that the landscapes they called home might have been slowly poisoning them with an invisible cascade of radioactive isotopes. As they came to understand how fission products had entered their homes and penetrated their bodies, reaching to their very genes, the feelings of rage and loss, along with the lawsuits and court battles that followed, divided the communities surrounding the Hanford plant into the sick and the healthy, those who had been offended and the perpetrators. On an individual level, the news led many to radically alter their biographies.

Trisha Pritikin was born in 1950 in Richland. Her father had been a safety engineer, a cautious man who never let her run behind the city's DDT fogging trucks. Pritikin considered her childhood in Richland blissful and healthy. Her father, ex-navy, had a speedboat and they would spend long summer days on the river, seeking the warm currents in the chilly Columbia to swim and float. While visiting relatives in Spokane in 1986, Pritikin read one of Steele's articles about Hanford contamination. Suddenly the delightfully warm river currents of her childhood memories transformed into unnaturally hot plumes from reactor outflow pipes that enfolded her youthful body in a sea of toxic effluent.

Perhaps, Pritikin wondered, Hanford's fission products explained the death of her infant brother, her mother's miscarriages and thyroid disease, her parents' early deaths from cancers, her own ongoing fertility problems, and why Pritikin felt awful most of the time with migraines, dizziness, gastrointestinal problems, extreme fatigue, and severe muscle contractions. Pritikin had her thyroid tested and learned it was in the final stages of shutting down. Once on thyroid medication, Pritikin finally managed to get pregnant, but she had a miscarriage and premature birth, which caused medical complications for her son.[2]

June Casey read the articles on Hanford in the spring of 1986 and also refashioned her life story. She tracked the Green Run in December 1949 to her freshman semester in college in Walla Walla. She remembered that when she returned home for the Christmas holidays her mother was horrified. Her daughter looked as if she had aged fifty years. Casey admitted she hadn't been feeling well. Her hair had been falling out in clumps, and she was tired all the time. The family doctor said he had never seen such a case of advanced hypothyroidism. Casey had never had health problems before. No one could figure it out. She took medication for her thyroid, lost all her hair, and wore a wig the rest of her life. She married and attempted to start a family but had a miscarriage, then a stillbirth, and finally in 1969 a son, John, born with neurological damage.[3]

When they won the lottery for Columbia Basin farmland, Juanita Andrewjeski and her husband, Leon, a veteran of the Korean War, were thrilled. They happily moved with their three children to eastern Washington in the early fifties. Andrewjeski didn't think much of it when, on the new farm, she had three miscarriages. It seemed to her that a lot of the neighboring women had similar troubles. Andrewjeski eventually had three more children. All six of her kids grew up on the farm, running and playing, helping in the fields on the high plateau above the Columbia River. I asked Andrewjeski to tell me the names of her children. She volunteered extra information to the list. "Bob was born in 1947. Janice in 1948. She's dead. Mark was born in 1953. He's dead. Jeannie was born in 1955. Krissy— she was 1957. Dead. I had Rod in '59. He's alive. Jeannie has liver problems."[4]

Leon Andrewjeski was diagnosed with heart disease in 1976, in his fifties. "All these big, strapping farmers were coming down with heart problems and cancers," Juanita told me over dinner in Richland. "It just didn't make sense." She continued, "Leon used to say he'd see guys out in the field in their white outfits, taking samples from our fields. It makes you wonder if you were a guinea pig."

Before news spread about Hanford's releases, Andrewjeski started marking up an emergency evacuation map distributed by the AEC. She placed red X's for the cancer cases and black dots for heart disease among her neighbors. Her map shows a lot of X's and dots.

The news that Hanford had let fly 700,000 curies' worth of radioactive iodine into the air currents and millions more curies into the water and ground led people even in Richland to question the plant management. "It's tragic," die-hard Hanford booster Sam Volpentest told the press, "that people were not told."[5]

Trying to contain the publicity problem, DOE Hanford manager Mike Lawrence at first followed the path of his predecessors. He denied everything. The federal government, he said, had spent "millions" monitoring public health and the environment, and there was, Lawrence asserted, no observable health effects caused by the plant's effluence.[6] When asked for the data supporting this claim, Hanford health physicists had to admit that despite the $2 billion spent over four decades on research on the health effects of ionizing radiation, this most basic question about the effects to public health from chronic radioactive contamination hung in the abyss of scientific uncertainty. Why, Spokane doctors demanded to know, if Hanford scientists intentionally released so much radioactive iodine, were there no epidemiological studies? Critics asked why DOE officials refused to hand over data on plutonium workers gathered by Thomas Mancuso in the sixties.[7] Angrily, critics charged Hanford managers with a "cover-up."[8]

The words "cover-up" and "guinea pig" appeared frequently in the years after Chernobyl, but the sadder truth was that plant researchers had nothing to hide. There were almost no studies that tracked the genetic and health effects of people living on landscapes bombarded for decades with low and in some places perilously high doses of ionizing radiation. At the time, the gold standard for research on the health effects of radiation was the Atomic Bomb Casualty Commission (ABCC) studies, carried out among bomb survivors in Japan. The American-funded ABCC study looked at survivors who had been hit by a large blast of gamma radiation from external sources.[9] The Hanford communities, however, had been exposed to long-term, usually low doses of alpha and beta rays from ingested radioactive particles, which scientists considered to be much more dangerous than gamma rays. By the nineties, the ABCC project had many critics who charged that the study overlooked radiation-related illnesses other than thyroid abnormalities and a few types of cancer, and greatly underestimated genetic effects and the rates of cancer among survivors. Critics also charged that the ABCC studies exaggerated the doses survivors received, doses that then set standards for permissible levels of exposure.[10] Research on Marshall Islanders, in the path of thermonuclear tests, also focused only on thyroid abnormalities and a few types of carcinomas. When the islanders complained they gave birth to babies that resembled "cats, rats and the insides of turtles," AEC sponsors suppressed requests for genetic studies.[11]

After Chernobyl, Lowell Sever, an epidemiologist at Battelle Labs, told the press, "There are no solid studies anywhere in the world that suggest an association between birth defects and low-level radiation exposure of the kind that occurred downwind of Hanford." Yet Sever himself had carried out a study of birth defects from 1968 to 1980 in the Hanford environs. His studies were only marginally useful because they omitted the years 1945 to 1957, the period of Hanford's heaviest pollution.[12] Even so, he found an increase in congenital neural tube disorders among offspring of workers and people surrounding the plant. Sever, however, discounted Hanford as a cause because similar birth defects were not found among atomic bomb survivors in Japan.[13] He suggested agricultural chemicals could be the culprit, but in the same years, epidemiologists and state officials rejected pesticides as the cause of cancer clusters in California, also for lack of evidence directly connecting pesticides to health problems.[14] Rather than it being a cover-up, John Gofman, the former medical director of Lawrence Livermore labs who clashed with the AEC in the sixties, was closer to the truth when he told Steele, "It's easy to have no observable health effects, when you never look."[15]

To address the dire need for answers, the Centers for Disease Control (CDC) recommended that the Department of Energy carry out studies of exposed populations around Hanford. Under pressure from lawsuits, DOE officials reluctantly funded a study, but they gave the important job of estimating the dose of radioactive isotopes to Battelle Northwest Labs, a longtime Hanford contractor.[16]

Researchers for the $18 million Hanford Environmental Dose Reconstruction (HEDR) study fed data with a person's age, sex, place of residence, and diet into computer models to come up with estimated doses for the years 1974–1980, when the Hanford plant was all but shuttered.[17] Those doses were then used as a baseline to calculate rates of thyroid illness in a second multimillion-dollar study funded by the CDC and conducted by researchers at the Fred Hutchinson Cancer Research Center in Seattle in the nineties. The two studies drew on assumptions derived from the ABCC studies, criteria that greatly narrowed the field of inquiry. Scientists searched for doses of radioactive iodine high enough to cause a few types of carcinomas and thyroid disease, based on estimates derived from Japanese survivors. Downwinders connected their sheep born without eyes to birth defects in their children, but the studies did not address genetic effects. Nor did it address other health problems that Russian scientists discussed in medical literature as chronic radiation syndrome.[18] Part of the hangover of the Cold War was that Russian science was often considered suspect because of its dependence on the state. Bruce Amundson, a senior scientist at Fred Hutchinson Cancer Research Center, made a trip in 1992 to Ozersk, where he was amazed to find the vast body of research on the Muslumovo cohort. "In our open society," he told Karen Dorn Steele, "we made a conscious decision not to study our offsite [exposed] population. In a closed society, the Soviets were able to carry on extensive, secret studies over the same period. They are way ahead of us in understanding what may have happened to their people."[19]

Finally, no one thought to question the synergistic impact of people exposed to both radioactive isotopes and the deadly sea of agricultural chemicals—herbicides, hormone-based weedkillers, and chlorinated organic hydrocarbons such as DDT—which had saturated the landscape since the fifties. After years of controlled laboratory experiments, in which research had been narrowly compartmentalized, scientists did not have the tools, nor even the questions, to address the multiplicity of poisons deposited in the eastern Washington environment.[20]

As the Hanford studies dragged on, five thousand people filed suits against federal contractors who had operated at Hanford. The plaintiffs, called "downwinders," waited anxiously for the results of the Hanford Thyroid Disease Study (HTDS), hoping it would provide determining evidence. But the case, which seemed so clear to plaintiffs living in areas where most people they knew were chronically ill, proved elusive. In the last decades of the twentieth century, lawyers representing manufacturers had created a highly restricted set of rules regarding the evidence necessary to prove damages from environmental contamination.[21] U.S. district judge Alan McDonald severely limited the number of eligible claimants in the downwinder case by ruling that in order to qualify, plaintiffs had to prove they had received a dose of Hanford radiation high enough to cause twice the number of cancers as would occur in the population at large.[22]

Nor was local justice blind. Judge McDonald owned a million dollars' worth of real estate in the Columbia Basin, and he displayed the kind of probusiness attitude that had long supported the regional defense industry. Telling the press, "The government's limited resources should be focused on [nuclear] cleanup and not diverted by litigation," Judge McDonald delayed and obstructed the case.[23] Ten years passed with no hearing, then another five years. The defense had good reasons to postpone the case. The federal government was committed to cover all legal fees for the five former contractors being sued. Corporate defense lawyers had no motivation to settle out of court or resolve the issue speedily to avoid expensive legal fees. By 2003, the Chicago law firm Kirkland and Ellis had racked up $60 million in legal fees, bankrolled by taxpayers.[24] Downwinders, on the other hand, did not have time or deep pockets. Many plaintiffs were older and sick. They struggled with expensive medical bills and their lawyers worried over the mounting legal costs.[25]

Finally, in January 1999, researchers from Fred Hutchinson announced to a hushed crowd in Richland the results of the thyroid disease study. They found, they said, no correlation between estimated doses and thyroid disease and cancer among the 3,193 people born between 1940 and 1946 they tracked in their study. Hanford, in other words, was not the reason people were sick.[26] An eruption of outrage, shouts, and tears followed the announcement, alternating with feelings of relief from property owners worried about their land values and a sense of vindication among Hanford boosters who had felt under siege for decades.[27]

Trisha Pritikin found herself wedged between these two emotional polarities. Her father, a former Hanford safety engineer, lay dying from an aggressive case of thyroid cancer. Gasping for air, he stopped speaking to his daughter because she was involved in the downwinder cause.[28] His defiant silence during his last days illustrates how, as sociologist Ulrich Beck postulated, resistance to understanding a threat grows with proximity. The people most severely affected by a hazard are often the ones who deny the peril most vehemently in order to keep on living, or, in Pritikin's father's case, to finish dying.[29]

In the late nineties, former Hanford workers started to speak about their health problems. Because workers wore film badges, their doses were easier to prove, and researchers began to try to reconstruct employees' doses. Of an estimated 250,000 workers, however, researchers could locate records for only 100,000 of them. Records for types of employees who were often the ones most directly exposed—women workers, part-timers, and subcontractors— were missing.[30] Meanwhile, X-rays of former workers found diminished lung capacities and pathologies at far higher rates than expected.[31] "Over the years, I used to think those whistle-blowers were all a bunch of nonsense, that they were just out for the money," said former worker Beulah "Boots" McCulley after her son, a Hanford worker, suffered a debilitating accident at the plant. "But [now] I think they deserve whatever they can get."[32]

During the heated debates over the question of Hanford contamination, the science of "experts" was often pitted against local knowledge wielded by farmers and "lay people." In angry meetings in Richland and Pasco, scientists pulled out charts and graphs and showed how it was not possible for people to have been harmed by the plant because "on average" they were well within permissible doses. Locals replied that what the scientists said made no sense, that in their communities they could pinpoint localities where most people had health problems. In response, the Seattle-based scientists discussed ions, rads, and isotopes. They clashed with people who wanted to talk about illnesses, tumors, and cancers of their loved ones. The dry, bristling scientists, who dismissed stories of families' illnesses as "anecdotal," reminded a lot of people of the arrogant scientists in Richland who the downwinders believed had caused their problems in the first place.[33] Meanwhile, the scientists and many Hanford veterans took the charges of illness as personal, moral accusations. Referring to the Green Run, one former health physicist asked me angrily, "Do you think those scientists would have released that iodine if they knew it would harm people?"[34] The diseases clustered in bodies, families, and communities were undeniably real, but scientists could identify no cause for them, which made Juanita Andrewjeski's map look like mere coincidence.

It is not that one form of knowledge, expert or local, was right and the other wrong. The two domains of knowledge reflected diverging interests. Both local and expert knowledge were limited, and because the record of ingested and ambient radioactive isotopes was nearly impossible to trace backward in time, both forms of knowledge were anecdotal and circumstantial. Most often, however, in court and congressional hearings, science wielded by experts was seen as "objective," while women such as Andrewjeski tallying up her sick children and neighbors on a dog-eared map were labeled "subjective" or "anecdotal." As such, the scientists' claims to truth were taken more seriously.

Yet science is a process of simplification in understanding complex processes. Radiation pathway studies were based on models, averages, and aggregate populations with a streamlined view of single isotopes entering the body through singular means. The spread of radioactive effluence and its integration into the environment, however, occurred not in the aggregate but at random points as air currents, river eddies, and groundwater followed particular patterns, not generalized ones. At the hot spots, bodies were drenched in fission products not at average levels but at great intensities.[35] Scientists breezing in from Seattle had only a cursory grasp of the hot spots.[36] Yet to determine where they were and to detect exposure would mean going over every square foot of the exposed 75,000-square-mile territory, measuring radioactivity in plants, roots, soil, groundwater, and air up to 2,000 feet above the surface. It would mean knowing the land the way children know the back lot or the way a farmer understands the nutrients of the soil, drainage patterns, the dips and turns in a field, and the vagaries of wind and weather. To do a thorough epidemiological study, the scientists would have had to get to know the population on intimate terms, not just the people living there but those who had moved or died; it would take knowing who had had a miscarriage, who was sick and with what, which couples had trouble with infertility, and which kids were just not right somehow. It would take the kind of knowledge people in extended families or close-knit communities possess. The Hanford scientists, sequestered by day on the nuclear reservation, often transplanted from elsewhere to Richland, where they were seen as arrogant and socially isolated, did not have that kind of knowledge.

40

Cassandra in Coveralls

Tom Bailie grew up on a dry-land farm in Mesa on territory the AEC opened in the fifties to the Columbia Basin Project. Known as "Mr. Downwinder," Bailie often served as an informal spokesman for the cause.[1] Bailie shows up in dozens of articles and almost every book about Hanford. Talking with him makes it easy to see why. Bailie has the gift of gab spiced with an uncanny knack for colorful sound bites. He also looks, dresses, and drawls just like a farmer on the Western range should, which makes for good copy. Because it takes historians a long time to research a story, I got to know Bailie well. Over the years, we became friends.

The first time I met Bailie, we climbed into his swather and rode up and down a row of alfalfa he was cutting for export to Japan. He told me that the television journalist Connie Chung had ridden in the same seat. I understood Bailie was offering me a photo op worthy of national TV. As he drove, Bailie kept up a monologue.

"When I was a kid, I used to love Buck Rogers, and then one day I looked out my window to see these men in space suits shoveling dirt from our front yard into little metal boxes. I was thrilled, but my mom panicked. She ran out and asked the scientists what was wrong. 'Nothing, ma'am,'" Bailie said, cupping his hand over his mouth to mimic a voice behind a mask, "'everything is just fine.'" The scientists then asked for the beaks and feet of the geese his father shot. And they went away.

"I finally realized," Bailie quipped on another day, "why me and my buddies are still going strong and the goody-two-shoes we went to school with are sick or gone."

"Why, Tom?"

"Because when their mothers told them to eat their milk and vegetables, they did, while me and my friends snuck off to the store and bought soda and Twinkies."

Bailie says when he ran for the state legislature in the 1980s, he campaigned among senior citizens ("because they vote") and noticed that while in some communities the old folks were in their nineties, still farming, his community and others had very few in that age group. Bailie asked his campaign manager, "Why we don't have any old folks?"

"They all died of cancer."

"Why is that?" he asked.

"I don't know."

Bailie asked the old people if they used pesticides. "Yes, we did until that communist, lesbian bitch Rachel Carson came along."

"See," Bailie pointed out to me, "everyone used DDT, so that didn't make the difference."

Bailie said he saw a pattern. The communities without elderly were in the hills; the ones with old people were in the valleys. When I gave Bailie a blank stare, he penciled out for me how the wind followed the contours of the land, funneling up the hillsides enclosing the valley.

I never knew what to make of Bailie's stories. They started at anecdotal and unscientific and escalated to outright fantastic. Driving in Pasco one day, he had me pull up near the rail yards to take a look at a low-slung, cement-block building abandoned behind cyclone fencing, the old Pasco slaughterhouse. "These guys in tan suits used to pull up in beige cars marked with consecutive plates. They were looking for our crooked lambs and calves." Bailie leaned closer to make sure I was listening, "Twenty percent of our livestock were malformed. The feds would go in, say something to the manager, and then come out with stainless-steel containers. They were collecting organs—like regular body snatchers!"

As Bailie tells it, he didn't always harbor suspicions of government agents in unmarked cars. He said he was once a freedom-loving, take-it-or-leave-it American patriot. During Vietnam he tried to enlist but he was rejected because of birth defects. Nonetheless, he spurned the hippies and peace movement and was proud of living alongside the plutonium plant in a community of like-minded people who knew the value of a strong defense. As Bailie learned about Hanford emissions, however, his former political certainties eroded. Gradually Bailie focused on the medical history of his extended family, a history in which his parents, aunts, uncles, and sisters came down with cancer, while Bailie, born with a hole in his chest, was told at age eighteen that he was sterile.[2] As a child he had long stays in the Kadlec Hospital in Richland, where he was treated in an iron lung for a mysterious paralysis. He remembered a strange blue light, the door to a ward guarded by soldiers, and being awakened to shouts. When he asked the nurse what was wrong, she hushed him, saying, "Go back to sleep. Those are just the men from Hanford."

As Bailie talked, I'd often have a vertiginous sensation that I'd entered the studio of a midnight talk radio show and wasn't allowed to leave. Bailie, some-times uncouth and usually inappropriate, would ramble from conjecture to rumor to conspiracy. His stories were hard to follow and much, much harder to believe. A couple of journalists have said as much, one calling him a "blowhard."[3] Bailie might just be the most oft-quoted unreliable narrator in American history.

Bailie knew he didn't sound very credible. "I'm a kid, they give us milkshakes and pass a meter over our stomachs; there's a *naval* base in landlocked Pasco. My friend's father who runs the train depot is really, my friend tells me, an FBI agent, and no one around here but me seems to think all this is strange."

I have often relied on the assumption that unreliable narrators are worth a good listen because there are some accounts most people don't believe, not because they are not true but because, as with the mythical prophet Cassandra, society is resistant to them. So I seek out unreliable narrators and then cross-check the facts.

Almost everything Bailie told me panned out.

In the early sixties, scientists did collect samples from farms in Benton and Franklin Counties, and took bioassays of wild game and livestock. They tested the drinking water and collected bovine thyroids from the slaughterhouses in Pasco, Moses Lake, and places as far away as Wenatchee.[4] From 1949, Hanford researchers also harvested organs of plutonium workers and neighboring farmers for research, and investigators funded by the AEC secretly gathered bones of children worldwide to measure radioactive fallout.[5] The Kadlec Hospital did have a guarded ward with thick cement-lined rooms to protect staff from the bodies of patients too radioactive to be near. Twinkies and poor nutrition might indeed have helped Bailie. Hanford studies showed that people who ate food purchased in grocery stores had lower counts of radioactive by-products.[6] Another study showed that pigs fed poor diets retained fewer radioactive isotopes than those on healthy diets.[7] Bailie's guess that farmers living on the hills had more exposure than those below also tracked with Hanford researchers' descriptions of plumes of radioactive iodine heading "upslope into the valleys."[8] Over the years, Bailie tipped me off about plant accidents and gave me mini lectures on topography, soil qualities, and the path of radioactive particles through the digestive tract. "But what do I know?" Bailie would conclude at the end of each monologue. "I'm just a dumb farmer."

Bailie had a point there, too. How did Bailie, with his high school education, bouncing around country roads in a battered Chevy, come to the same conclusions as an army of Hanford researchers supported by multimillion-dollar budgets? Bailie was obsessed with the downwinders' cause. He carried around big, messy folders of clippings and documents, and he put that knowledge to work alongside his farmer's understanding of local history, geography, geology, and weather, mixed in with a good deal of gossip, rumor, family lore, and coffee-shop conjecture.

Bailie often landed right in the middle of the battles over Hanford's health effects. He ran for public office and talked to every reporter who called him until his wife just about divorced him. A lot of fellow farmers wished Bailie would shut up before their land and crops were stigmatized as radioactive and lost all value. As the community divided into people who backed the plutonium plant and those who suspected they had been poisoned, Bailie became a lightning rod. Lots of people stopped talking to him, including friends and family members. He had trouble extending his credit lines at the town bank and he lost his farm. Tom came to be a reviled miscreant in his community because in his obstinate refusal to stop talking he pointed out how some truths, visible to the naked eye, had been overlooked, while silences had begotten certainties.

The 1990s health studies were supposed to provide answers and bring peace to the quarrelling communities of the Columbia Basin, but they only raised more questions and provoked more bitterness.[9] After Fred Hutchinson researchers concluded that there was no connection between Hanford radiation doses and downwinders' thyroid diseases, Trisha Pritikin and Tim Connor, a former HEAL activist, convinced officials at the Centers for Disease Control to review the study. The CDC review found that the researchers overstated their conclusions. The reviewers emphasized that the subject population had three times the number of cases of thyroid disease than would be expected. Other reviewers found that the first HEDR dose reconstruction study generally underestimated the population's doses.[10] This made sense when downwinders discovered that lawyers from the firm Kirkland and Ellis had sat in on original meetings of the controversial HEDR dose reconstruction study in order to design the study for "litigation defense."[11] Downwinder lawyers, meanwhile, learned that Judge McDonald, who had for a decade obstructed and delayed the downwinders' lawsuit, owned an orchard in Ringold, directly across the river from Hanford. Acknowledging that if a jury should determine Hanford hazardous, his agricultural holdings could decline in value, McDonald had to recuse himself.[12] Finally, in 2000 the federal government agreed to pay up to $150,000 each in compensation to former Hanford employees after a study showed elevated numbers of twenty-two different cancers among nuclear workers. Downwinders grew exasperated to see that while they had waited sixteen years for trials that ended in a draw, workers got a payout.[13] Elusive justice, medical research that had been tampered with, contradictory verdicts—for many downwinders, the whole process felt like a fix.

Rather than concede defeat, however, the downwinders did something really remarkable. They took the job of medical proof into their own hands and created a new kind of people's epidemiology. Joining forces with doctors, scientists, and social justice advocates, the downwinders devised a health survey that they distributed to friends, neighbors, family members—anyone who might have ingested radioactive isotopes. In the survey, they asked for the kind of local knowledge about patterns in family health, diet, landscape, and wind that might contribute to localized exposure to radiation. Analyzing the results from eight hundred complete surveys, they compared disease rates with those among a control population and found the downwind population six to ten times more likely to have thyroid disease and other illnesses. Although the community-based study conflicted with the government-funded studies, the results had a great deal in common with the conclusions of the studies Hanford scientists had carried out on animals over the years and which Russian scientists had conducted on Techa River populations.[14] The downwinders' epidemiology also validated people's knowledge about their health and their landscapes, which felt good after a decade of being told they were wrong and ignorant.[15]

In 2009, Bailie took me to his Connell High School reunion. The class of 1968 was meeting at Michael's Cafe in downtown Connell, which is less a town than a strip off the highway with a hotel, a state prison, a food processing plant, and a string of mobile homes. Twenty-five years had passed since locals had learned of Hanford's dangerous emissions. Even so, as we went in, Bailie said it wouldn't be a good idea to mention thyroid or health problems: "People don't want to talk about that." Bailie looked nervous. In the nineties, most of his fellow classmates had derided his downwinder activism. I think he expected to get dressed down for bringing along a snooping historian.

While Bailie went off to greet some old friends, I sat down at a table prepared to make small talk, but I didn't get a chance. Pat rolled up in a wheelchair. She told me she has multiple sclerosis, as does her sister. She said she used to pick peaches down in Ringold, across the river from the plant, and attributes her MS to Hanford. Linda chimed in that her mother had troubles with her thyroid. Her father, always slim and active, had had heart problems very young. Crystal (thyroid and lung cancer, never smoked) said she had not bothered with that downwinder business when she was going through her own health problems, but when her daughter got cancer and had fertility problems, that made her angry. Gwen (thyroid disease) didn't look well. Her husband had to heave her from her walker to a chair. Gwen's parents had moved to eastern Washington from California after they had won the Bureau of Land Management lottery in the early fifties. The classmates all grew up on farms that had been established on land opened for irrigation just downwind from Hanford. They talked about the green books the scientists gave them to record every bushel of wheat and pound of potatoes. All that crazy detail, they laughed.

Bailie joined us and, turning to Gwen, asked, "Remember how your mother used to say she didn't feel well because of the water? And your father used to say, 'You are crazy, woman! That is a twelve-hundred-foot artesian well.' Remember that?" Bailie continued, "No one knew then that we shared an aquifer with Hanford's leaking waste tanks. We all drank from that aquifer." Getting agitated, Bailie pulled out a dinner napkin and drew a line marking a country road, then made an X to mark the location of a farm owned by a family named Holmes. "She got bone cancer. The girls both had thyroid problems." Bailie's pen turned a 45-degree angle and stopped to indicate another farm. "She drowned her deformed baby in the bathtub and then committed suicide." Bailie's pen paused again. "She had leukemia, and up there the baby was born with no head."[16] Bailie's pen stopped at Gwen's farm. Gwen's mother died of leukemia in her forties. Her father also died of cancer. Gwen had a lifetime thyroid condition (and died a few years after the reunion). "That's what we used to call the death mile."

A man walked up to our table. He'd had a few drinks. His eyes were red and his speech was slurred. The man said Bailie was full of bull, that he had grown up on a farm downwind and he was fine. "We have plenty of eighty-seven-year-olds around here." Bailie nodded, abnormally silent. The other classmates stared at their laps. I was surprised, as I'd never known Bailie to back down from a debate. Bailie later explained that his critic was having serious health problems. "I couldn't argue," Bailie said. "I felt sorry for him."

Before Bailie and other downwinders started talking, there had been no real debate about Hanford's health effects. Instead there was an appearance of scientific debate that produced an often calculated confusion and uncertainty. Before the downwinders, there was no debate because there was no public record of sick people. Sick Hanford workers, sick farmers, and sick neighbors suffered silently for years without realizing they possibly shared a fate with thousands of others. In the years after Chernobyl, as the downwinders talked, met, campaigned, and traveled to Japan, Ukraine, and Ozersk, they came to learn of tens of thousands of people like themselves. The sick bodies of self-proclaimed downwinders helped to map the invisible geographies of contamination, hidden for decades in the interior West. Using their bodies as evidence, the downwinders pointed to the gaping contradiction that while the nuclear reservation was considered contaminated enough to require a cleanup that cost more than $100 billion, the people living next to the reservation were thought to be unharmed. In their frustration at the walls put up by experts long used to zoning and compartmentalizing territory and information, the downwinders created alternative ways to produce knowledge. By telling and retelling their medical tragedies, the downwinders undermined established certainties and inspired a new understanding that science, in order to produce convincing results, needs to emerge from laboratories to grasp the specificities and complexities of environments, locales, and the bodies that inhabit them.

41

Nuclear Glasnost

In the Soviet Union, the Chernobyl accident ignited what had been a small Russian environmental movement into a massive conflagration.[1] Emerging groups of greens demanded to know more about their country's nuclear past. In June 1989, officials from the Ministry of Medium Machine Building released a thick pamphlet about the 1957 Kyshtym explosion. With this official acknowledgment, newspaper stories followed, sparked by a keen interest in the long-cloistered Maiak plant.[2] Soon after, American congressional representatives and scientists began visiting the top-secret plant for on-site inspections. The news of the nuclear accident in the Urals, on top of the Chernobyl catastrophe, struck a keen blow to the credibility of the Communist Party leadership but did little to rock the establishment in the closed city of Ozersk. City residents, worried about the future of their pretty lakeside city, supported their plant and party leaders. For them, the end of the Cold War was an ominous threat.

In the nearby city of Cheliabinsk, long a bastion of conservatism and patriotism, an antinuclear movement took shape. One of its founders was Natalia Mironova. She was an unlikely leader for an antinuke movement because her life had never been severed from the Soviet nuclear industry. She grew up in postwar East Germany, where her parents worked for the Soviet Vismuth Company, an agency founded to mine uranium abroad for the Soviet weapons industry. As an adult, she worked as an energy engineer, making visits to nuclear power plants. In Cheliabinsk, she grew interested in plans to build a new nuclear power plant in the southern Urals and went to an early antinuke demonstration. There Mironova noticed a man holding a poster that read "No More Mutants." She questioned him and learned that he was from a village called Muslumovo.[3] He had an eye-popping story to tell about a nuclear disaster of which no one had ever heard. It turned out that this grade-school-educated farmer knew a great deal about the health effects of fission products.

Mironova, a tall, good-looking brunette, morphed naturally into her role as the Hanoi Jane of defense-industry-dependent, no-foreigners-allowed Cheliabinsk province. She described to me the day that solidified her transformation from loyal Soviet public servant to just-as-loyal protestor. Because in 1989 there was no public debate about the proposed new reactor, she and several others reserved a large hall and invited the director of the Maiak plutonium plant and the province's governor to a debate. In the crowded auditorium, Mironova made her case, using as evidence the Chernobyl explosion and the 1957 accident at the Maiak plant. In response, the Maiak director, Victor Fetisov, did not defend his plant's safety record, but instead berated Mironova in the usual way of party bosses accustomed to monopolizing the podium. He said she was not an expert, knew nothing about the Maiak plant, and so had no right to speak on the topic. Suddenly from the crowd an elderly man stood up. The man said Mironova was right, that he lived on the Techa River and had seen what radiation does to fish, animals, and people, and that it was time the authorities started to tell the truth about it. The crowd roared in support, silencing the powerful plant director.

Mironova smiled telling me that story. Besting the big corporate boss, correcting what she felt was wrong, moving a crowd to defy those who carelessly wielded power—it felt good. Mironova founded a group called the Movement for Nuclear Safety, which grew in the next few years with grants from foreign donors and help from scores of volunteers. The nonprofit rented an office in Cheliabinsk. Mironova won election in 1990 to the province's Council of People's Deputies on an antinuke platform, and she became chair of a new Committee for Radiation Safety. Boris Yeltsin, then a dissident political rival of Mikhail Gorbachev, aided the cause by disclosing information about the dangerous practices of chemical and nuclear weapons production in the Urals. Yeltsin had been party boss of the neighboring Sverdlovsk province, home to several closed nuclear and military installations. He let slip insider information that helped to further discredit the Communist Party, which he was intent on overturning.[4] Mironova's group started a campaign to cease the importation of nuclear waste into the province for processing and storage. They managed to stop, for the time being, plans for the new reactor.[5] Later in 1991, as the first Russian president, Yeltsin founded the Ministry of Environmental Protection and signed Russia's first comprehensive environmental law. Yeltsin opened Cheliabinsk to foreigners and appeared in Muslumovo, where he announced that the people living on the radiated Techa would finally be moved.

The Movement for Nuclear Safety held seminars and served as advisors for village groups to petition the Russian government for compensation for what they charged were radiation-related health problems. Mironova's group worked to declassify the records of the formerly secret health clinics in Cheliabinsk (FIB-4) and Ozersk. Amazingly, in 1995, the Yeltsin administration issued an official apology for the wrongful exposure of Soviet citizens.[6] That statement opened the door, in 1996, to a family's lawsuit against the Maiak plant in an Ozersk court. The parents successfully sued for damages for the genetic deformities of their child, who belonged to the third generation to live on the trace.[7] This was an incredible admission in Russian courts of the link between Maiak's fission products and genetic damage. The success of the lawsuit speaks to the importance of historical context in the weighing of scientific evidence. The first cases heard after the fall of Communism reflected the general suspicion by which many Russians viewed their government, past and present.

These events of the early nineties were breathtaking. The environmental injustices of the past would be redressed. Russia would become a state governed by law, not by command, whim, and backroom deals. Mironova called the period from 1991 to 1993 "the glory years of Russian democracy."

Checking herself, she added, "They were the only years."

For Mironova democracy ended in 1993 when President Boris Yeltsin defeated rebellious parliamentarians not with laws and debate but with soldiers and tanks in Moscow streets. With democracy escorted at gunpoint out of the Russian White House, new elections swept in a class of powerful and often corrupt businessmen and former party bosses.[8] Yeltsin supported nuclear glasnost as long as it bolstered his larger goal of propelling the Communist Party from power. Once that mission was accomplished, antinuclear protestors and victims' groups gradually got in the way. Civic groups such as the Movement for Nuclear Safety, which had backed perestroika and helped end communist rule, were largely displaced.

The bellwether cases threatened to expose the Maiak plant and the Russian government to phenomenal liabilities. Another case of a boy who was pressed into service as a liquidator after the 1957 accident was successful in a regional court but was overturned in a superior court. In the following years, petitioners lost case after case. In 1998, Yeltsin issued an order enabling all government industries to classify information they considered sensitive. This decree was probably a response to lobbying from the energy and defense industries, including the renamed Russian atomic energy agency, Rosatom.[9] In 1999, agents from the Federal Security Bureau (FSB), the KGB's larger and more powerful heir, accused several scientists reporting on nuclear safety of passing state secrets. Some scientists received obscene phone calls and death threats.[10] Without public scrutiny, powerful ministers were again free to act without fear of charges of corruption, malfeasance, or environmental damage.

Freedom from public scrutiny was the last thing the aging, ailing Maiak plant needed in the years when budgets grew skimpy, wages went unpaid, and the best and brightest Russian technicians and scientists were abandoning Russian nuclear installations for jobs abroad.[11] Meanwhile, the infrastructure in Russia was failing; miners went on strike, the supply of electricity grew sporadic, and workers left their unpaid jobs untended. The Maiak plant piled up dozens of accidents in the nineties, each one issuing more radioactive pulp, liquids, and aerosols.[12] On September 9, 2000, for example, the region's electrical power grid crashed, and the backup generators of the plant's remaining nuclear reactors failed to kick in. For forty-five heart-stabbing minutes, plant operators raced to fix the generators. They got them running two minutes before the reactors over-heated, which would have caused explosions eclipsing Chernobyl.[13]

Alarmed, more groups formed in the Urals under the umbrella of "greens." They targeted the Maiak plant, which was the only place in the country process-ing spent nuclear fuels, a process that results in a great deal of radioactive waste, all of which was mounting behind aging dams holding highly radioactive effluent in open reservoirs above the Techa River. Protestors floated down the radiated Techa in rubber rafts and placed vats of irradiated fish in front of the governor's office. They took disturbing pictures of children with twisted limbs and blurred minds. They snapped photos of handheld dosimeters registering 800 micro-roentgens at roadside food stands.[14] The activists claimed that there was an underground radioactive lake moving toward the Cheliabinsk watershed. The Maiak leadership, sons of the plant's founding fathers, fought back. They charged the greens with costing the province $2 million annually in litigation and com-pensation, another $50 million in lost plant revenue, and hundreds of local jobs, and that none of this economic sabotage was accidental. They asserted the envi-ronmental activists worked for foreign governments in order to undermine Russia's economic and defensive potential.[15]

In Cheliabinsk, Mironova tracked what she called the death of Russian de-mocracy with the start of harassment of the Movement for Nuclear Safety. Offi-cial harassment began in small measures in the mid-nineties with lawsuits and charges of tax violations, mixed with press accounts smearing the group as op-portunists, running a front to net Western grants. Movement activists cam-paigned for funds to evacuate villages on contaminated land, for compensation for families with medical problems, and for money for prescriptions, prostheses, and medical monitoring programs. Maiak and Rosatom officials learned how to battle the news stories featuring crippled kids and their sickly parents. Govern-ment spokespeople claimed victims' groups were "welfare cases," ne'er-do-wells, intent on soaking the government for subsidies.[16] Other people not directly connected with the nuclear industry emerged to affirm this interpretation.

In Cheliabinsk, I talked to Vladimir Novoselov, a historian who wrote several books on the Maiak plant. Novoselov grew up in Ozersk. In the mid-nineties, Maiak directors asked him to write the first authorized history of plutonium production in the Urals. He and former first party secretary Vitalii Tol'stikov were given enviable access to the plant archives. They penned an account critical of the plant's record of accidents and environmental degradation. Their work gave plant veterans courage to speak to other authors about what they had witnessed.[17]

Novoselov and Tol'stikov's second book on the same topic, published two years later, contained far more muted criticism and a lot of apologetics. When I asked Novoselov why, he admitted that the plant management had warned him that if he wrote another unsympathetic book about the plant, it would be the end of his career as a professor. That was enough for Novoselov. Just before we met, he had given an interview to a local newspaper in which he stated that because the Techa region is so poor, the locals' only source of income is compensation claims as radiation victims. The real cause of illness, he said, was not radiation but alcoholism; the reason for the unusually high rates of birth defects was inbreeding.[18]

Novoselov, a son of Ozersk, understood how to play by the rules in the nuclear security state. His "expert" testimony went a long way toward discrediting villagers' claims. Novoselov wrote his first book when many liberals hoped that the Russian government would right the wrongs of the Soviet state and in so doing show that "we live in a state governed by law."[19] By the time he wrote his second book, that hope had all but vanished.

42

All the King's Men

The Russian economy crashed in 1998. Many professional and blue-collar workers sank into poverty as hyperinflation annihilated savings and salaries, which the bankrupt Russian government often failed to pay anyway. University professors took in laundry. Middle-class professionals sold candy bars on the streets. The radiation biologist Natalia Manzurova, too, hit rock bottom. In 1992, she returned from four years' work in the Chernobyl zone as an invalid. For much of the nineties she was bedridden and nursed by her teenage daughter. By the time she recovered, the Cold War was long over and there were few jobs for nuclear scientists. With no income or savings, Manzurova haunted trash bins, looking for scraps and returnable bottles.

"I was raised to believe that you gave everything to your government and in return your government would always provide for you," Manzurova told me one afternoon outside an oncology ward, where she was waiting for surgery on a tumor. "Now I realize how wrong I was."

In carefully managed elections in 2000, Vladimir Putin won office. He promised to restore the broken economy, end the conflict with the rebellious Chechen minority, and rebuild Russia as a world power. Soon after he gained power, bombs exploded in several Russian cities, and Putin pledged a war on terrorism, which included a large-scale hunt for spies and saboteurs, accompanied by new restrictions on civil rights and freedom of information and a harsher approach to political opponents. Putin's policies were greatly aided by rising oil prices. As state revenues increased, the Russian economy revived. With it, Putin's popularity grew. No one wanted to return to taking in laundry; no one wanted hostage-taking terrorists in their school.

On September 11, 2001, burglars broke into the offices of the Movement for Nuclear Safety and stole the computers' memory containing the groups' financial statements and correspondence. In the months that followed, Russian federal tax officials brought charges against the group for tax evasion. Local prosecutors filed suits for violations of obscure laws. Young people right out of law school joined the movement and formed a whole wing of the office. But the pressure did not cease. Tax officials tried to bankrupt Mironova personally. The group had to stop taking funds from foreign foundations. They could not pay their rent and moved to members' private apartments. By 2002, the nonprofit had to disband.[1]

The Urals' environmental movement could have ended there, but Mironova had trained other activists and inspired them. One protégé was Natalia Manzurova. In 2005, Manzurova founded a nonprofit in Ozersk to focus on the environmental contamination of the Maiak plant. This was a brave act. Residents of Ozersk had largely sat out the glasnost era. They didn't topple statutes or cast out party magnates. The double fence surrounding the closed city remained. When polled about opening the closed city in 1989 and 1999, residents voted overwhelmingly to keep their gates, guards, and pass system for fear of an influx of criminal riffraff. Half of the polled nuclear scientists said they would move away if the city was opened.[2] As the nineties wore on, there were only a few perceptible changes in Ozersk: the hammer-and-sickle emblem was swapped out for the Russian Federation tricolor, and signs of capitalism materialized in the form of open-air traders who stood shifting from one foot to another in the infinite Siberian frost.

Nadezhda Kutepova was one of those traders. She sold women's lingerie from a sidewalk card table. Before the economy crashed, Kutepova had gone to university and married a local cop in one of those disposable late-Soviet-era marriages, producing one child before the divorce. Like many young people in Russia in the nineties, she cast about in the uncertain post-Soviet economic and political landscape until one day, as she tells it, she went to a lecture. The lecturer told the audience what Kutepova had not known before: that her town produced plutonium for Soviet nuclear weapons, that there had been major Chernobyl-like accidents in the city before she was born, and that the surrounding landscape of jeweled lakes and birch and pine forests was highly contaminated with long-lasting radioactive isotopes. This information sent Kutepova into shock.

As a child, Kutepova had overheard her mother, an oncologist, mention in phone conversations patients who had been "overexposed" or "radiated," but the words had never registered. Her father had always told her—credibly, to her mind—that his factory made candy wrappers. Learning that he worked at a plutonium plant and had cleaned up one of its major accidents explained his early and painful death from cancer. The shock of her gullibility propelled Kutepova into action. She founded a women's organization in Ozersk and soon after joined up with Manzurova. The two women made a powerful team. Manzurova, schooled in physics and biology, was sober, quiet, and hardworking; Kutepova, trained in sociology and law, was talkative, energetic, and equally hardworking.

Women showed up at the new nonprofit and told Manzurova and Kutepova stories that tested their credulity. Villagers described being pressed into service to clean up radiated territories in the fifties and sixties. Some of the women had been pregnant at the time; others had been schoolchildren. By Russian law, "liq-uidators" of environmental catastrophes qualify for special medical subsidies and pensions. Because Soviet law forbade employment of children and pregnant women in hazardous conditions, however, their work had not been recorded in the labor books that Soviet citizens carried. Having no records, the ailing vil-lagers had been denied status as liquidators and with it compensation for med-ical benefits and treatment. Kutepova and Manzurova agreed to represent the women in Russian courts, although previous cases had failed.[3] As word got around, more people, mostly villagers, appealed to Kutepova and Manzurova to represent them in court for funds to move out of villages such as Tatarskaia Karabolka and Muslumovo.

I sometimes went along with Kutepova as she traveled to villages and took depositions. The villagers would come to appointments armed with records of estimated exposure and health problems that extended through several genera-tions. These farmers with village educations knew a great deal about the uptake of radioactive isotopes in soils, plants, and human organs. They had dug through local archives and complicated scientific texts to determine exposure levels. They laid out the differences between internal and external doses. They explained how the half-lives of isotopes are responsible for the limited usefulness of full-body scans. They described in poignant and telling detail the medical complica-tions of their family members. The villagers educated Kutepova and then me, teaching from the painful lessons they had learned living a "natural experiment" on a radioactive landscape. With this information, Kutepova filed suits in local courts for compensation.

Kutepova and Manzurova have mostly represented villagers outside Ozersk. I asked why few Ozersk residents asked for their help. Kutepova replied:

> This is the kind of mentality we have in a closed city. Villagers on the outside, although they do not have the same kind of education and some are illiterate, are not taken in by the state and corporate PR cam-paigns. Nor do they have anything to fear losing if they joined a lawsuit to claim compensation. The atomic industry functions by spoiling those who work in the nuclear installations with federal subsidies. They spend a lot on public relations, targeting Ozersk. These residents believe the myths that PR officials spread about the safety of the plant and the "exaggerated" legacy of contamination. Maybe I am one of those people, too. I still live here [in Ozersk], after all, with my kids.[4]

Kutepova remains in gated Ozersk because the schools, health care, and public services are excellent. She considered moving to neighboring Kyshtym, but the schools there are poor, municipal budgets meager, the crime rate high.[5]

So far, Kutepova has lost her cases in the regional courts and in the provincial courts. She expects to lose in the Russian Supreme Court as well, but she plans to take her cases to the European Court of Human Rights in Strasbourg, France. The Russian Parliament ratified the European Convention on Human Rights in 1998, after which the European Court's decisions have the status of domestic Russian law.[6]

In 2005, FSB officials called in Kutepova to question her about "nuclear secrets" she allegedly imparted to a CIA agent at an international conference. At the time, Kutepova was working with a Moscow-based sociologist, Olga Tsepilova, on a planned poll in Ozersk. They never carried out the poll because FSB agents arrested Tsepilova when she arrived in Ozersk.[7] In 2008, FSB agents attempted to search Kutepova's home and office without search warrants. In 2009, tax inspectors charged her nonprofit, Planet of Hopes, with tax evasion.[8] If she had been found guilty, Kutepova, a single mother of four, would have faced stiff fines and a thirty-year prison sentence. After a furious court battle, a Cheliabinsk arbitrage court acquitted Kutepova of the fraud charges, but she soon encountered other difficulties. The FSB tried to recruit a colleague as an informant. They threatened to fire her boyfriend, a nuclear engineer, from the plant, and guards at the gates of Ozersk routinely harassed the infamous "activist," trying to withdraw her pass to the city.

With each setback, Kutepova doggedly sought protection in Russian law. She filed charges of police harassment against the FSB for attempting to search her home without a warrant. She requested to see the regulations empowering guards at the city gates to seize residents' passes. It turns out there were no regulations, just a long-established practice and a quota for guards to confiscate a certain number of passes each month. Kutepova threatened to file a suit, and officials ended the illegal practice. After an investigation, the FSB suspended the officer who had ordered the unwarranted search of Kutepova's home.

If I were keeping score, that would be two points for Kutepova.

Kutepova is by far the most litigious person I have ever met. She listed merrily the scores of lawsuits she had brought against officialdom in "Absurdovo." In her cheerful, indefatigable way, she is accomplishing in this citadel of the nuclear weapons complex what Soviet dissidents set out to do in the sixties—to force officials to abide by their own laws.

For many years, residents of Ozersk have described Kutepova in the local press and in online listservs as a troublemaking opportunist, a possible spy against the nation, and a certain traitor to the community. But the usefulness of independent observers keeping an eye on the plant has become increasingly apparent. In 2004, independent investigators discovered that the background radiation coming from the Techa River doubled and then tripled in just a few months. State inspectors found a leak in a dam holding back a radioactive reservoir in which plant engineers were dumping radioactive waste. It had been known for a decade that the dam was weak. The federal government had issued funds to fix it, but plant managers had taken no action.[9]

The Maiak general director, Vitalii Sadovnikov, claimed he didn't have the 350 million rubles to rebuild the dam, but the Maiak plant had registered 1.4 billion rubles in profits, plus another 4.3 billion rubles in income from foreign partners and sales of hard currency during the years of dumping, 2001–04. Sadovnikov had also invested heavily in fashionable Moscow offices and paid himself 1.7 million rubles in bonuses.[10] In 2005, a regional district attorney brought criminal charges against Sadovnikov for environmental contamination. The Rosatom director suspended Sadovnikov, but Sadovnikov, a well-connected businessman and regional parliamentarian, managed to get the charges dropped.[11] Sadovnikov returned to his position.

One point for Maiak.

In 2007, however, when an accident occurred at the Maiak plant, Kutepova alerted the press to the gross discrepancies between Sadovnikov's official reporting of the accident and workers' accounts on the city listserv, which described a much larger event. This time Sadovnikov's cover-up gave Rosatom officials justification to finally fire Sadovnikov.[12]

One point for Kutepova.

As the years go by, the relationship between the plant and Kutepova has become less adversarial. In 2010, a Maiak manager showed Kutepova and Manzurova a barn at the former experimental research station where lab animals, for decades dosed with radioactive substances for various tests, had been kept in cages. The animals were long gone, but the barn and cells were so radioactive the manager did not know what to do with them. He explained that if they demolished the massive building, it would send radioactive dust directly toward Ozersk. He asked the two women what he should do.

Kutepova laughed at that, "Here they were, asking *us* what to do!"

Manzurova came up with a solution—wrapping the building before demolition. Inspired by this incident, Manzurova and Kutepova started making other suggestions. They proposed that the Russian government construct a sarcophagus over the contaminated Techa River along a sixty-mile stretch to contain its radiating mud bottom, stem floods, and keep unsuspecting bystanders away. They have plans, too, for the evacuation of Muslumovo and Karabolka, decades overdue.

I puzzled over this question of who is addressing the hazards of the plutonium complex. It is not all the king's men. Rather, two women—marginally employed, working out of their apartments, Manzurova fighting health problems, Kutepova raising small children—are the ones shouldering the massive detritus of the Soviet nuclear weapons complex. Why has it been left to them and their village collaborators to suggest solutions and legal parameters for cleaning up the fifty-year disaster of the gargantuan plutonium plant?

As Manzurova and Kutepova and the villagers who support them enter this history, suddenly the usual narrative of the Soviet atomic program—which features brilliant scientists leading thousands of willing minions—makes less sense. The powerful Ministry of Medium Machine Building, which became first Minatom and then Rosatom, with all its money, political access, heavy machinery, labs, scientists, foreign subsidies, and army of civil servants, cannot somehow manage the problems the program created. That is because to seriously address the vortex of public health, environmental, economic, and genetic problems revolving around the plant would mean first to see them. For decades, the powerful men leading the Soviet nuclear industry have been empowered only to see and hear certain things. Legislating against "panic" and "exaggeration," they failed to focus on the fifty-year waste management crisis and the tens of thousands of people who have been exposed on heavily contaminated landscapes. This blindness made it possible over the years to cut the budgets for monitoring and waste management while continuing to dump into the environment.

Instead the plant leadership focused on subsidies from Moscow and, in the post-Soviet period, from abroad. In that respect, charges of a welfare mentality have been improperly directed at victims' groups. The welfare mentality flourished for decades in nuclear Ozersk. Plant managers and residents learned to ask early and often for funds, which were diverted from waste management, plant safety, and emergency evacuation projects and went instead to ensure prosperity for the chosen plutonium people. Their affluence bought a great deal of silence and submission, which further empowered the spread of radioactive isotopes.

Plant managers say they will continue to dump radioactive waste into open reservoirs until at least 2018. This dumping represents a kind of perpetual environmental debt, a debt that villagers have been underwriting with their health and well-being for generations, but which in the last two decades Russian and American taxpayers have also been paying for in the form of cleanup funds. Perhaps, from the perspective of those in Ozersk, that was a good strategy. The contaminated environment secures a future of subsidies for the one-industry city, even after the desire for plutonium has evaporated. The most immediate plan calls for another fifty years of amelioration projects to contain several hundred million curies' worth of radioactive waste. When the projects are completed, plant managers confirm that the plant will require "long-term monitoring, control, and maintenance."[13] If nothing else, the twenty-four-thousand-year half-life of plutonium guarantees job security.

As for Kutepova, Manzurova, and their collaborators, if the meek are to inherit the earth, they must first save it. The activists are necessary because, as history has shown, Moscow bosses cannot trust Maiak bosses to manage radioactive waste without supervision. In the climate of what President Putin has dubbed the Russian "nuclear renaissance," a plan to build scores of nuclear power reactors for domestic use and export abroad, there is little ideological room for Rosatom officials to dwell on the topic of contaminated landscapes. From the perspective of Rosatom, it is easier to let the women carp about it and, with their meager resources, occasionally catch unscrupulous plant managers. Maiak managers, in turn, occasionally toss obstacles such as tax evasion and espionage charges into the activists' path to slow them down. Then Russian politicians can call it "controversy" or "democracy," knowing they hold the cards.

Well, almost all the cards. For Kutepova, Manzurova, and their village collaborators are not backing down. Thanks to the post-Chernobyl decade of nuclear glasnost, the story is out, and others are starting to see it.

While I was staying in Kyshtym, I was visited by Louisa Surovova from Ozersk, who, like Manzurova, was formerly a nuclear biologist working on the trace. She told me she had wanted her husband to come for our interview because he had stories to tell, but he stayed home because he was afraid. I asked Surovova why she had come.

She said that in the late eighties, before the research station closed, she wrote a scientific paper stating that radioactive contamination would harm especially the third generation of offspring. The article was just science, she said, a prognosis based on research data, until recently, when her granddaughter, at age ten, came down with a severe case of Crohn's disease, which is very rare in children. Surovova attributes her granddaughter's disease to genetic mutations caused by her own childhood exposure to radiation. Watching her grandchild's disease develop, Surovova grasped that her scientific prediction had come to haunt her own biography. That recognition forced her to see other families with sick children. In the months before our meeting, Surovova had begun to collaborate with Manzurova and Kutepova.

One point for Kutepova.

43

Futures

At the small Richland airport, in a converted hangar, I visited Sergei Tolmachev, the director of the United States Transuranium and Uranium Registries (USTUR). The registry is unique, the agency's promotional material claims, as the only national repository of irradiated body parts in the United States.[1] Tolmachev showed me shelves with large boxes holding jars of dissolved tissue and bone. Each box was encoded with numbers indicating the human donor and type of organ—liver, kidney, heart, gallbladder. Across from the shelves were large freezers containing body parts that lab technicians had yet to dissolve, ash, and measure for deposition of radioactive isotopes. I was starting to comprehend the magnitude of this work. I was also beginning to feel queasy.

In the operating chamber, Carlos Mendez, the lab's part-time pathologist, was snapping on rubber gloves to dissect a man's arm.[2] He separated the arm into upper and lower sections, placing skin, fatty tissue, bone and muscle in discrete piles. A second, undissected hand lay on the table, palm upward, in the gesture of a meditating yogi. Tolmachev explained that the Department of Energy had severely cut the repository's budget in the last few years, so they had dropped from a few dozen employees to just four. In Mexico, Mendez had worked as a doctor, but he had crossed the U.S. border illegally and done some time in jail, so he was not certified to practice medicine in the United States. Mendez watched my discomfort at his work with an angry scowl. I got the feeling he preferred to operate on the living rather than the irradiated dead.

Mendez picked up an electric saw. As Tolmachev and I left the room, the sound of the saw's teeth biting into bone trailed us. Tolmachev told me that to save money, the registry's former director was asked to take an early retirement. Pointing at his chest, Tolmachev said in his charming Russian accent, "So I am new, cheap director." Tolmachev has a degree in nuclear chemistry from a classified wing of a Moscow institute. In the Soviet Union, he worked in a "box," a restricted military research facility. Because of security regulations, Tolmachev never dreamed he would visit the United States, let alone direct an American nuclear installation.

Tolmachev's lanky figure biking through Richland signifies the marriage, after a protracted, quarrelsome courtship, of the American and Soviet nuclear complexes. In the nineties, American military strategists, who had long cultivated fear of Soviet warheads, suddenly shifted to greater anxieties that the Soviet weapons complex would collapse, scattering nuclear materials and weapons scientists across the globe. Anxiously, DOE officials hired Russian nuclear scientists and American lawmakers funneled taxpayers' money to Russian nuclear facilities.[3]

The wedding of the Russian and American nuclear weapons' superstars was a small affair, with restricted media access. Without the brinkmanship and gruff posturing of the Cold War era, the public has largely been anesthetized into believing the nuclear arms race is over. On the contrary, we are in the midst of what anthropologist Hugh Gusterson calls the "second nuclear age." Nuclear research and development were better funded in 2011 than at the height of the Cold War. In the United States, the federal government plans to spend $700 billion on nuclear weapons development in the coming decade. Not to be outgunned, the Russian government projects an outlay of $650 billion to revamp its military, much of that to be spent on strategic nuclear weapons.[4] As Lawrence Korb, a former Pentagon official in the Reagan administration, put it, "The Cold War is over, and the military-industrial complex has won."[5]

Tolmachev, though trained in the enemy's camp, makes a great DOE employee. He is a firm believer in medical containment. He said his repository of irradiated body parts shows the benign effects of plutonium in a body up to certain tolerances, which he believes should be raised. All his donors, he said, were in their seventies and eighties when they died. "They lived forty years with plutonium in their bodies and they were fine," Tolmachev told me. "You journalists," he added, shaking his head, "get it all wrong."[6]

The effects of radiation on health remain highly controversial, as the continuing debate about the effects of Chernobyl demonstrates; estimates of deaths from the accident range from thirty-seven to a quarter of a million. The controversy is not surprising. I have argued in this history that highly controlled medical research on the effects of radioactive isotopes on human bodies manufactured knowledge, doubt, and dissent in a way that created a gulf of opinions. But there also existed a strange lack of curiosity. In Cold War America, the National Cancer Institute (NCI) did not address the question of radiation and cancer. When I asked Allen Rabson, former deputy director of the NCI, why, he replied, "Radiation as a carcinogen was cut and dried, but to pursue that would have meant to leave science and go political."[7] Instead, NCI officials left the field of medical research to AEC officials, who were highly invested in promoting the safety of nuclear power.

AEC officials took over funding of the first Atomic Bomb Casualty Commission studies in Japan in order to keep "misleading" and "unsound" reports "to a minimum."[8] In subsequent studies, researchers used the compromised ABCC studies to set standards. The standards helped to focus research narrowly on a limited dose range and a few cancers, while mostly ignoring other possible effects of radioactive isotopes in bodies. Creating universal standards is how science works. Yet in the case of the plutonium disasters, this model largely failed.

Tolmachev embodies the global citizen armed with mobile knowledge, transcending boundaries, free-flowing in the most modern ways.[9] Yet the modern cult of mobility glosses over the grounded, local quality of the risks modern technologies created in the specific environmental terrain surrounding the plutonium plants. Medical studies conducted near Hanford and Ozersk, relying on averages and estimates from standards calculated in other places, often failed to take into account the native and site-specific ways radioactive isotopes integrated with earthly and bodily landscapes. Despite decades of practice, researchers still do not have tools sensitive enough to determine any but the most obvious health effects.

Meanwhile, the nuclear industry worked to relativize radioactive contamination. Since the fifties, the American public has heard a great deal about background radiation in cities such as Denver, though background radiation has little to do with the hazards of ingesting fission products.[10] In the United States, the public also heard that nuclear power is "green." In the sixties nuclear power saved fish that might have perished in hydroelectric dams. In the twenty-first century, nuclear power promises a carbon-free future. Soon after the Fukushima catastrophe, public relations agents dusted off a five-year-old report on the dangers of the coal industry.[11] It was a maneuver that had been seen before: the Atomic Energy Commission first trumpeted the hazards of coal mining in the mid-sixties.[12] In the decades before the Fukushima disaster, the Japanese government and corporate proponents of nuclear power censored the Chernobyl disaster from textbooks and spent millions on advertising an image of nuclear safety. Meanwhile, Japanese power companies glossed over accidents, doctored safety reports, and failed to purchase emergency equipment for fear of alerting workers to the dangers of the industry.[13]

Another important way to neutralize the plutonium disasters has been to naturalize them. In the last decade, officials have repurposed the Hanford, Maiak, and Chernobyl territories as wildlife preserves. The Chernobyl zone, open to tourism, features a breathtakingly beautiful terrain of forest, lakes, and streams. Journalists and scientists describe it as teeming with wildlife.[14] Tim Mousseau, an evolutionary biologist, however, tracked birds in the Chernobyl zone and found a zone of ecological calamity. Even in areas of moderate contamination, 18 percent of the birds he followed had deformities; 40 percent of male barn swallows were sterile, and the total number of swallows was depressed by 66 percent. Mousseau could not find in the hot spots bumblebees, butterflies, spiders, or grasshoppers. Whole zones in the Zone are dead.[15] In eastern Washington, the territory around the Hanford reservation is promoted as the last stand of original shrub-sage habitat in the Columbia Basin, yet periodically deer and rabbits wander from the preserve and leave radioactive droppings on Richland's lawns.[16] In the eighties, the Hanford Reach gained recognition as the last free-flowing stretch of the Columbia River. Watching the river's water level rise and fall with demands for electricity in Portland while measuring the radioactivity of mulberry trees on the graded gravel banks of the Columbia makes calling the Hanford segment "wild" a reach indeed.[17]

Wine production is the Columbia River region's latest diversification project. Tourists are encouraged to tour tasting rooms located in a large arc around the Hanford Nuclear Reservation, which figures as an unnamed area on the vineyard tour map. As I tasted a few wines, I mentioned to a vintner that a lot of Wanapam Indians down the road are sick with cancers and that a Centers for Disease Control study found local Indians had a one in fifty chance of getting cancer, in part because of their traditional diet of Columbia River fish.[18] If the study is correct, I asked, how did she feel about growing wine grapes so near the mothballed plutonium plant? She testily replied that the Indians had a lot of problems with alcoholism, inbreeding, and poor diet. I had heard that before.[19]

I sympathized with the vintner. It is difficult to face the problem of nuclear waste because it is so outsized. The 2011 Fukushima meltdowns ranked with Chernobyl in terms of disaster because the managers of the TEPCO power company had no place to store an enormous volume of nuclear waste other than in pools right next to the power reactors, an act akin to stowing gunpowder in a gasworks.[20] Russian officials promoting the "nuclear renaissance" are competing with a Hitachi-GE partnership to export nuclear reactors to developing countries in Asia and the Middle East.[21] A big part of Russia's sales pitch is the promise to reprocess spent fuel, using its existing, already overtaxed Maiak reprocessing plant as the dumping ground. Geographer Shiloh Krupar describes nuclear wastes as among the living dead. Like zombies in a 1950s film—there is no way to kill them, no way to keep them from returning.[22]

Plutonium production required undemocratic and unsafe decisions and policies that Americans and Soviets found politically palatable because the world's first plutonium plants germinated distinct communities—of affluent permanent employees separate from temporary, migrant laborers—on zoned-off landscapes that effectively rendered invisible not only the massive nuclear installations but also the environmental and health problems they produced. Plutopia's spatial compartmentalization appeared natural because it mirrored divisions in Soviet and American society between free and unfree labor, between majority white and minority nonwhite populations, and often between those people thought to be safe and those left in the path of radiation. Twentieth-century history centers around the inscription of race and class on the landscape. In the plutonium zones, these boundaries ensured state secrets and subordinate workers increasingly concerned with preserving the consumer privileges and social welfare benefits granted them inside their limited-access cities.

Plutopia was locally popular also because it served up an ever-expanding economy delivering a continually increasing volume of consumer goods for an endlessly rising standard of living made possible by government subsidies for select workers. Residents of plutopia displayed a fantastic faith in scientific progress and economic efficiency. Many understood their city's universal, classless affluence as the materialization of the American dream or Communist utopia, an affirmation that their national ideology was correct. Self-assurance and confidence bred patriotism, loyalty, submission, and silence.

These special communities proliferated as ably as the nuclear technologies that spawned them. Communities like Richland were reproduced in California, Texas, Georgia, Idaho, and New Mexico. Communities like Ozersk repeated across the Urals, Kazakhstan, Siberia, and parts of European Russia. It was a compelling model with global prospects. In the 1960s, by means of secret treaties, American CIA agents and advisors helped install GE-brand nuclear power plants in Japan. GE engineers had initially designed Fukushima's first reactors for American nuclear submarines. The reactors' design was intended for an environment with lots of water and few people—suitable for conditions under the ocean's surface, but less so for civilian uses on land.[23] As a consequence, Japanese nuclear developers appropriated the practice of choosing poor, remote coastal regions for nuclear reactors and then subsidizing the new "nuclear villages" handsomely. Japanese nuclear villagers, in turn, welcomed reactors because they meant jobs and tax revenue, new community centers, schools, theme parks, swimming pools, even free diapers.[24]

As the arms race fed nuclear communities, the culture and way of life they produced nurtured the arms race. The highly subsidized nuclear communities built up a "universe of contentment" that was very hard, politically, to dismantle. President Eisenhower was the first to complain of the obdurate qualities of the military-industrial complex. The former five-star general conceded he was no match for it. President Barack Obama, coming to power on a no-nukes platform, has not been able to curb the acceleration in nuclear weapons spending.[25] In the Soviet Union, Khrushchev tried to deescalate nuclear arms production. So did Mikhail Gorbachev. Both Soviet leaders were unseated bodily, with the help of military and security officers.

The plutonium cities were the product not merely of technology and science but of the larger cultures that created them. They led their countries in creating new kinds of communities to harbor nuclear families. And the residents of plutopia were not alone in their desires. Outside plutopia, Americans and Soviets also rushed to acquire physical and financial security in the form of residential space in exclusive, ethnically segregated, federally subsidized communities. Over time, plutopia came to be seen not as the anomalous, isolated social configurations they were, but as aspirational versions of bedroom communities sprawling across the American and Soviet landscapes. In this way, spatial practices normalized and rendered invisible the massive nuclear installations hidden in plain sight.

I have never entered Ozersk, but from photos, it looked to be a very pleasant version of many Soviet-era cities, with high-rises, broad streets, and vast, empty public squares. I did spend a good deal of time in Richland, where I was overcome with its unexceptionality. Richland did not look or feel like a Cold War outpost. It hardly differed from most American postwar suburbs. I felt at home cruising along Richland's wide, straight arteries past big-box stores set back from the roads to make room for acres of parking. It rarely occurred to me that the arteries were evacuation routes, the parking lots served as firebreaks, and the windowless shopping centers doubled as shelters in case of disaster.[26]

Richland and Ozersk are easily recognizable because in these citadels of plutonium, at ground zero of nuclear Armageddon, people who had choices made the same kind of trade-offs of consumer and financial security for civil rights and political freedoms as did their fellow citizens nationally. The invisible zoning of territory along sliding hierarchies of value gave the lie to American and Soviet promises of equal opportunity and mobility, but it was a lie that was easy to overlook. Exchanging long-held goals of universal equality for the exclusivity of a "few good men" or "the chosen people," Americans and Russians came to shun accessible, equitable public housing in favor of limited-access cities, in the Soviet case, and monoclass suburbs, in the American context. In creating these zones of affluence, the boosters of defense and progress also created zones of blight and environmental sacrifice where the boosters themselves would never live. By the twenty-first century, large numbers of Russians and Americans sought to live in gated communities, which had become, ironically, not a stigma of incarceration but an object of desire, the end of the pioneering nuclear family's mobile, half-century quest for security, health, and happiness.[27]

The generals, engineers, and scientists who created Hanford and Ozersk made assurances of safety and security by relying on clear boundaries between nations and ideologies, between leaders and public, between nuclear zones and non-nuclear zones, between bodies and environments, humans and animals, nature and culture.[28] These distinctions were largely fictional. Just as American and Soviet ideas about the family and prosperity morphed into one another while scientific discoveries crossed enemy lines, so, too, fission products migrated from industrial to residential zones, from soils into food, from air to lungs to bloodstream, bone marrow, and finally DNA, so bodies themselves now serve as nuclear waste repositories. Yet even as fiction, borders have their uses. Compartmentalization, Generals Groves and Beria knew, was an important technology for keeping secrets and suppressing bad news. Sorting ideas into "Communist" and "capitalist" kept people from questioning their superiors for fear of being labeled enemies. Histories subdivided into "science," "environment," "culture," and "architecture" also fail to unwrap the larger story of how scientific research, cultural currents, urbanization trends, territorial zoning, policy, and finance created the landscapes we call home at the same time they produced warheads, missiles, and defense budgets increasingly impossible to shoulder.

The nuclear weapons complexes created militarized landscapes that blossomed far beyond the immediate territories of the plutonium plants. In the American case, Tri-Cities boosters lobbied for and won highways, bridges, dams, schools, housing, and farm subsidies to ensure national security and self-sufficiency. These allocations also provided financial security for citizens who felt increasingly entitled as a result of Cold War promises of universal affluence. Across the militarized American terrain rolled multiple wars: the war on poverty, the war on drugs, the war on cancer, and the war on terror. These wars were fought with the usual martial technologies: bulldozers to clear blighted cities, classified financial security maps, chemical weapons adapted for both food production and chemotherapy, nuclear technologies modified for food preservation and medicine, and weapons of surveillance turned on native populations.

Taken as a complex, the militarized landscape also helped produce health epidemics. In the United States from 1950 to 2001, the overall age-adjusted incidence of cancer increased by 85 percent. Childhood cancer, once a medical rarity, has become the most common disease killer of American children.[29] Cancer rates are just the end of a continuum of American health problems that include diabetes, heart disease, asthma, and obesity. These social-cultural-economic problems written on the body are also etched across American communities, one-quarter of which are within four miles of a Superfund site filled with plastics, chemical solvents, pesticides, nuclear waste, and all the other unwanted detritus of consuming societies.[30] This too is the house plutonium built.

Russia scores no better. From 1960 to 1985, cancer rates in the Soviet Union grew from 115 to 150 per 100,000 people.[31] By the mid-1990s, Russia's death rate had reached its highest peacetime level in the twentieth century. Only a third of Russian infants are born healthy. Russia annually falls at the bottom among nations in a host of categories, including life expectancy, fertility, and infant mortality.[32] As Soviet leaders raced to keep up with American weapons manufacturers, they created landscapes where the promises of Communist utopia were shifted to citizens living in well-supplied cities requiring registration passes, while villagers legally immobilized in the hinterland languished in poverty and poor health.

The history of Richland and Ozersk begs telling because the problems of plutopia are not going away. The 2011 meltdown of three reactors in Fukushima followed the patterns established in the plutonium zones: military designs scant on safety, accidents followed by denial and minimization, delayed evacuations to zones also contaminated, and the deployment of short-term, minimally paid "jumpers" to do the dirtiest work.[33] In the United States, Russia, and Japan, where private corporations reap profits from nuclear production, federal governments have indemnified nuclear enterprises against nuclear accidents, in this way socializing the financial risk of nuclear adventure.

Fortunately, the story does not end there. Faced with mute and paralyzed leaders, Japanese citizens did what villagers in the Urals and farmers in eastern Washington did: they took matters into their own hands. They organized through social media networks and community organizations, bought Geiger counters and tested their food, air, and soil to produce their own, detailed maps of contamination. With this knowledge they forced corporate executives and government officials to take action too.[34]

I embarked on this project because I wanted to learn about the pioneers of the nuclear security state. I considered the citizens of plutopia, on the frontier of the nuclear arms race, to be the cultural founders of the early twenty-first century, in which people around the globe have become subjects of surveillance—antiterrorist, financial, and medical. As citizens of the new millennium, we are listened to, watched, and tracked, sometimes by means of our own joyful collusion.

In the course of writing this book, however, I came to know people who could not share a meal with me because of medical dietary restrictions. I met individuals who lifted their shirts to show me the cross-hatching of scars left from multiple surgeries. Watching these courageous people who insisted on asking questions, sought their own answers, and spoke even when their supervisors attempted to silence them, I came to visualize a different kind of nuclear pioneer. This group is on the march. Some are wearing protective jumpsuits and masks; some are thin and pale. Others are children. A few shuffle with oxygen tanks or roll along in wheelchairs. As conflicts over resources, wealth, and power merge with struggles over risk, health, and safety, these people are defining a new kind of citizenship in which they demand, in addition to their political and consumer rights, biological rights. Along with freedom from want and tyranny, they insist on freedom from risk and contamination. These determined people are, in other words, any number of us, as we are all citizens of plutopia.

Archives and Abbreviations

AOKMR   Arkhivnyi otdel Kunashakskogo munitsipal'nogo raiona, Kunashak, Russia

BPC   Boris Pash Collection, Hoover Institution, Stanford University, Palo Alto, California

CBN   *Columbia Basin News*

CREHST   Columbia River Exhibition of History, Science and Technology, Richland, Washington

DOE Germantown   AEC Secretariat Files, 1958–1966, Department of Energy, Germantown, Maryland

DOE Opennet   Department of Energy Opennet System, online, https://www.osti.gov/opennet

EOL   Ernest O. Lawrence Papers, Special Collections, Bancroft Library, University of California, Berkeley

FCP   Fred Clagett Papers, Special Collections, University of Washington, Seattle

FTM   Frank T. Matthias Diary, Department of Energy Public Reading Room, Richland, Washington

GWU   George Washington University, National Security Archive, www.gwu.edu/~nsarchiv/radiation/dir/mstreet/commeet/meet5/brief5/tab_fbr5f3m.txt

HMJ   Henry M. Jackson Papers, Special Collections, University of Washington, Seattle

HML   Hagley Museum and Library, Wilmington, Delaware

JPT   James P. Thomas Papers, Special Collections, University of Washington, Seattle

LKB   Leo K. Bustad Papers, Washington State University Special Collections

NAA	National Archives, Atlanta, Georgia
NARA	National Archives and Records Administration, College Park, Maryland
NYT	*New York Times*
OGAChO	Ob'edinennyi Gosudarstvennyi Arkhiv Cheliabinskoi Oblasti, Cheliabinsk, Russia
PRR	Department of Energy Public Reading Room, Richland, Washington
RPL	City of Richland History Collection, Richland Public Library, Richland, Washington
RT	Private collection of Robert Taylor, Pasco, Washington
SPI	*Seattle Post-Intelligencer*
SR	*The Spokesman Review*
TCH	*Tri-City Herald*
UWSC	Special Collections, University of Washington, Seattle

Notes

Introduction

1. R. E. Gephart, *Hanford: A Conversation About Nuclear Waste and Cleanup* (Columbus, OH: Battelle Press, 2003), 5.25. Estimates at the Maiak plant are far greater, at 1 billion curies; Vladislav Larin, "Neizvestnyi radiatsionnye avarii na kombinate Maiak," www.libozersk.ru/pbd/mayak/link/160.htm (accessed March 19, 2012).
2. Yoshimi Shunya, "Radioactive Rain and the American Umbrella," *Journal of Asian Studies* 71, no. 2 (May 2012): 319–31.
3. Jack Metzgar, *Striking Steel: Solidarity Remembered* (Philadelphia: Temple University Press, 2000), 7, 156.
4. John M. Findlay and Bruce William Hevly, *Atomic Frontier Days: Hanford and the American West* (Seattle: University of Washington Press, 2011), 84.
5. T. C. Evans, "Project Report on Mice Exposed Daily to Fast Neutrons," July 18, 1945, NAA, RG 4nn-326-8505, box 54, MD 700.2, "Enclosures."
6. Adriana Petryna, *Life Exposed Biological Citizens After Chernobyl* (Princeton, NJ: Princeton University Press, 2002).
7. The most recent publications include Gabrielle Hecht, *Being Nuclear: Africans and the Global Uranium Trade* (Cambridge, MA: MIT Press, 2012); Richard Rhodes, *Twilight of the Bombs: Recent Challenges, New Dangers, and the Prospects for a World Without Nuclear Weapons* (New York: Vintage, 2011); Findlay and Hevly, *Atomic Frontier Days*; Jonathan Schell, *The Seventh Decade: The New Shape of Nuclear Danger* (New York: Metropolitan Books, 2007); Sharon Weinberger and Nathan Hodge, *Nuclear Family Vacation: Travels in the World of Atomic Weaponry* (New York: Bloomsbury, 2008); Max S. Power, *America's Nuclear Wastelands* (Pullman: Washington State University Press, 2008); V. N. Kuznetsov, *Zakrytye goroda Urala* (Ekaterinburg: Akademiia voenno-istoricheskikh nauk, 2008).

Chapter 1: Mr. Matthias Goes to Washington

1. FTM, December 16–22, 1943.
2. Katherine G. Morrissey, *Mental Territories: Mapping the Inland Empire* (Ithaca, NY: Cornell University Press, 1997), 32–35.
3. Michele Stenehjem Gerber, *On the Home Front: The Cold War Legacy of the Hanford Nuclear Site* (Lincoln: University of Nebraska Press, 1992), 22.
4. Paul C. Pitzer, *Grand Coulee: Harnessing a Dream* (Pullman: Washington State University Press, 1994), 341–43.
5. Ibid., 116.
6. FTM, January 5, 1943.

7. Robert S. Norris, *Racing for the Bomb: General Leslie R. Groves, the Manhattan Project's Indispensable Man* (South Royalton, VT: Steerforth Press, 2002), 214.

8. FTM, December 17 and 22, 1943.

9. As quoted in Gerber, *On the Home Front*, 12. See also George Hopkins to Nichols et al., December 26, 1942, NAA, RG 326-8505, box 41, 600.03 "Location"; "Completion Report: Hanford Engineer Works, Part I," April 30, 1945, NAA, RG 326-8505, box 46, 400.22 "General."

10. D. W. Meinig, *The Great Columbia Plain: A Historical Geography, 1805–1910* (Seattle: University of Washington Press, 1968), 6.

11. John M. Findlay and Bruce William Hevly, *Atomic Frontier Days: Hanford and the American West* (Seattle: University of Washington Press, 2011), 60.

12. Patricia Nelson Limerick, "The Significance of Hanford in American History," in *Terra Pacifica: People and Place in the Northwest States and Western Canada* (Pullman: Washington State University Press, 1998), 53–70.

13. Ted Van Arsdol, *Hanford: The Big Secret* (Vancouver, WA: Ted Van Arsdol, 1992), 13–15; Peter Bacon Hales, *Atomic Spaces: Living on the Manhattan Project* (Urbana: University of Illinois Press, 1997), 47–70.

14. FTM, March 26, 1943.

15. Norris, *Racing for the Bomb*, 217–21.

Chapter 2: Labor on the Lam

1. "Photographs and Films from the Hanford Engineer Works E. I. Du Pont de Nemours & Company," HML.

2. Estimates of construction workers from 1943 to 1945 range from 94,000 to 132,000. Harry Thayer, *Management of the Hanford Engineer Works: How the Corps, DuPont and the Metallurgical Laboratory Fast Tracked the Original Plutonium Works* (New York: ASCE Press, 1996), 93.

3. "The Manhattan District History," PRR, HAN 10970, 58.

4. "Daily Employment During Construction Period" and "Total Daily Terminations," HML, acc. 2086, folder 20.13; "Completion Report: Hanford Engineer Works, part I," April 30, 1945, NAA, RG 326-8505, box 46, folder 400.22 "General."

5. Crawford Greenewalt Diary, vol. 3, August 7, 1943 and January 13, 1944, HML.

6. "Semi-Monthly Report for Hanford Area," August 5, 1943, and "Progress Reports," 3/43–12/43, NAA, RG 326-8505, box 46, MD 600.914.

7. Groves to Nichols, telegram, November 16, 1943, Groves to Ackart, November 19, 1943, NAA, RG. 326-8505, box 41, MD 600.1, "Construction and Installation."

8. Nell Macgregor, "I Was at Hanford," 17, UWSC, acc. 1714-71, box 1, folder 1; Ted Van Arsdol, *Hanford: The Big Secret* (Vancouver, WA: Ted Van Arsdol, 1992), 24.

9. Peter Bacon Hales, *Atomic Spaces: Living on the Manhattan Project* (Urbana: University of Illinois Press, 1997), 103.

10. Michele Stenehjem Gerber, *On the Home Front: The Cold War Legacy of the Hanford Nuclear Site* (Lincoln: University of Nebraska Press, 1992).

11. Van Arsdol, *Hanford*, 37.

12. Groves, "Memorandum," November 9, 1943, NAA, RG 326-8505, box 52, MD 624, "Housing."

13. Hales, *Atomic Spaces*, 117–25.

14. See "Field Progress Report, Part f, Maps and Plans," March 31, 1944, NAA, RG 326-8505, box 46, 600.914.

15. James W. Parker Memoirs, HML, acc. 2110, 5.

16. Pap A. Ndiaye, *Nylon and Bombs: DuPont and the March of Modern America* (Baltimore: Johns Hopkins University Press, 2007), 167.

17. S. L. Sanger and Robert W. Mull, *Hanford and the Bomb: An Oral History of World War II* (Seattle: Living History Press, 1989), 96.

18. Parker Memoirs, 15.

19. Sanger, *Hanford and the Bomb*, 96.
20. Ibid, 93.
21. Van Arsdol, *Hanford*, 44.
22. Ibid.
23. T. B. Farley, "Protection Security Experience to July 1, 1945," October 2, 1945, PRR, HAN 73214; Leslie R. Groves, *Now It Can Be Told: The Story of the Manhattan Project* (New York: Da Capo Press, 1983), 139; Hales, *Atomic Spaces*, 177.
24. Sanger, *Working on the Bomb*, 140.
25. Farley, "Memorandum," October 2, 1945, PRR, HAN 73214, 17; W. B. Parsons, "Surveillance Logs," October 4, 1944, NAA, RG 326 8505, box 103.
26. Mary Catherine Johnson-Pearsall, Alumni Sandstorm (on-line archives), www.alumnisandstorm.com, November 9, 1998.
27. Sanger, *Working on the Bomb*, 138–39.
28. "Richland, Atomic Capital of the West," *Bosn's Whistle*, November 16, 1945.
29. Sanger, *Working on the Bomb*, 95. On infestations, see Matthias to Friedell, October 16, 1944, NAA, RG 326-87-6, box 16, folder "Telegrams."
30. "Total Daily Terminations," HML, acc. 2086, folder 20.13.

Chapter 3: "Labor Shortage"

1. George Q. Flynn, *The Mess in Washington: Manpower Mobilization in World War II* (Westport, CT: Greenwood Press, 1979), 165.
2. Cindy Hahamovitch, "The Politics of Labor Scarcity: Expediency and the Birth of the Agricultural 'Guestworkers' Program," Center for Immigration Studies, December 1999.
3. Harry Thayer, *Management of the Hanford Engineer Works: How the Corps, DuPont and the Metallurgical Laboratory Fast Tracked the Original Plutonium Works* (New York: ASCE Press, 1996), 27.
4. In 1944, Matthias did commandeer 150 steamfitters from the army. Peter Bacon Hales, *Atomic Spaces: Living on the Manhattan Project* (Urbana: University of Illinois Press, 1997), 185.
5. Carl Abbott, *The Metropolitan Frontier: Cities in the Modern American West* (Tucson: University of Arizona Press, 1993), 20.
6. John M. Findlay and Bruce William Hevly, *Atomic Frontier Days: Hanford and the American West* (Seattle: University of Washington Press, 2011), 27.
7. FTM, February 18, 1944.
8. FTM, September 23, 1943.
9. FTM, February 26, 1944.
10. Ibid.
11. Matthias, "Field Progress Report," March 31, 1944, NAA, RG 326-8505, box 46, 600.914, "Progress Reports HEW."
12. "Spanish-American Program," April 1944, HML, acc. 2086, folder 20.13.
13. Robert Bauman, "Jim Crow in the Tri-Cities, 1943–1950," *Pacific Northwest Quarterly*, Summer 2005, 126.
14. Memo from Richland Human Rights Commission to Richland City Council, August 6, 1969, Human Rights Commission folder, RPL.
15. Otto S. Johnson, "Manpower Meant Bomb Power," September 1945, HML, acc. 2086, folder 20.13.
16. James W. Parker Memoirs, HML, acc. 2110, 1.
17. Pap A. Ndiaye, *Nylon and Bombs: DuPont and the March of Modern America* (Baltimore: Johns Hopkins University Press, 2007), 121.
18. Administration Personnel, HML, acc. 2086, folder 20.10. Manhattan Project officials also introduced segregation to Oak Ridge, Tennessee. Russell B. Olwell, *At Work in the Atomic City: A Labor and Social History of Oak Ridge, Tennessee* (Knoxville: University of Tennessee Press, 2004), 21.

19. Flynn, *Mess in Washington*, 149–71.
20. Church, "HEW Policy Recommendations," April 17, 1943, NAA, RG 326-8505, box 42, f. 600.18, "HEW Operations."
21. Draft, "Federal Prison Industries Operating Contract," 1947, RT; FTM, June 10, June 11, July 5, August 4, 1943.
22. Herbert Taylor, March 24, 1944, RT.
23. Draft, "Federal Prison Industries Operating Contract."
24. Herbert Taylor, March 24 and April 2, 1944, RT.
25. Draft, "Federal Prison Industries Operating Contract."
26. Frank T. Matthias, "Hanford Comes of Age," January 1946, HML, acc. 2086, folder 20.10.
27. James W. Parker Memoirs, HML, acc. 2110, 9.
28. Nell Macgregor, "I Was at Hanford," 17, UWSC, acc. 1714–71, box 1, folder 1, 8–9.
29. Otto S. Johnson, "Manpower Meant Bomb Power," September 1945, HML, acc. 2086, folder 20.13; Thayer, *Management of the Hanford Engineer Works*, 82.
30. Bradley Seitz to E. H. Marsden, January 8, 1944, NAA, RG 326-8508, box 54, MD 700.2.
31. FTM, December 18–22, 1943.
32. Macgregor, "I Was at Hanford," 54.
33. FTM, December 18–22, 1943.
34. Yuletide Festival programs, December 1944, HML, Matthias Photo Collection.
35. "Completion Report," April 30, 1945, NAA, RG 326-8505, box 46, 400.22, "General."
36. Groves to E. G. Ackart, November 16, 1943, Groves to Nichols, November 16, 1943, NAA, RG 326-8505, box 41; MD 600.1, "Construction and Installation"; E. DeRight to Nichols, December 14, 1943, NAA, RG 326-8505, box 46, MD 600.914, "Progress Reports HEW."

Chapter 4: Defending the Nation

1. Buck to Matthias, April 1944, HML, acc. 20.15.
2. "WACs Visit Indian Tribe," *Sage Sentinel*, April 28, 1944, 1.
3. Click Relander, *Drummers and Dreamers: The Story of Smowhala the Prophet and His Nephew Puck Hyah Toot, the Last Prophet of the Nearly Extinct River People, the Last Wanapams* (Caldwell, ID: Caxton Printers, 1956), 51–55.
4. FTM, September 15, 1943.
5. The Wanapam lands were ceded to the United States by treaty without their consent or participation. Andrew H. Fisher, *Shadow Tribe: The Making of Columbia River Indian Identity* (Seattle: University of Washington Press, 2010), 83.
6. FTM, April 2, 1944.
7. Ibid.
8. Author interview, Rex Buck, May 8, 2008, Priest Rapids Dam, Washington.
9. FTM, April 2, 1944; Norman G. Fuller to Matthias, September 20, 1945, HML, acc. 2086, folder 20.15.
10. FTM, September 15, 1943, April 2, 1944.
11. Fred Foster to Matthias, September 4, 1945, HML, acc. 2086, folder 20.15.

Chapter 5: The City Plutonium Built

1. FTM, March 2, 1943.
2. "Memorandum of Conference," April 1, 1943, NAA, RG 326-8505, box 60, folder "Meetings DuPont."
3. See Peter Bacon Hales, *Atomic Spaces: Living on the Manhattan Project* (Urbana: University of Illinois Press, 1997), 120–26 and Charles O. Jackson, *City Behind a Fence: Oak Ridge, Tennessee, 1942–1946* (Knoxville: University of Tennessee Press, 1981), 71.
4. T. B. Farley, "Protection Security Experience to July 1, 1945," October 2, 1945, PRR, HAN 73214.

5. Hardy Green, *The Company Town: The Industrial Edens and Satanic Mills That Shaped the American Economy* (New York: Basic Books, 2010), 56.

6. Hales, *Atomic Spaces*, 99.

7. FTM, June 24, 1943.

8. FTM, June 21 and 27, 1945.

9. Crawford Greenewalt Diary, July 8 and January 9, 1943, HML.

10. FTM, June 28, 1945; Greenewalt Diary, March 1944, HML.

11. Wendy L. Wall, *Inventing the "American Way": The Politics of Consensus from the New Deal to the Civil Rights Movement* (New York: Oxford University Press, 2008), 51.

12. Pap A. Ndiaye, *Nylon and Bombs: DuPont and the March of Modern America* (Baltimore: Johns Hopkins University Press, 2007), 118–19; Barton J. Bernstein, "Reconsidering the 'Atomic General': Leslie R. Groves," *Journal of Military History* 67 (July 2003): 895.

13. Robert F. Burk, *The Corporate State and the Broker State: The du Ponts and American National Politics, 1925–1940* (Cambridge, MA: Harvard University Press, 1990), 295–96.

14. During World War II, DuPont produced 70 percent of all explosives manufactured in the United States. Ndiaye, *Nylon and Bombs*, 111; quote from 152.

15. Matthias, "Notes on Village," April 17, 1943; Church Sawin to Daniel Haupt, April 17, 1943; Matthias to Yancey, April 19 and April 23, 1943, HML, acc. 2086, folder 20.63.

16. Matthias to Yancey, April 23, 1943; Yancey to Matthias, April 24, 1943, April 26, 1943; "Memo from General Groves to Matthias," April 27, 1943, HML, acc. 2086, folder 20.63.

17. "Conference Notes," April 1, 1943, Wilmington, NAA, RG 326 8505, box 183, f MD 319.1, "Report—Hanford Area"; FTM, June 24 and September 11, 1943, and October 12, 13, and 15, 1944.

18. DuPont sought a higher percentage of high-end houses than other projects. Travis to Nichols, April 19, 1943, NAA, RG 326-8505, box 52, folder MD 624, "Housing."

19. FTM, October 18, 1943.

20. As quoted in Hales, *Atomic Spaces*, 96.

21. On the costs of low-density sprawl in the 1930s–1940s, see Kenneth T. Jackson, *Crabgrass Frontier: The Suburbanization of the United States* (New York: Oxford University Press, 1985), 131, and Howard L. Preston, *Automobile Age Atlanta: The Making of a Southern Metropolis, 1900–1935* (Athens: University of Georgia Press, 1979).

22. "History of the Project," vol. 1, PRR, HAN 10970, and FTM, November 16, 1943.

23. *Richland Villager*, February 6, 1947.

24. K. D. Nichols, *The Road to Trinity* (New York: Morrow, 1987), 107–8.

25. J. S. McMahon, "Village Administration Experience," July and August 1946, PRR, HAN 73214, Bk.-17.

26. DuPont to District Engineer, "Monthly Report, August 1944," September 20, 1944, NAA, RG 326-8505, box 182, MD 319.1, "Reports—DuPont."

27. Wall, *Inventing the "American Way,"* 49–55.

28. Jack Metzgar, *Striking Steel: Solidarity Remembered* (Philadelphia: Temple University Press, 2000), 156; Joan Didion, *Where I Was From* (New York: Knopf, 2003), 115.

29. On depictions of Richland as "middle class," see Paul John Deutschmann, "Federal City: A Study of the Administration of Richland," M.A. thesis, University of Oregon, 1952, 301–5, and John M. Findlay and Bruce William Hevly, *Atomic Frontier Days: Hanford and the American West* (Seattle: University of Washington Press, 2011), 98.

30. Groves to Philip Murray, CIO, April 19, 1946, Robert Norris Papers, box 41, folder "Labor Relations," Hoover Institution Archives, Palo Alto, CA. For labor conflict and unions as intelligence targets, see FTM, September 8, 1944, and W. B. Parsons, "List of Unions," May 23, 1944, NAA, RG 326 8505, box 103, Folder, "Policy Books of Intelligence Division."

31. James W. Parker Memoirs, HML, acc. 2110, and "Monthly Report, July 1944," August 18, 1944, NAA, RG 326-8505, box 182, MD 319.1, "Reports—DuPont."

32. Peter Bacon Hales, "Building Levittown: A Rudimentary Primer," University of Illinois at Chicago, http://tigger.uic.edu/~pbhales/Levittown/building.html.

33. FTM, June 24, 1943.
34. FTM, June 28 and July 15, 1943, February 7, 8, and 9, 1944.
35. M. T. Binns, "Housing Experience to July 1, 1945," August 3, 1945, in "Village Operations, Part I," PRR, acc. 3097, 4–6, 9–10.
36. FTM, December 21, 1943.
37. FTM, August 18, 1943.
38. Hales, *Atomic Spaces*, 194.
39. As quoted in Michele Stenehjem Gerber, *On the Home Front: The Cold War Legacy of the Hanford Nuclear Site* (Lincoln: University of Nebraska Press, 1992), 61.
40. Leroy Arthur Sheetz, "Richland—the Atomic City," *Christian Science Monitor*, January 18, 1947; *Business Week*, December 18, 1948, 65–70; George W. Wickstead, "Planned Expansion for Richland, Washington," *Landscape Architecture* 39 (July 1949): 167–75.
41. Carl Abbott argues that Richland was a new kind of American community, neither a classic company town nor a manifestation of the Green City movement that created New Deal towns such as Greenbelt, Maryland. Abbott, "Building the Atomic Cities: Richland, Los Alamos, and the American Planning Language," in Bruce Hevly and John M. Findlay, eds., *The Atomic West* (Seattle: University of Washington Press, 1998), 90–115.
42. Among many histories detailing the history of postwar American suburbs, see Robert O. Self, *American Babylon: Race and the Struggle for Postwar Oakland* (Princeton, NJ: Princeton University Press, 2003); Elaine Tyler May, *Homeward Bound: American Families in the Cold War Era* (New York: Basic, 1999); Amanda I. Seligman, *Block by Block: Neighborhoods and Public Policy on Chicago's West Side* (Chicago: University of Chicago Press, 2005); Beryl Satter, *Family Properties: Race, Real Estate, and the Exploitation of Black Urban America* (New York: Metropolitan, 2009). Among important correctives to the Lewis Mumford–inspired critique of the suburbs as solely elite and white, see Matthew D. Lassiter, *The Silent Majority: Suburban Politics in the Sunbelt South* (Princeton, NJ: Princeton University Press, 2006) and Kevin Michael Kruse and Thomas J. Sugrue, eds., *The New Suburban History* (Chicago: University of Chicago Press, 2006).

Chapter 6: Work and the Women Left Holding Plutonium

1. Groves to Area Engineer, HEW, September 1, 1944, NARA, RG 77, entry 5, box 41.
2. Pap A. Ndiaye, *Nylon and Bombs: DuPont and the March of Modern America* (Baltimore: Johns Hopkins University Press, 2007), 136.
3. W. O. Simon, "Census Survey Tabulation," August 16, 1944, HML, box 2, folder 20.63.
4. P. W. Crane, "Technical Department Functions and Organization to July 1, 1945," PRR.
5. Author interview with Joe Jordan, May 17, 2008, Richland, WA.
6. G. W. Struthers, "Procurement and Training of Non-Exempt Personnel," September 6, 1945, PRR.
7. "Questions Asked of Dr. Chet Stern in Conference by Dr. Warren," NAA, June 24, 1943, RG 326-8505, box 12, I.E.2, "General Correspondence"; Nichols to Daniels, April 24, 1943, NAA, RG 326-66A-1405, box 9, folder 600.1, "Hanford."
8. "Completion Report," April 30, 1945, NAA, RG 326-8505, box 46, folder 400.22, "General."
9. "HW Radiation Hazards for the Reactor Safeguard Committee," July 27, 1948, PRR, HW 10592.
10. "Minutes of Richland Community Council," meeting no. 20, May 9, 1949, Richland Public Library, 1.
11. Struthers, "Procurement and Training."
12. Ted Van Arsdol, *Hanford: The Big Secret* (Vancouver, WA: Ted Van Arsdol, 1992), 64.
13. Williams to Groves, August 29, 1944, and De Right to Williams, August 25, 1944, NAA, RG 326-8505, box 55, MD 729.3 "Radiation, Book 1."
14. Struthers, "Procurement and Training."
15. Ibid.; Peter Bacon Hales, *Atomic Spaces: Living on the Manhattan Project* (Urbana: University of Illinois Press, 1997), 117.

16. Author interview with Marge DeGooyer, May 16, 2008, Richland, WA.

17. Ruth Howes and Caroline L. Herzenberg, *Their Day in the Sun: Women of the Manhattan Project* (Philadelphia: Temple University Press, 1999), 142.

18. Laurie Williams, "At Hanford Plutonium Lab, She Could Really Cook," *TCH,* October 31, 1993, C8.

19. Author interview, DeGooyer.

20. Michele Stenehjem Gerber, *On the Home Front: The Cold War Legacy of the Hanford Nuclear Site* (Lincoln: University of Nebraska Press, 1992), 45; Ian Stacy, "Roads to Ruin on the Atomic Frontier: Environmental Decision Making at the Hanford Reservation, 1942–1952," *Environmental History* 15, no. 3 (July 2010): 415–48.

21. Howes and Herzenberg, *Their Day in the Sun,* 142, 195.

Chapter 7: Hazards

1. Barton C. Hacker, *The Dragon's Tail: Radiation Safety in the Manhattan Project, 1942–1946* (Berkeley: University of California Press, 1987), 44, 52–53; J. Samuel Walker, *Permissible Dose: A History of Radiation Protection in the Twentieth Century* (Berkeley: University of California Press, 2000), 9.

2. Walker, *Permissible Dose,* 7–8.

3. Stafford Warren, "Case of Leukemia in Mr. Donald H. Johnson," February 7, 1945, NAA, RG 326-8505, box 54, MD 700.2, "Enclosures."

4. Christopher Sellers, "Discovering Environmental Cancer: Wilhelm Hueper, Post–World War II Epidemiology, and the Vanishing Clinician's Eye," *American Journal of Public Health* 87, no. 11 (November 1997): 1824–35; Devra Lee Davis, *The Secret History of the War on Cancer* (New York: Basic Books, 2007), 97–102.

5. Robert Proctor, *Cancer Wars: How Politics Shapes What We Know and Don't Know About Cancer* (New York: Basic Books, 1995), 36–44.

6. Hacker, *The Dragon's Tail,* 53.

7. Greenewalt to Compton, April 2, 1943, reference in Stone to Compton, April 10, 1943, NAA, RG 326-8505, box 55, MD 729.3, "Radiation Book 1."

8. "Questions Asked of Dr. Chet Stern in Conference by Dr. Warren," June 24, 1943, NAA, RG 326-8505, box 12, I.E.2, "General Correspondence."

9. Peter Bacon Hales, *Atomic Spaces: Living on the Manhattan Project* (Urbana: University of Illinois Press, 1997), 284.

10. Hymer Friedell to Nichols, February 14, 1945, NAA, RG 326-8505, box 54, MD 700.2, "Essays and Lectures."

11. Bradley Seitz, "Manhattan District Health Program," January 8, 1944, 1944, NAA, RG 326-8508, box 54, MD 700.2.

12. Hales, *Atomic Spaces,* 281.

13. Nichols to DuPont Co., October 30, 1943, and Traynor to Williams (draft), October 30, 1943, NAA, RG, 326-8505, box 55, MD 729.3, "Radiation Book 1"; Compton to Warren, October 28, 1944, box 54, MD 700.2, "Essays and Lectures."

14. Davis, *The Secret History,* 21, 31.

15. R. E. Rowland, *Radium in Humans: A Review of U.S. Studies* (Argonne, IL: Argonne National Lab, 1994), 25; Eileen Welsome, *The Plutonium Files: America's Secret Medical Experiments in the Cold War* (New York: Dial Press, 1999), 49–50, 66. See also Ross M. Mullner, *Deadly Glow: The Radium Dial Worker Tragedy* (Washington, DC: American Public Health Association, 1999).

16. Robley Evans, "Protection of Radium Dial Workers and Radiologists from Injury by Radium," *Industrial Hygiene and Toxicology* 25, no. 7 (September 1943): 253–69.

17. Williams to Nichols, October 7, 1943, NAA, RG 326-8505, box 55, MD 729.1, "Radiation, Book 1."

18. H. M. Parker, "Status of Health and Protection at the Hanford Engineer Works," in *Industrial Medicine on the Plutonium Project* (New York: McGraw-Hill 1951), 476–84.

19. David Goldring, "Draft of Report," October 2, 1945, NAA, RG 326-87-6, box 15, "Miscellaneous."
20. Greenewalt to Nichols, April 14, 1943, and Greenewalt to Compton, April 14, 1943, NAA, RG 326-8505, box 54, MD 700.2 "Fish Research."
21. Crawford Greenewalt Diary, January 22, 1943, HML; William Sapper, "Conference with Ichthyologist," NAA, June 12, 1943, RG 326-8505, box 60, "Meetings and Conferences."
22. Williams, "Radioactivity Health Hazards—Hanford," June 26, 1944, NAA, RG 326-8505, box 55, MD 729.1, "Radiation, Book 1."
23. Nichols to Groves, September 8, 1944, NARA, RG 77 5, box 83, "General Correspondence 1942–1948."
24. Hamilton to Dr. Herman Hilberry, September 29, 1944, EOL, reel 43 (box 28), folder 40.
25. Welsome, *Plutonium Files*, 27, 29–30.
26. William Moss and Roger Eckhardt, "The Human Plutonium Injection Experiment," *Los Alamos Science* 23 (1995): 194.
27. J. G. Hamilton, "Review of Research upon the Metabolism of Long-life Fission Products October 1, 1942–April 30, 1943," July 13, 1943, EOL, reel 43 (box 28), folder 40.
28. Hales, *Atomic Spaces*, 291.
29. R. S. Stone to J. G. Hamilton, 1943, EOL, reel 43 (box 28), folder 40; Williams to Nichols, October 7, 1943; Marsden to Nichols, October 18, 1943; Traynor to Williams (draft), October 25, 1943.
30. Hamilton, "A Review"; Hamilton to Compton, July 28 and October 6, 1943; Hamilton to Stone, October 7, 1943; Hamilton, "A Brief Review of the Possible Applications of Fission Products in Offensive Warfare," May 27, 1943, EOL, reel 43 (box 28), folder 40.
31. H. J. Curtin to R. L. Doan, October 19, 1943, EOL, reel 43 (box 28), folder 40.
32. Ibid.
33. Hamilton, "Survey of Work Done by the 48-A Group at Berkeley," April 24, 1945, EOL, reel 43 (box 28), folder 41.
34. Hamilton, "A Review."
35. Williams, "Radioactivity Health Hazards—Hanford," June 26, 1944, NAA, RG 326-8505, box 55, MD 729.1, "Radiation, Book 1."
36. J. E. Wirth, "Medical Services of the Plutonium Project," in *Industrial Medicine on the Plutonium Project* (New York: McGraw-Hill, 1951), 32.
37. Greenewalt Diary, January 29, 1944, HML; Norwood to Stone, September 9, 1944, NAA, RG 326-8505, box 54, MD 700.2, "Medical Correspondence"; Compton to Hamilton, April 8, 1944, EOL, reel 43 (box 28), folder 40.
38. Stone to Norwood, October 25, 1944, NAA, RG 326-8505, box 54, MD 700.2, "Medical Correspondence."
39. Compton to Warren, October 28, 1944, NAA, RG 326-8505, box 54, MD 700.2, "Essays and Lectures."
40. Stone to Hamilton, December 13, 1943, EOL, reel 43, (box 28), folder 40.
41. Hamilton to Stone, January 4, 1944, EOL, reel 43, (box 28), folder 40.

Chapter 8: The Food Chain

1. "Radiation Hazards," September 1, 1944, NAA, RG 326-8505, box 55, MD 729.3.
2. Hamilton, "Decontamination Studies with the Products of Nuclear Fission," 1944; Hamilton, "Progress Report for March 1945," April 6, 1945, EOL, reel 43 (box 28), folders 40–41.
3. Stone to Hamilton, January 30, 45, EOL, reel 43 (box 28), folder 41; Finkel and Brues, "The Shift of Strontium 89 from the Mother to the Fetus and Young," NAA, RG 326-87-6, box 24, "Summary Medical Research Program." My thanks to Harry Winsor for help interpreting these documents.
4. Wirth, "Medical Services," 32.
5. Eileen Welsome, *The Plutonium Files: America's Secret Medical Experiments in the Cold War* (New York: Dial Press, 1999), 68, 79.

6. Fast neutrons are neutrons with velocities much greater than the speed of sound. They are produced with every fission, but in conventional reactors are moderated or slowed down. My thanks to Harry Winsor for this explanation.

7. T. C. Evans, "Project Report on Mice Exposed Daily to Fast Neutrons," July 18, 1945, NAA, RG-326-8505, box 54, MD 700.2, "Enclosures."

8. Stone to Norwood, "Exposures Exceeding Tolerance," October 25, 1945, NAA, RG 326-8505, box 54, MD 700.2, "Essays and Lectures."

9. "Chronic Radiation Program," NAA, RG 326-87-6, box 24, "Summary Medical Research Program."

10. On immune disorders, see Peter Bacon Hales, *Atomic Spaces: Living on the Manhattan Project* (Urbana: University of Illinois Press, 1997), 290.

11. Susan Lindee, *Suffering Made Real: American Science and the Survivors at Hiroshima* (Chicago: University of Chicago Press, 1994), 62.

12. Ibid., 59.

13. "Experiments to Test the Validity of the Linear R-Dose/Mutation Rate Relation at Low Dosage," n.d., NAA, RG 326-87-6, box 24, "Summary Medical Research Program."

14. Ibid. The report was probably written by Donald Charles, who worked during the war at the University of Rochester. See Lindee, *Suffering Made Real*, 65.

15. Herman Muller, "Time Bombing Our Descendants," *American Weekly*, November 1946.

16. Crawford Greenewalt Diary, January 22 and February 12, 1943, HML.

17. See Hales, *Atomic Spaces*, 144–48, for a fuller discussion of Hanford's meteorology studies.

18. FTM, February 21, 1944; Matthias to Groves, October 24, 1960, HML, acc. 2086, 20.92.

19. Jacobson and Overstreet to Stone, February 15, 1944, EOL, reel 43 (box 28), folder 40.

20. Hamilton to Hilberry, September 29, 1944, and Hamilton, "Progress Report for month of December 1944," January 4, 1945, EOL, reel 43 (box 28), folder 40.

21. Hamilton to Hilberry.

22. Michele Stenehjem Gerber, *On the Home Front: The Cold War Legacy of the Hanford Nuclear Site* (Lincoln: University of Nebraska Press, 1992), 147.

23. Ibid., 162; "Monthly Report, December 1956," DOE Opennet, HW-47657; "Release of Low-Level Aqueous Wastes," DOE Germantown, RG 326/1359/7, 6–7.

24. Jacobson and Overstreet to Stone, November 15, 1944; Hamilton to Stone, November 15, 1944, EOL, reel 43, (box 28), folder 40.

25. Jacobson and Overstreet, "Absorption and Fixation of Fission Products and Plutonium by Plants," June 1945, EOL, reel 43 (box 28), folder 40.

26. Stone to Hamilton, April 28, 1944, EOL, reel 43 (box 28), folders 41.

27. H. M. Parker to S. T. Cantril, July 10, 1954, PRR, HW-7-1973.

28. On the symbolic importance of salmon, see Richard White, *The Organic Machine* (New York: Hill and Wang, 1995).

29. As a classified report stated, "It is necessary to remove the government from any claim as to the injury to the large and lucrative salmon industry in the Columbia River." "Summary Medical Research Program," NAA, 326-87-6, box 24, "Fish Program."

30. Greenewalt to Nichols, April 14, 1943, NAA, RG 326-8505, box 54, MD 700.2, "Fish Research."

31. Untitled photos and "Hanford Thayer to Warren," August 19, 1944, NAA, RG 326-8505, box 54, MD 700.2, "Fish Research."

32. Hanford Thayer, "Fisheries Research Program," March 12, 1945, NAA, RG 326-8505, box 54, MD 700.2, "Fish Research"; "Fish Program," RG 326-87-6, box 24, "Summary Medical Research Program."

33. Hanford Thayer to Warren, July 18, 1944, NAA, RG 326-8505, box 54, MD 700.2, "Fish Research."

34. Within the retention basins, where effluent cooled before returning to the river, the radioactivity of the water measured from 1.8 to 2.4 rads in 1945. H. M. Parker to S. T. Cantril, September 11, 1945, PRR, HW 7-2346.

35. Hanford Thayer, "Site W Hazards to Migratory Fishes," May 22, 1943, NAA, RG 326-87-6, box 24, G-36.

36. Warren to Harry Wensal, OSRD, October 5, 1943, and C. L. Prosser and K. S. Cole, "Biological Research: Fish," 1944, NAA, RG 326 8505, MD 700.2, "Fish Research."

37. Richard Foster, "Weekly Report, 146 Building," October 14, 1945, NAA, RG 326-8505, box 54, MD 700.2, "Fish Research."

38. Gerber, *On the Home Front*, 117–8.

39. Another kill occurred October 11, 1945. Foster, "Weekly Report, 146 Building."

40. On classification of medical studies, see Russell B. Olwell, *At Work in the Atomic City: A Labor and Social History of Oak Ridge, Tennessee* (Knoxville: University of Tennessee Press, 2004), 119–20.

Chapter 9: Of Flies, Mice, and Men

1. Parker to Cantril, PRR, HW-7-1973; J. E. Wirth, "Medical Services of the Plutonium Project," in *Industrial Medicine on the Plutonium Project* (New York: McGraw-Hill, 1951), 20.

2. Roger Williams, "Radioactivity Health Hazards—Hanford," June 26, 1944, NAA, RG 326-8505, box 55, MD 729.1, "Radiation, Book 1."

3. Russell B. Olwell, *At Work in the Atomic City: A Labor and Social History of Oak Ridge, Tennessee* (Knoxville: University of Tennessee Press, 2004), 52–53.

4. Stone, "Exposures Exceeding Tolerance," October 25, 1945, NAA, RG. 326-8505, box 54, MD 700.2, "Essays and Lectures."

5. On these differences of opinion, see Hymer Friedell, "Comment on Tolerance Values for Radium and Product," May 11, 1945, EOL, reel 43 (box 28), folder 41.

6. Matthias, "Reports for Week Ending 29 April," May 12, 1944, NAA, RG 326-87-6, box 15, "Teletypes and Telegrams," and "Obstetrical and Gynecological Statistics from Discharged Patients," March, April, May, June (etc.) 1945, RG 326-8505, box 54, MD 701, "Medical Attendance."

7. R. S. Stone, "General Introduction," in *Industrial Medicine*, 14.

8. Foster, "Fish Life Observed in the Columbia River on September 27, 1945," NAA, RG 326-8505, box 54, MD 700.2, "Fish Research."

9. Lauren Donaldson, "Fisheries Inspection on the Columbia River in the Area Above Hanford, Washington, October 25 and 26, 1945," NAA, RG 326-8505, box 54, MD 700.2, "Fish Research."

10. Olwell, *At Work in the Atomic City*, 49–63.

11. Warren to Groves, "Report on Beri-Beri," NAA, RG 326-8505, box 54, MD 700.2, "Reports, Book 1."

12. Matthias to Warren, 1945, NAA, RG 326-87-6, box 16, "HEW Reports."

13. P. C. Leahy to District Engineer, January 7, 1946, NAA, RG 326-8505, box 54, MD 702, "Medical Examinations."

14. Robert Fink to Friedell, December 5 and December 27, 1945; Hamilton to Stone, July 7, 1946, NAA, RG 326-8505, box 54, MD 700.2, "Essays and Lectures"; Peter Bacon Hales, *Atomic Spaces: Living on the Manhattan Project* (Urbana: University of Illinois Press, 1997), 273–300.

15. Hamilton, "Progress Report for the Month of June and October 1945," EOL, reel 43 (box 28), folder 41.

16. Hales, *Atomic Spaces*, 284.

17. Testimony of Elmerine Whitfield Bell, granddaughter of Elmer Allen or "Cal-3," US Advisory Committee on Human Radiation Experiments, March 15, 1995, George Washington University, National Security Archive.

18. Stone to Norwood, October 25, 1944, NAA, RG 326-8505, box 54, MD 700.2, "Medical Correspondence."

19. W. B. Parsons, "Employment of Barbadians and Jamacians," November 23, 1944, NAA, RG 326 8505, box 103.

20. Williams to Groves, August 24, 1944, and Yancey to Matthias, August 1, 1944, NARA, RG 77, entry 5, box 41.

21. R. E. De Right to Roger Williams, August 25, 1944, NAA, RG 326 8505, box 55, MD 729.3, "Radiation Book 1."

22. Groves to Williams, August 26, 1944, Groves to Williams, September 7, 1944, Williams to Groves, August 30, 1944, and Warren to Groves, "Trial Evacuation of Site 'W,'" September 13, 1944, NARA, RG 77, entry 5, box 41; Friedell, "Comment on Tolerance Values."

23. Hymer Friedell to Morris E. Daily, May 10, 1944, and J. N. Tilley, "Richland Medical Plan," February 19, 1945, NAA, RG 326 8508, box 54, MD 700.2, "Specimens"; MD 701, "Medical Attendance."

24. R. L. Richards to Nichols, April 10, 1944, NAA, RG 326 8505, box 54, MD 701, "Medical Attendance."

25. H. M. Parker, "Report on Visit to Site W by G. Failla," July 10, 1945, PRR, HW-7-1973.

26. Groves to Naylor, September 24, 1945 NARA, RG 77, entry 5, box 83.

27. "Memorandum for the Chief, Military Intelligence, December 12, 1945," NARA, RG 77, box 85, folder "Goudsmit."

28. Williams, "Radioactivity Health Hazards—Hanford," June 26, 1944, NAA, RG 326 8505, box 55, MD 729.1, "Radiation, Book 1."

29. DuPont, "Monthly Report, May 1943," June 5, 1943, NAA, RG 326 8505, box 182, MD 319.1, "Reports—DuPont"; Groves to Nichols, telegram, November 16, 1943, Groves to Ackart, November 19, 1943, NAA, RG 326 8505, box 41, MD 600.1, "Construction and Installation"; DeRight to Nichols, December 14, 1943, RG 326 8505, box 46, MD 600.914, "Progress Reports HEW"; Greenewalt Diary, vol. 3, January 13, 1944, HML.

30. Williams, "Radioactivity Health Hazards—Hanford."

31. Hamilton to Compton, April 24, 1945, EOL, reel 43 (box 28), folder 41.

32. Joshua Silverman, "No Immediate Risk: Environmental Safety in Nuclear Weapons Production, 1942–1985," PhD diss., History Department, Carnegie Mellon University, 2000, 60.

33. H. M. Parker to S. T. Cantril, July 10, 1945, PRR, HW-7-1973.

34. H. M. Parker, "Radiation Exposure Data," February 8, 1950, PRR, HW-19404.

35. Silverman, "No Immediate Risk," 96.

36. Matthias to Groves, October 24, 1960, HML, acc. 2086, 20.92.

37. Parker, "Report on Visit to Site W."

38. The time required to remove half of the materials deposited in the body varied from a few seconds to two weeks depending on the isotope. R. S. Stone, "General Introduction," in *Industrial Medicine on the Plutonium Project* (New York: McGraw-Hill, 1951), 11.

39. Williams to Nichols, April 12, 1945 and Warren to Wirth, Norwood, and Groves, February 10, 1945, NAA, RG 326 8505, box 54, MD 700.2, "Enclosures." For other contested fatalities, see Fred A. Bryan, "Transmittal of Blood Smears," February 14, 1946, in the same file, and Olwell, *At Work in the Atomic City*, 118.

40. "Insurance Agreement Covering the Hanford Engineer Works," June 17, 1943, JPT, 5433-001, 11.

41. For assertions of the Manhattan Project's excellent medical record, see, for example, Wirth, "Medical Services of the Plutonium Project," 19–35.

Chapter 10: The Arrest of a Journal

1. N. I. Kuznetsova, "Atomnyi sled v VIET," in *Istoriia Sovetskogo atomnogo proekta: Dokumenty, vospominaniia, issledovaniia* (Moscow: Ianus-K, 1998), 64.

2. Yuli Khariton and Uri Smirnov, "The Khariton Version," *Bulletin of the Atomic Scientists*, May 1993, 22.

3. Kuznetsova, "Atomnyi sled," 62–81.

4. Alexandr Kolpakidi and Dmitrii Prokhorov, *Imperiia GRU: Ocherki istorii Rossiiskoi voennoi razvedki* (Moscow: Olma Press, 2001), 2:174.

5. Alexander Vassiliev, "Black Notebook #35," in *The Vassiliev Notebooks: Cold War International History Project Virtual Archive*, www.cwihp.org. See also Allen Weinstein and Alexander Vassiliev, *The Haunted Wood: Soviet Espionage in America—the Stalin Era* (New York: Random House, 1999), 37; John F. Fox Jr., "What the Spiders Did: U.S. and Soviet Counterintelligence Before the Cold War," *Journal of Cold War Studies* 11, no. 3 (Summer 2009): 206–24.

6. Alexander Vassiliev, "Yellow Notebook #4," in *The Vassiliev Notebooks*, 5–6; Michael R. Dohan, "The Economic Origins of Soviet Autarky 1927/28–1934," *Slavic Review* 35, no. 4 (December 1976): 603–35.

7. Weinstein and Vassiliev, *Haunted Wood*, 28.

8. Kolpakidi and Prokhorov, *Imperiia GRU*, 174.

9. Weinstein and Vassiliev, *Haunted Wood*, 67.

10. Max Holland, "I. F. Stone: Encounters with Soviet Intelligence," *Journal of Cold War Studies* 3 (Summer 2009): 159.

11. L. D. Riabev, *Atomnyi proekt SSSR: Dokumenty i materialy*, vol. I, bk. 1 (Moscow: Nauka, 1999), 22, and vol. II, bk. 6 (2006), 754–62.

12. Riabev, *Atomnyi proekt SSSR*, vol. I, bk. 1, 239–40.

13. Campbell Craig and Sergey Radchenko, *The Atomic Bomb and the Origins of the Cold War* (New Haven, CT: Yale University Press, 2008), 44.

14. Riabev, *Atomnyi proekt*, vol. I, bk. 1, 242–43.

15. Anatoli A. Iatskov, "Atom i razvedki," *Voprosi istorii estestvoznaniia i tekhniki* 3 (1992): 105.

16. Riabev, *Atomnyi proekt*, vol. I, bk. 1, 244–45.

17. V. Chikov, "Ot Los-Alamosa do Moskvy," *Soiuz* 22, no. 74 (May 1991): 18.

18. Riabev, *Atomnyi Proekt*, vol. I, bk. 2, 259.

19. G. N. Fleurov, a physicist, wrote Stalin about his suspicions of an A-bomb project in the West. B. V. Barkovskii, "Rol' razvedki v sozdanii iadernogo oruzhiia," in *Istoriia Sovetskogo atomnogo proekta: Dokumenty, vospominaniia, issledovaniia* (Moscow: Ianus-K, 1998), 87–134.

20. Riabev, *Atomnyi proekt*, vol. I, bk. 1, 265–66.

21. Vassiliev, "Yellow Notebook #1," 192.

22. Ibid.

23. E. A. Negin, *Sovetskii atomnyi proekt: Konets atomnoi monopolii* (Nizhnii Novgorod: Izd-vo Nizhnii Novgorod, 1995), 59.

24. Riabev, *Atomnyi proekt*, vol. I, bk. 1, 244–45.

25. Ibid., 276.

26. Campbell and Radchenko, *The Atomic Bomb*, 51; Jeffrey Richelson, *Spying on the Bomb: American Nuclear Intelligence from Nazi Germany to Iran and North Korea* (New York: Norton, 2006), 64.

27. Riabev, *Atomnyi proekt*, vol. I, bk. 1, 276–79, 363–64.

28. Negin, *Sovetskii atomnyi proekt*, 59.

29. Riabev, *Atomnyi proekt*, vol. I, bk. 1, 348–50.

30. Ibid., 276–79.

31. Ibid., 368–73.

32. J. Dallin, *Soviet Espionage* (New Haven, CT: Yale University Press, 1955), 457; Conant to Gromyko, July 16, 1942, NARA, RG 227 169, box 33, "B-2000 Russia."

33. Campbell and Radchenko, *Atomic Bomb*, 12; Kai Bird and Martin Sherwin, *American Prometheus: The Triumph and Tragedy of J. Robert Oppenheimer* (New York: Knopf, 2005), 164.

34. Vassiliev, "Yellow Notebook #4," 116–18.

35. Anonymous, "Diary of Visits to Germany, March-July, 1945 (private)," BPC, box 2, folder 7, 22; "Memorandum for the Chief," December 12, 1945, NARA, RG 77, box 85, "Goudsmit," 96; "Memorandum to the Chief of Staff," April 23, 1945, and John Lansdale, "Capture of Material," July 10, 1946, Hoover Institute, Robert Norris Papers, box 38. On Soviet knowledge of American actions in their zone, see *Atomnyi proekt*, vol. II, bk. 2, 339, and

Pavel Oleynikov, "German Scientists in the Soviet Atomic Project," *Nonproliferation Review* 7, no. 2 (2000): 4–5. On German scientists to the US, see Linda Hunt, *Secret Agenda: The United States Government, Nazi Scientists, and Project Paperclip, 1945 to 1990* (New York: St. Martin's Press, 1991), 20.

Chapter 11: The Gulag and the Bomb

1. Vladimir Gubarev, "Professor Angelina Gus'kova," *Nauka i zhizn'* 4 (2007): 18–26; E. A. Negin, *Sovetskii atomnyi proekt: Konets atomnoi monopolii* (Nizhnii Novgorod: Izd-vo Nizhnii Novgorod, 1995), 64.
2. G. A. Goncharov and L. D. Riabev, *O sozdanii pervoi otechestvennoi atomnoi bomby* (Sarov: RFIATS-VNIIEF, 2009), 44–45.
3. I. Afanas'ev and V. A. Kozlov, *Istoria Stalinskogo gulaga: Konets 1920-kh-pervaia polovina 1950-kh godov* (Moscow: ROSSPEN, 2004), 1:30.
4. V. P. Nasonov and B. L. Vannikov, *B. L. Vannikov: Memuary, vospominaniia, stat'i* (Moscow: TSNIIatominform, 1997), 89–90; Mikhail Vazhnov, *A. P. Zaveniagin: Stranitsy zhizni* (Moscow: PoliMEdia, 2002), 9–11.
5. Nasonov and Vannikov, *B. L. Vannikov*, 92. For an alternative version, see Negin, *Sovetskii atomnyi proekt*, 61–62.
6. Resolution no 9887 was passed on August 20, 1945. Abram Isaakovich Ioirysh, *Sovetskii atomnyi proekt: Sudby, dokumenty, sversheniia* (Moscow: IUNITI-DANA, 2008), 187.
7. Arkadii Kruglov, *Kak sozdavalas atomnaia promyshlennost' v SSSR* (Moscow: TSNIIatominform, 1994), 54; Michael Gordin, *Red Cloud at Dawn: Truman, Stalin, and the End of the Atomic Monopoly* (New York: Farrar, Straus and Giroux, 2009), 85, 99.
8. V. Vachaeva, *A. P. Zaveniagin: K 100-letiiu so dnia rozhdeniia* (Noril'sk: Nikel, 2001), 25–26; Vazhnov, *A. P. Zaveniagin*, 6.
9. Career NKVD officers included Sergei Kruglov, Victor Abakumov, Vasilii Chernyshev, and Pavel Meshik. A. Volkov, "Problema no. 1," *Istoriia otechestvennykh spetssluzhb*, http://shieldandsword.mozohin.ru/index.html.
10. O. V. Khlevniuk, *The History of the Gulag: From Collectivization to the Great Terror* (New Haven, CT: Yale University Press, 2004), 182.
11. A. B. Suslov, *Spetskontingent v Permskoi oblasti, 1929–1953 gg* (Ekaterinburg: Ural'skii gos. universitet, 2003), 118–21, 125.
12. Oleg Khlevniuk, "Introduction," in *Istoriia stalinskogo Gulaga*, 3:46–47.
13. V. A. Kozlov and O. V. Lavinskaia in *Istoriia stalinskogo Gulaga*, 6:59–64.

Chapter 12: The Bronze Age Atom

1. V. Chernikov, *Osoboe pokolenoe* (Cheliabinsk: V. Chernikov, 2003), 19.
2. L. D. Riabev, *Atomnyi proekt SSSR: Dokumenty i materialy*, vol. I, bk. 1 (Moscow: Nauka 1999), 46.
3. Paul R. Josephson, *Red Atom: Russia's Nuclear Power Program from Stalin to Today* (New York: W. H. Freeman, 2000), 89.
4. O. V. Khlevniuk, *The History of the Gulag: From Collectivization to the Great Terror* (New Haven, CT: Yale University Press, 2004), 35.
5. A. P. Finadeev, *Togda byla voina, 1941–1945: Sbornik dokumentov i materialov* (Cheliabinsk: n.p., 2005), 65.
6. Wilson T. Bell, "The Gulag and Soviet Society in Western Siberia, 1929–1953," PhD diss., University of Toronto, 2011, 246, 306.
7. Riabev, *Atomnyi proekt*, vol. II, bk. 2, 354–55, 358; V. Chernikov, *Za zavesoi sekretnosti ili stroitel'stvo No. 859* (Ozersk: V. Chernikov, 1995), 17.
8. Mark Bassin, "Russian Between Europe and Asia: The Ideological Construction of Geographical Space," *Slavic Review* 50, no. 1 (Spring 1991): 1–17.

9. Rapoport, "Prikaz," January 14, 1946, and February 12, 1946, OGAChO, 1619/2c/43, 2, 3.
10. Chernikov, *Za zavesoi*, 17.
11. Ibid., 8.
12. Donald A. Filtzer, *Soviet Workers and Late Stalinism: Labour and the Restoration of the Stalinist System After World War II* (Cambridge: Cambridge University Press, 2002), 22.
13. V. N. Kuznetsov, *Zakrytye goroda Urala* (Ekaterinburg: Akademiia voenno-istoricheskikh nauk, 2008), 86–87.
14. Boris Khavkin, "Nemetskie voennoplennye v SSSR i Sovetskie voennoplennye v Germanii," *Forum noveishei vostochnoevropeiskoi istorii i kul'tury* no. 1 (2006): 2.
15. Chernikov, *Za zavesoi*, 25.
16. V. Novoselov and V. S. Tolstikov, *Taina "Sorokovki"* (Ekaterinburg: Ural'skii rabochii, 1995), 65.
17. Efim P. Slavskii, "Kogda strana stoila na plechakh iadernykh titanov," *Voenno-istoricheskii zhurnal* 9 (1993): 13–23.
18. "Vypolnenie proizvodstvennogo plana" (1949), OGAChO, 1619/1/363, 1.
19. As of January 1, 1947, 9,000 prisoners were on site; the ChMS NKVD enterprise wanted to send in more prisoners but could not for lack of housing. Four months later, the ChMS NKVD enterprise had 13,000 prisoners, 7,000 special settlers, and 8,200 POWs, only 47 percent of whom were healthy enough to work. See Kazverov to A. N. Komorovskii, 1946, OGAChO, R-1619/2/48, 46–59 and 80–91.
20. Rapoport, "O resul'tatakh [*sic*] proverki lagernogo uchastka," March 15, 1946, OGAChO, 1619/2/44, 42–43; Rapoport, "Rasporiazhenie," February 28, 1946, OGAChO, 1619/2/43, 36.
21. Kuznetsov, *Zakrytye goroda*, 90.
22. Rapoport, "Vsem nachal'nikam podrazdelenii ChMS i SY 859," September 16, 1946, OGAChO, R-1619/2/51, 6–8.
23. Kazverov to A. N. Komarovskii, 1946, OGAChO, R-1619/2/48, 46–59.
24. Richard Rhodes, *Dark Sun* (New York: Simon and Schuster, 1995), 276.
25. Mikhailov nachal'nikam laguchastkov ChMS MVD, January 18, 1947, OGAChO, R-1619/2/51, 5–6.
26. Chernikov, *Za zavesoi*, 84.
27. Kazverov to Komarovskii, 46–59.
28. Zakharov Beloborodovu, 1949, OGAChO, 288/42/33, 4–15.
29. Rapoport, "O merakh uvelicheniia potoka posylok," October 31, 1946; Liutkevich Rapoportu, December 11, 1946, and Divbunov, "Po prevlecheniiu posylok-peredach," December 11, 1946, OGAChO, R-1619/2/45, 25–26, 31–32.
30. Filtzer, *Soviet Workers*, 41–43; S. Kruglov, "O razdelenii stroitel'stva no. 859," October 11, 1946, OGAChO, 161/2/41, 10–14.

Chapter 13: Keeping Secrets

1. V. N. Novoselov and V. S. Tolstikov, *Atomnyi sled na Urale* (Cheliabinsk: Rifei, 1997); N. V. Mel'nikova, *Fenomen zakrytogo atomnogo goroda* (Ekaterinburg: Bank kul'turnoi informatsii, 2006); Vladimir Gubarev, *Belyi arkhipelag Stalina* (Moscow: Molodaia gvardiia, 2004).
2. Rapoport, "Prikaz," April 13, 1946, OGAChO, 1619/2c/43, 42–43; Rapoport, "Ob organizatsii otpravki rabsily spetsposelentsev v SU-859," July 16, 1946, OGAChO, 1619/2/43, 66.
3. A. B. Suslov, *Spetskontingent v Permskoi oblasti, 1929–1953 gg* (Ekaterinburg: Ural'skii gos. universitet, 2003), 130.
4. Kazverov, "Dokladnaia zapiska za IV kvartal 1946 goda" and "Dokladnaia zapiska za 1 kvartal 1947 goda," OGAChO, R-1619/2/48, 46–59, 80; Saprikin, "V sviazi s postupleniem novogo spets. kontingenta," June 17, 1946, 1619/2/43, 66–67; Rapoport, "Ob organizatsii

laguchastka no. 9 pri Stroiupravlenii no. 859, ChMS MVD," May 27, 1946, 1619/2/43, 63–64. On dangerous prisoners, see V. N. Kuznetsov, *Zakrytye goroda Urala* (Ekaterinburg: Akademiia voenno-istoricheskikh nauk, 2008), 61.

5. Rapoport, "Prikaz o sniatii s ucheta," July 26, 1946, OGAChO, 1619/23/48, 86–87; Rapoport, "Vsem nachal'nikam," 6–8. See also Lynne Viola, *The Unknown Gulag: The Lost World of Stalin's Special Settlements* (New York: Oxford University Press, 2007), 95.

6. Kazverov to Komarovskii, 1946, OGAChO, R-1619/2/48, 46–59.

7. Rapoport, "Prikaz ob organizatsii shtrafnoi kolonnii zakliuchennykh," February 26, 1946, OGAChO, 1619/2/434, 27; Rapoport, "O meropriiatiiakh dal'neishego usileniia okhrany," April 22, 1946, OGAChO, 1619/2/44, 54–57.

8. Kazverov, "Dokladnaia zapiska," 1947, 46–59.

9. "Mikhailov nachal'nikam laguchastkov ChMS, MVD," January 18, 1947, OGAChO, R-1619/2/51, 5–6.

10. V. Chernikov, *Za zavesoi sekretnosti ili stroitel'stvo No. 859* (Ozersk: V. Chernikov, 1995), 145.

11. Riabev, *Atomnyi proekt SSSR*, vol. II, bk. 4 (Moscow: Fizmatlit, 2004), 198.

12. Chernikov, *Za zavesoi*, 145.

13. Ibid., 130, 145.

14. Alexandr Isaevich Solzhenitsyn, *One Day in the Life of Ivan Denisovich* (New York: Signet Classics, 2008).

15. Rapoport, "O resul'tatakh [*sic*] proverki," March 15, 1946, OGAChO, 1619/2/44: 42–43; "Prikaz po upravleniiu Cheliabmetallurgstroiia MVD SSSR," April 5, 1946, OGAChO, 1619/1/39, 256; "O merakh usileniia rezhima," April 947, OGAChO, R-1619/2/50, 53–54.

16. Kuznetsov, *Zakrytye goroda*, 13.

17. V. Novoselov and V. S. Tolstikov, *Taina "Sorokovki"* (Ekaterinburg: Ural'skii rabochii, 1995), 124.

18. Vladyslav B. Larin, *Kombinat "Maiak"—Problema na veka* (Moscow: KMK Scientific Press, 2001), 199; "Protokol no. 1, politotdela Bazy no. 10," January 5, 1949, OGAChO, P-1137/1/15, 1–5.

19. "Dokladnaia zapiska," no earlier than November 1946, OGAChO, P-288/1/141, 12.

20. Ibid., 13.

21. Ibid.

22. Rapoport, "O vvedenii vremennogo propusknogo rezhima na stroitel'stve MVD SSSR no. 859," July 23, 1946, OGAChO, R-1619/2/44, 79–80; I. P. Zemlin and I Gashev, *Desant polkovnika P. T. Bystrova* (Ozersk: Po Maiak, 1999), 16. Regulations for special military regime zones were first created in 1934. Irina Bystrova, *Voenno-promyshlennyi kompleks SSSR v gody kholodnoi voiny: Vtoraia polovina 40-kh-nachalo 60-kh godov* (Moscow: IRI RAN, 2000), 16.

23. Chernikov, *Za zavesoi*, 39; Rapoport, "Rasporiazheniia po Cheliabmetallurgstroiu MVD SSSR," October 17, 1946, OGAChO, 1619/1/39, 300.

24. Novoselov and, Tolstikov *Taina "Sorokovki,"* 126; Ia. P. Dokuchaev, "Ot plutoniia k plutonievoi bombe," in *Istoriia Sovetskogo atomnogo proekta: Dokumenty, vospominaniia, issledovaniia* (Moscow: IAnus-K, 1998), 279–312.

25. David Holloway, *Stalin and the Bomb: The Soviet Union and Atomic Energy, 1939–1956* (New Haven, CT: Yale University Press, 1994), 185.

26. Rapoport, "Prikaz nachal'nikam," September 19, 1946, OGAChO, 1619/1/39, 146.

27. D. Antonov Beloborodovu, August 21, 1949, OGAChO, 288/42/35; Rapoport, "O zapreshchenii zakupki produktov," September 5, 1946, OGAChO, 1619/2/43, 79–80.

28. Rapoport, "Rasporiazhenie," February 28, 1946, and "Ob organizatsii 7 laguchastka," August 1, 1946, OGAChO, 1619/2/43, 36, 75. On lack of an urban plan, "Protokol no. 3 partiinogo sobraniia partorganizatsii UKSa, Bazy no. 10," June 3, 1949, OGAChO, P-1167/1/4, 35–39.

29. "Protokol no. 3, zakrytogo partiinogo sobraniia partorganizatsii," January 27, 1948, OGAChO, 1142/1/4, 1–7.

Chapter 14: Beria's Visit

1. E. A. Negin, *Sovetskii atomnyi proekt: Konets atomnoi monopolii* (Nizhnii Novgorod: Izd-vo Nizhnii Novgorod, 1995), 67.
2. Michael Gordin, *Red Cloud at Dawn: Truman, Stalin, and the End of the Atomic Monopoly* (New York: Farrar, Straus and Giroux, 2009), 153.
3. Iu. I. Krivonosov, "Okolo atomnogo proekta," in *Istoriia Sovetskogo atomnogo proekta: Dokumenty, vospominaniia, issledovaniia* (Moscow: IAnus-K, 1998), 354.
4. N. V. Mel'nikova, *Fenomen zakrytogo atomnogo goroda* (Ekaterinburg: Bank kul'turnoi informatsii, 2006), 26.
5. Ibid., 24.
6. A. V. Fateev, *Obraz vraga v Sovetskoi propaganda, 1945–1954 gg.* (Moscow: RAN, 1999), 70.
7. Ibid., 63.
8. Vladislav Zubok, "Stalin and the Nuclear Age," in *Cold War Statesmen Confront the Bomb* (New York: Oxford University Press, 1999), 58.
9. L. D. Riabev, *Atomnyi proekt SSSR: Dokumenty i materialy*, vol. II, bk. 6 (Moscow: Nauka, 2006), 236–37, 246–47, 248, 302, 350–52.
10. Riabev, *Atomnyi proekt*, vol. II, bk. 3 (Moscow: Nauka, 2002), 128, 199, 214.
11. "Mikhailov nachal'nikam laguchastkov ChMS MVD," January 18, 1947, OGAChO, R-1619/2/51, 5–6; Riabev, *Atomnyi proekt*, vol. 1, bk. 1, 195.
12. Kruglov, "O razdelenii stroitel'stva no. 859," October 11, 1946, OGAChO, 1619/2/41, 10–14.
13. V. Chernikov, *Za zavesoi sekretnosti ili stroitel'stvo No. 859* (Ozersk: V. Chernikov, 1995), 44–45.
14. V. Novoselov and V. S. Tolstikov, *Taina "Sorokovki"* (Ekaterinburg: Ural'skii rabochii, 1995), 132–33, 142.
15. Riabev, *Atomnyi proekt*, vol. II, bk. 2, 488–89; bk. II, vol. 3, 199, 260–61.
16. Francis Sill, "Manhattan Project: Its Scientists Have Harnessed Nature's Basic Force," *Life*, August 20, 1945.
17. Novoselov and Tolstikov, *Taina Sorokovki*, 132.
18. Gordin, *Red Cloud at Dawn*, 82.
19. "O narusheniiakh zemlepol'zovaniia kolkhozov," August 24, 1946, OGAChO, 274/20/10, 38; "O stroitel'stve baraka," August 8, 1946, OGAChO, R 274/20/10, 34.
20. "Chertezh zemel'nykh uchastkov," April 5, 1947, and "Soveshchania u nachal'nika stroitel'stva no 859," May 7, 1947, OGAChO, R 274/20/18, 120–22.
21. V. A. Kozlov, *Massovye besporiadki v SSSR pri Khrushcheve i Brezhneve: 1953—nachalo 1980-kh gg* (Moscow: Rosspen, 2010); Iu. N. Afanas'ev et al., eds., *Istoriia stalinskogo Gulaga*, vol. 7.
22. Anatoli A. Iatskov, "Atom i razvedki," *Voprosi istorii estestvoznaniia i tekhniki* 3 (1992): 103–32; B. V. Barkovskii, "Rol' razvedki v sozdanii iadernogo oruzhiia," in *Istoriia Sovetskogo atomnogo proekta*, 87–134.
23. David Holloway, *Stalin and the Bomb: The Soviet Union and Atomic Energy, 1939–1956* (New Haven, CT: Yale University Press, 1994), 56, 185.
24. Alexander Vassiliev, "Yellow Notebook #1," in *The Vassiliev Notebooks: Cold War International History Project Virtual Archive*, www.cwihp.org, 23, 39, 146; Vizgin, *Istoriia Sovetskogo atomnogo proekta*, 120.
25. Vassiliev, "Yellow Notebook #1," 287, 79.
26. Weinstein and Vassiliev, *Haunted Wood*, 208.
27. Novoselov and Tolstikov, *Taina "Sorokovki,"* 137.

28. Riabev, *Atomnyi proekt*, vol. 1, bk. 1, 188–89; vol. II, bk. 3, 203–7.
29. Kotkin, *Magnetic Mountain: Stalinism as a Civilization* (Berkeley: University of California Press, 1995); Katherine A. S. Siegel, *Loans and Legitimacy: The Evolution of Soviet-American Relations, 1919–1933* (Lexington: University Press of Kentucky, 1996), 128–30; Lennart Samuelson, *Plans for Stalin's War Machine: Tukhachevskii and Military-Economic Planning, 1925–1941* (New York: St. Martin's, 2000), 16.
30. James R. Harris, *The Great Urals: Regionalism and the Evolution of the Soviet System* (Ithaca, NY: Cornell University Press, 1999), 156, 167.
31. L. P. Sokhina, *Plutonii v devich'ikh rukakh: Dokumental'naia povest' o rabote khimiko-metallurgicheskogo plutonievogo tsekha v period ego stanovleniia, 1949–1950 gg* (Ekaterinburg: Litur, 2003), 32.
32. Chernikov, *Za zavesoi*, 48–49.
33. Leonid Timonin, *Pis'ma iz zony: Atomnyi vek v sud'bakh tol'iattintsev* (Samara: Samarskoe knizhnoe izd-vo, 2006), 12; Nikolai Rabotnov, "Publitsitsika—Sorokovka," *Znamia*, July 1, 2000, 162; author interview with Natalia Manzurova, August 11, 2009, Cheliabinsk.
34. Riabev, *Atomnyi proekt*, vol. II, bk. 3, 316–18.
35. Timonin, *Pis'ma iz zony*, 10.
36. Novoselov and Tolstikov, *Taina "Sorokovki,"* 140.
37. "Protokol no. 3, zakrytogo partiinogo sobraniia partorganizatsii Zavodoupravelniia," January 27, 1948, OGAChO, 1142/1/4, 1–7.
38. V. N. Kuznetsov, *Atomnyi proekt za koliuchei provolokoi* (Ekaterinburg: Poligrafist, 2004), 24.

Chapter 15: Reporting for Duty

1. N. V. Mel'nikova, *Fenomen zakrytogo atomnogo goroda* (Ekaterinburg: Bank kul'turnoi informatsii, 2006), 37.
2. L. D. Riabev, *Atomnyi proekt SSSR: Dokumenty i materialy*, vol. 1, bk. 1, (Moscow: Nauka, 1999), 176–77, 226–28, 250–52; B. Muzrukov, "Ob otkomandirovanii na Bazu-10," April 24, 1950, OGAChO, 288/42/40, 32–33.
3. V. Chernikov, *Za zavesoi sekretnosti ili stroitel'stvo No. 859* (Ozersk: V. Chernikov, 1995), 57–59. Saranskii exaggerated the completeness of the background checks. For a corrective, see "Zasedanie partiinogo aktiva," January 8, 1953, OGAChO, P-1137/1/48, 84–85.
4. N. I. Ivanov, *Plutonii, A Bochvar, Kombinat "Maiak"* (Moscow: VNII neorganicheskikh materialov, 2003), 8.
5. V. Novoselov and V. S. Tolstikov, *Taina "Sorokovki"* (Ekaterinburg: Ural'skii rabochii, 1995), 125.
6. Vladimir Gubarev, "Professor Angelina Gus'kova," *Nauka i zhizn'* 4 (2007): 18–26.
7. Nikolai Rabotnov, "Publitsistika—Sorokovka," *Znamia*, July 1, 2000, 162.
8. Author interview with Natalia Manzurova, August 11, 2009, Cheliabinsk.
9. For evidence of travel outside the zone, see "Sobranie partiinogo aktiva politotdela bazy-10," April 19, 1951, OGAChO, P-1137/1/31, 68–70, and Fokin, "Ob'iasnenie po delu o kraze moikh dokumentov," July 25, 1952, OGAChO, 288/42/51, 96–97.
10. "O neotlozhnykh meropriiatiakh," August 1946, OGAChO, 1619/2/41, 9; Riabev, *Atomnyi proekt*, vol. 1, bk. 1, 203–7, 210, 282, 320–21.
11. Sekretariu Cheliabinskogo Obkoma VKP/b F. N. Dadonovu, July 19, 1948, OGAChO, 288/43/30, 4.
12. "Postanovlenie Politotdela Bazy no. 10," June 29, 1949, OGAChO, P-1167/1/15, 76–81; Beloborodov, April 7, 1948, and "Uralets, nachal'nik ob'ekta B, MVD Matveistevy," April 24, 1948, OGAChO, 288/42/29, 2–10.
13. Meshik Beloborodovu, June 7, 1949, OGAChO, 288/42/34, 16; Riabev, *Atomnyi Proekt*, vol. II, bk. 3, 324–25 and bk. 4, 198.
14. Riabev, *Atomnyi proekt SSSR*, vol. 1, bk. 1, 250–52.

15. Novoselov and Tolstikov *Taina "Sorokovki,"* 100.

16. Chernikov, *Za zavesoi,* 77.

17. Dol'nik Beloborodovu, May 27, 1949, and Likhachev Beloborodovu, June 7, 1949, OGAChO, 288/42/35.

18. "Ob'iasnenie," 1946, OGAChO, 1619/1/161, 23, 44.

19. Gubarev, "Professor Angelina Gus'kov," 18–25.

20. Novoselov and Tolstikov, *Taina "Sorokovki,"* 143–46.

21. "Stenogramma sobraniia partiinogo aktiva, politotdela no. 106," January 30, 1952, OGAChO, P-1137/1/38, 67–69.

22. Riabev, *Atomnyi proekt,* vol. 1, bk. 1, 206.

23. Mel'nikova, *Fenomen,* 51–54.

24. Chernikov, *Za zavesoi,* 90.

25. Arkadii Kruglov, *Kak sozdavalas atomnaia promyshlennost' v SSSR* (Moscow: TSNIIatominform, 1994), 66.

26. Novoselov and Tolstikov, *Taina "Sorokovki,"* 107.

27. Riabev, *Atomnyi proekt,* vol. II, bk. 4, 431, 633–37.

28. Chernikov, *Za zavesoi,* 131.

29. Ibid., 116, 96; "Ob usilenii bditel'nosti i rezhima sekretnosti," March 26, 1953, OGAChO, P-1137/1/48, 79–85; "Protokol #8, Biuro Ozerskogo Gorkom," October 2, 1956, OGA-ChO, 2469/1/4, 1–12. On cleansing the larger Soviet nuclear bomb installations of Jews, see Mikhail Vazhnov, *A. P. Zaveniagin: Stranitsy zhizni* (Moscow: PoliMEdia, 2002), 95–97.

30. Chernikov, *Za zavesoi,* 57.

31. "Protokol no. 32, zasedeniia biuro Ozerskogo Gorkoma," September 24, 1957, OGAChO, 2469/1/121, 173.

32. "Spisok kandidatov i deputatov," 1957, OGAChO, 2469/1/120.

33. Novoselov and Tolstikov, *Taina "Sorokovki,"* 100–101.

34. Chernikov, *Za zavesoi,* 116.

35. V. V. Alexeev, ed., *Obshchestvo i vlast': 1917–1985,* vol. 2: *1946–1985* (Cheliabinsk: UrO RAN, 2006), 93; L. P. Kosheleva, E. Iu Zubkova, and G. A. Kuznetsova, *Sovetskaia zhizn,' 1945–1953* (Moscow: Rosspen, 2003), 198.

36. "O vyselenii iz osobo-rezhimnoi zony," February 8, 1948; "Ob otselenii grazhdan iz rezhimnoi zony ob'ekta no. 859" (1948), OGAChO, 23/1/22, 4–5, and "Tkachenko Beloborodovu," March 2, 1948, OGAChO, 288/43/21, 1–2.

37. Sh. Khakimov, *Neizvestnaia deportatsiia* (Cheliabinsk: Kniga, 2006), 16. For petitions asking to return, see OGAChO, R-274/20/30, 50–53, 66–67, 78, and 87.

38. Tkachenko T. Smorodinskomu, OGAChO, 288/42/34, 5–6.

39. "Ganichkin Belobordovu," April 4, 1946, OGAChO, 288/42/34, 7.

40. "Protokol no. 9, zakrytogo bioro Kyshtymksogo Gorkoma," May 15, 1948, OGAChO, 288/42/34, 6–9.

41. David R. Shearer, *Policing Stalin's Socialism: Repression and Social Order in the Soviet Union, 1924–1953* (New Haven, CT: Yale University Press, 2009); Gijs Kessler, "The Passport System and State Control over Population Flows in the Soviet Union, 1932–1940," *Cahiers du monde russe* 42, nos. 2–4 (April–December 2001): 478–504.

42. On tense relations between Kyshtym and Ozersk, see Viktor Riskin, "'Aborigeny' atomnogo anklava," *Cheliabinskii rabochii,* April 15, 2004.

Chapter 16: Empire of Calamity

1. L. D. Riabev, *Atomnyi proekt SSSR: Dokumenty i materialy,* vol. II, bk. 2 (Moscow: Nauka, 2000), 83–85.

2. Ibid., 451–56.

3. Paul R. Josephson, *Red Atom: Russia's Nuclear Power Program from Stalin to Today* (New York: W. H. Freeman, 2000), 88–90.

4. Vladyslav B. Larin, *Kombinat "Maiak"—Problema na veka* (Moscow: KMK Scientific Press, 2001), 77.

5. Vladimir Gubarev, *Belyi arkhipelag Stalina* (Moscow: Molodaia gvardiia, 2004), 302–3.

6. Riabev, *Atomnyi proekt*, vol. II, bk. 4, 459–60.

7. Ibid., 461–62; V. Novoselov and V. S. Tolstikov, *Taina "Sorokovki"* (Ekaterinburg: Ural'skii rabochii, 1995), 149–53.

8. Sergei Parfenov, "Kaskad zamedlennogo deistviia," *Ural* 8, no. 3 (2006).

9. Alexei Mitiunin, "Natsional'nye osobennosti likvidatsii radiatsionnoi avarii," *Nezavisimaia gazeta*, April 15, 2005.

10. Parfenov, "Kaskad."

11. Ia. P. Dokuchaev, "Ot plutoniia k plutonievoi bombe," in *Istoriia Sovetskogo Atomnogo Proekta: Dokumenty, Vospominaniia, Issledovaniia* (Moscow: IAnus-K, 1998): 291.

12. Larin, *Kombinat*, 27–28.

13. Ibid., 87–88.

14. Vladimir Gubarev, "Glavnii ob'ekt derzhavy: po stranitsam 'Atomnogo proekta SSSR,'" *Vsiakaia vsiachina: bibliotechka raznykh statei*, May 2010, http://wsyachina.com.

15. Riabev, *Atomnyi proekt*, vol. II, bk. 4, 425; L. P. Sokhina, *Plutonii v devich'ikh rukakh: Dokumental'naia povest' o rabote khimiko-metallurgicheskogo plutonievogo tsekha v period ego stanovleniia, 1949–1950 gg* (Ekaterinburg: Litur, 2003), 40–42. On late recognition of internal ingestion, see. A. K. Gus'kova, *Atomnaia otrasl' strany glazami vracha* (Moscow: Real'noe vremia, 2004), 101.

16. Nikolai Rabotnov, "Publitsitsika—Sorokovka," *Znamia*, July 1. 2000, 165.

17. L. P. Sokhina, "Trudnosti puskogovo perioda," in *Nauka i obshchestvo, istoriia Sovetskogo atomnogo proekta (40e–50-e gody)* (Moscow: Izdat, 1997), 138; Novoselov and Tolstikov, *Taina "Sorokovki,"* 160.

18. Larin, *Kombinat*, 83.

19. Riabev, *Atomnyi proekt*, vol. II, bk. 4, 338–39.

20. Vladyslav B. Larin, "Mayak's Walking Wounded," *Bulletin of the Atomic Scientists*, September/October 1999, 23.

21. Larin, *Kombinat*, 85–87.

22. Ibid.

23. Dokuchaev, "Ot plutoniia," 291.

24. Larin, *Kombinat*, 86.

25. Ibid., 87.

26. "Zasedanie partiinogo aktiva," July 6, 1951, OGAChO, P-1137/1/31, 162–68.

27. Larin, *Kombinat*, 47.

28. Sokhina, "Trudnosti," 139–40.

29. Larin, *Kombinat*, 113.

30. Mikhail Gladyshev, *Plutonii dlia atomnoi bomby: Direktor plutonievogo zavoda delitsia vospominaniiami* (Cheliabinsk: n.p., 1992), 6.

31. Sokhina, *Plutonii*, 97.

32. Ibid., 71–74.

33. Larin, "Mayak's Walking Wounded," 22, 24.

34. Sokhina, "Trudnosti," 144.

35. Sokhina, *Plutonii*, 92–93.

36. Novoselov and Tolstikov, *Taina "Sorokovki,"* 148–49.

37. Larin, *Kombinat*, 26.

38. Sokhina, *Plutonii*, 37–38.

39. V. N. Novoselov and V. S. Tolstikov, *Atomnyi sled na Urale* (Cheliabinsk: Rifei, 1997), 148.

Chapter 17: "A Few Good Men" in Pursuit of America's Permanent War Economy

1. As quoted in Paul John Deutschmann, "Federal City: A Study of the Administration of Richland," M.A. thesis, University of Oregon, 1952, 20.
2. "Memorandum to the File," April 24, 1946, PRR, HAN 73214, Bk.-17.
3. "Transcript of Press Conference," April 18, 1949, HMJ, 58/3; Walter Williams to Carroll Wilson, October 7, 1947, NARA, RG 326 67A, box 71, 600.1 (HOO); Rodney P. Carlisle with Joan M. Zenzen, *Supplying the Nuclear Arsenal: American Production Reactors, 1942–1992* (Baltimore: Johns Hopkins University Press, 1996), 56–57.
4. On the government monopoly, see "Terms of Reference for Security Survey Panel," 1950, NARA, RG 326 8505, box 6, "Security and Intelligence, 1947–1963."
5. Carlisle, *Supplying*, 162.
6. "Managers' Data Book," JPT, 5433-001, box 25; J. Gordon Turnbull, *Master Plan of Richland, Washington* (1948), 56.
7. Author interview with Ralph Myrick, August 19, 2008, Kennewick, WA; Richland Community Council Minutes, meeting #21, June 20, 1949, RPL.
8. Charles Edward Wilson is not to be confused with Charles Erwin Wilson, CEO of GM and later the secretary of defense under Eisenhower.
9. Charles E. Wilson, "For the Common Defense: A Plea for a Continuing Program of Industrial Preparedness," *Army Ordnance* XXVI, no. 143 (March-April 1944): 285–287.
10. Ibid., 287.
11. Garry Wills, *Reagan's America: Innocents at Home* (New York: Doubleday, 1987), 281.
12. W.F. Tompkins to V. Bush, March 28 and May 9, 1944; Bush to Tompkins, April 27, 1944; Colonel Rising, "Memorandum," July 24, 1944, NARA, OSRD General Records, RG 227, entry 13, box 36.
13. "Research Board for National Security" (Draft), July 24, 1944, NARA, RG 227, 13, box 36, 2; Charles Vanden Bulck, "Remarks for Presentation at the Atomic Industrial Forum Symposium," March 3, 1958, NARA, RG 326 A1 67A, box 8, folder 10.
14. M. A. Tuve, "Suggested Pattern for Stable Contracting Agencies," September 12, 1944, and Tuve to Bush, September 29, 1944, NARA, RG 227 13, box 36.
15. Tuve, "Suggested Pattern."
16. Ibid., 12.
17. GE's contracts with the Atomic Energy Commission eventually totaled $4.2 billion from 1946 to 1975. INFACT, *Bringing GE to Light: How General Electric Shapes Nuclear Weapons Policies for Profits* (Philadelphia: New Society Publishers, 1988), 97–98.
18. See Peter Galison and Bruce William Hevly, *Big Science: The Growth of Large-Scale Research* (Stanford, CA: Stanford University Press, 1992).
19. Caroll L. Wilson to Robert Oppenheimer, February 10, 1948, NARA, RG 326 1A 67A, box 71, 600.1 (HOO); Richard Rhodes, *Dark Sun* (New York: Simon and Schuster, 1995), 280.
20. "Directive for Type B Production Unit Areas," November 6, 1947; "Kellex Contract with General Electric," September 23, 1947, and Fred C. Schlemmer to Carroll Wilson, July 18, 1947, NARA, RG 326 67A, box 71, 600.1 (HOO).
21. Walter J. Williams, October 8, 1948, JPT, 5433-1, box 24; "Excerpts from Delbert Meyer's Thesis on History of Tri-Cities" (1959), CREHST, Acc. 2006.1 box 2, folder 6.1.
22. *Villager*, October 21, October 28, and November 4, 1948.
23. Schlemmer to Wilson, July 14, 1947, NARA, RG 326 67A, box 71, 600.1 (HOO); "Russ Atom Blast to Speed New Projects at Hanford," *TCH*, October 11, 1949, 1.
24. "Report on Building Project 234–35," NARA, RG 326 67A, box 71, 600.1 (HOO).
25. Williams to Wilson, July 16, 1947, NARA, RG 326 67A, box 71, folder 600.1 (HOO).
26. "Report by Falk Architectural Consultants," September 4, 1949, NARA, 326, entry 67A, box 71, folder 600.1.
27. A decade later, neighboring Kennewick built a high school for a million dollars. *TCH*, October 10, 1958, 1.
28. "Report on Building Project 234-35," RG 326-1A-67A, box 71, folder, 600.1 (HOO).

29. David Inglis, "Atomic Profits and the Question of Survival," *Bulletin of the Atomic Scientists* IX, no. 4 (1953): 118; Christopher Drew, "Pentagon Changes Rules to Cut Cost of Weapons," *NYT*, September 15, 2010.

30. Roy B. Snapp, August 15, 1951, NARA, RG 326-67A, box 8, folder 10-14, 23.

31. Paul John Deutschmann, "Federal City: A Study of the Administration of Richland," MA thesis, University of Oregon, 1952, 149–50; *TCH*, 16, October , 1949, 1–2. See also Michael Gordin, *Red Cloud at Dawn: Truman, Stalin, and the End of the Atomic Monopoly* (New York: Farrar, Straus and Giroux, 2009), 250–51.

32. *TCH*, October 15, 1950, 4.

33. *TCH*, September 1 and October 4, 1949, 4, 1; October 30, 1950, 4.

34. *Villager*, October 28, 1948, 1; *TCH*, September 30, 1949, 24; *CBN*, May 4 and 11, June 2, July 3, 1950, 1.

35. On the role of boosters in national defense, see Roger W. Lotchin, *Fortress California, 1910–1961: From Warfare to Welfare* (Urbana: University of Illinois Press, 2002).

36. Maria E. Montoya, "Landscapes of the Cold War West," in Kevin J. Fernlund, ed., *The Cold War American West, 1945–1989* (Albuquerque: University of New Mexico Press, 1998), 15–16.

37. Patricia Nelson Limerick, "The Significance of Hanford in American History," in *Terra Pacifica: People and Place in the Northwest States and Western Canada* (Pullman: Washington State University Press, 1998), 53–70.

38. *TCH*, October 24, 1950, 1. See *CBN*, October 18, 1952, 2.

39. See *CBN*, September 2 and June 7, 1950, 1. For examples of justifications for federal projects in southeastern Washington, see Glenn Lee Papers, Washington State University Libraries, series ten, eleven, and fourteen.

40. Gordin, *Red Cloud*, 43.

41. On Richland as indefensible, see Harold D. Anamosa, "Passive Defense Survey," May 7, 1953, NARA, RG 326 67B, box 154, folder 9, 4–7; W. F. Libby to Douglas McKay, March 25, 1955, NARA, RG 326 67B, box 154, folder 11.

42. Paul Loeb, *Nuclear Culture: Living and Working in the World's Largest Atomic Complex* (Philadelphia: New Society, 1986), 70.

43. Thomas W. Evans, *The Education of Ronald Reagan: The General Electric Years and the Untold Story of His Conversion to Conservatism* (New York: Columbia University Press, 2006), 92–99; Wendy Wall, *Inventing the "American Way": The Politics of Consensus from the New Deal to the Civil Rights Movement* (New York: Oxford University Press, 2008), 207.

44. "Monthly Report of Hanford District Civil Defense," July 9, 1951, CREHST, 2006.001 box 1, folder 3.1.

45. *TCH*, September 1, 1950, 7.

46. *CBN*, December 13, 1952, 1; *TCH*, November 17, 1957, 1; *TCH*, November 27, 1957, 1.

47. On the patriotic consensus, see Russell B. Olwell, *At Work in the Atomic City: A Labor and Social History of Oak Ridge, Tennessee* (Knoxville: University of Tennessee Press, 2004), 3.

Chapter 18: Stalin's Rocket Engine

1. Jeffrey Richelson, *Spying on the Bomb: American Nuclear Intelligence from Nazi Germany to Iran and North Korea* (New York: Norton, 2006), 93.

2. Paul R. Josephson, "Atomic-Powered Communism: Nuclear Culture in the Postwar USSR," *Slavic Review* 55, no. 2 (Summer 1996): 297–324; V. P. Vizgin, "Fenomen 'kul'ta atoma' v CCCP (1950–1960e gg.) in *Istoriia Sovetskogo atomnogo proekta* (Moscow: IAnus-K, 1998), 439–40.

3. Yuli Khariton and Uri Smirnov, "The Khariton Version," *Bulletin of the Atomic Scientists*, May 1993, 27–29.

4. Riabev, *L. D. Riabev, Atomnyi proekt SSSR: Dokumenty i materialy*, vol. II, bk. 6 (Moscow: Nauka, 2006), 748, and bk. 4, 755.

5. "Sobranie partiinogo aktiva politotdela bazy-10," April 19, 1951, OGAChO, P-1137/1/31, 31–39.

6. On Moscow as aspiration, see Vera Dunham, *In Stalin's Time: Middleclass Values in Soviet Fiction* (Cambridge: Cambridge University Press, 1976), 49.

7. Vladimir Bokin and Marina Kamys, "Posledstviia avarii na kombinate 'Maiak,'" *Ekologiia* 4, April 2003.

8. Riabev, *Atomnyi proekt*, vol. II, bk. 4, 379–80, 570–71; Elena Zubkova, *Russia After the War: Hopes, Illusions and Disappointments, 1945–1957* (Armonk, NY: Sharpe, 1989), 86.

9. As quoted in David Holloway, *Stalin and the Bomb: The Soviet Union and Atomic Energy, 1939–1956* (New Haven, CT: Yale University Press, 1994), 148.

10. Dunham, *In Stalin's Time*, 4.

11. Author interview with Vladimir Novoselov, June 26, 2007, Cheliabinsk, Russia.

12. Holloway, *Stalin and the Bomb*, 186.

13. Anita Seth, "Cold War Communities: Militarization in Los Angeles and Novosibirsk, 1941–1953," PhD diss., Yale University, 2012, 161–224.

14. "Postanovlenie Politotdela Bazy no. 10," June 29, 1949, OGAChO, P-1167/1/15, 76–81.

15. Wendy Goldman, *Terror and Democracy in the Age of Stalin: The Social Dynamics of Repression* (Cambridge: Cambridge University Press, 2007) 45–47, 116; Jeffrey J. Rossman, *Worker Resistance Under Stalin: Class and Revolution on the Shop Floor* (Cambridge, MA: Harvard University Press, 2005); Donald A. Filtzer, *Soviet Workers and De-Stalinization: The Consolidation of the Modern System of Soviet Production Relations, 1953–1964* (Cambridge: Cambridge University Press, 1992), 155.

16. Elena Zubkova, et al., *Sovetskaia zhizn', 1945–1953* (Moscow: Rosspen, 2003), 81–82, 625; A. V. Fateev, *Obraz vraga v Sovetskoi propaganda, 1945–1954 gg.* (Moscow: RAN, 1999), 178–79.

17. "Svodki, Cheliabinsksogo Obkoma," March 5, 1948, OGAChO, P-288/12/194, 3–5.

18. M. E. Glavatskii, ed., *Rossiia, kotoruiu my ne znali, 1939–1993* (Cheliabinsk: Iuzhnoe-ural'skoe knizhnoe izdatel'stvo, 1995), 59–62.

19. Kosheleva, *Sovetskaia zhizn'*, 209.

20. "O vypolnenii postanovleniia biuro obkoma," September 18, 1948, OGAChO, 288/42/29; "Postanovlenie politotdela bazy no 10," March 30, 1954, OGAChO, P-1137/1/15, 32–41.

21. "O khode zhilishchnogo stroitel'stva," June 18, 1948, OGAChO, 288/42/29; V. Chernikov, *Za zavesoi sekretnosti ili stroitel'stvo No. 859* (Ozersk: V. Chernikov, 1995), 211.

22. "Protokol no. 3," June 3, 1949, OGAChO, P-1167/1/4, 35–39.

23. Riabev, *Atomnyi proekt*, vol. II, bk. 3, 393–94; "Morkovin Beloborodovu," September 15, 1949, and "Dol'nik Beloborodovu," no earlier than October 1949, OGAChO, 288/42/35.

24. "Protokol no. 1, politotdela Bazy no. 10," January 5, 1949, OGAChO, 1137/1/15, 1–5.

25. "Uralets Beloborodovu," September 30, 1949, OGAChO, 288/42/35.

26. "Protokol no. 10, politotdela Bazy no. 10," April 1, 1949, OGAChO, 1137/1/15: 76–81; "Semenov Beloborodovu," September 29, 1949, OGAChO, 288/42/35.

27. "Zasedanie partiinogo aktiva," April 19, 1951, OGAChO, 1137/1/31, 27.

28. "Protokol no. 2," October 10, 1956, OGAChO, 2469/1/2, 9–10.

29. "Protokol No. 3, zasedanie biuro Ozerskogo gorkoma KPSS," August 29, 1956, OGAChO, 2469/1/3, 45–55.

30. "Protokol no. 3," 35–39.

31. "Sobranie partiinogo aktiva," April 10, 1952, OGAChO, P-1137/1/38, 171–72.

32. Riabev, *Atomnyi proekt*, vol. II, bk. 4, 248–50; Chernikov, *Za zavesoi*, 80, 30.

33. "Sobraniia partiinogo aktiva," 142–46.

Chapter 19: Big Brother in the American Heartland

1. "Utopian Life Only a Mirage in Atom Town," *Chicago Tribune*, July 24, 1949, 4.

2. "The Atom: Model City," *Time*, December 12, 1949.

3. David Stevens, Rex E. Gwinn, Mark W. Fullerton, and Neil R. Goff, "Richland, Washington: A Study of Economic Impact," 1955, CREHST, 2006.001, 1, Folder 3.1.

4. Stevens et al., "Richland, Washington"; "JCAE Hearings 'Free Enterprise in Richland," June 23, 1949, HMJ, acc. 3560-2/58/29.

5. R. W. Cook, July 27, 1951, NARA, RG 326 67B, box 8, folder 10-4.

6. George W. Wickstead, "Planned Expansion for Richland, Washington," *Landscape Architecture* 39 (July 1949): 174.

7. "Atomic Cities' Boom," *Business Week*, December 18, 1948, 65–70.

8. *TCH*, February 11, 1950.

9. Paul Nissen, "Editor's Life at Richland Wasn't an Easy One!" part II, *TCH*, October 25, 1950, 1–2.

10. "Exhibit D, Villagers, Inc. Balance Sheet," July 10, 1945, PRR, Han 73214, bk. 17.

11. Nissen, "Editor's Life," part II, 2.

12. Nissen, "Editor's Life," part III, October 26, 1950, 1.

13. Ibid.

14. Author interview, Annette Heriford, May 18, 2008, Kennewick, WA.

15. W. B. Parsons, "List of Unions," May 23 1944, NAA, 326 8505, box 103, folder "Policy Books of Intelligence Division."

16. Robert Michael Smith, *From Blackjacks to Briefcases: A History of Commercialized Strikebreaking and Union Busting in the United States* (Athens: Ohio University Press, 2003).

17. Carleton Shugg to General Manager, September 10, 1947, NARA, RG 326 67A, box 16, folder 231.4.

18. Claude C. Pierce Jr., "Reorganization of the Intelligence and Security Division," September 6, 1945, NAA, RG 326 8505, box 103, folder "Policy Books of Intelligence Division"; "AEC Security Costs," December 4, 1953, NARA, RG 326 67B, box 154, folder 11.

19. Stevens et al., "Richland, Washington," 55; "JCAE Hearings 'Free Enterprise in Richland,'" June 22, 1949, HMJ, 3560-2/58/29.

20. David Witwer, "Westbrook Pegler and the Anti-Union Movement," *Journal of American History* 92, no. 2 (September 2005): 527–52.

21. F. A. Hayek, *The Road to Serfdom* (Chicago: University of Chicago Press, 1994).

22. "Minutes of Richland Community Council," meetings no. 20 and 21, May 9 and June 20, 1949, RPL.

23. *TCH*, October 26, 1950, 1; Carroll Wilson to Joint Commission on Atomic Energy, April 11, 1947, NARA, RG 326 67A, box 39, folder 352.9.

24. Deutschmann, "Federal City," 143, 268–74; *TCH*, January 12, 1950, 1.

25. *TCH*, October 24, 1950, 1.

26. Prout to Fred Schlemmer, December 20, 1948, and Schlemmer, "AEC Rental Rates," January 26, 1949, NARA, RG 326 67A, box 57, folder 480.

27. *TCH*, February 1, 8, 17, and 25, 1950.

28. R. W. Cook, July 27, 1951, NARA, RG 326 67B, box 8, folder 10-4; Gordon Dean to Brien McMahon, February 19, 1951, HMJ, acc. 3560-2/58/58-26.

29. "Minutes of Richland Community Council," meeting no. 26, November 14, 1949, RPL; Sumner Pike to Estes Kefauver, February 23, 1951, NARA, RG 326 67B, box 8, folder 9.

30. J. A. Brownlow to Brien McMahon, February 15, 1951, HMJ, acc. 3560-2/58/58-26; "Richland Community Council," September 20, 1952, RPL; K. E. Fields to Oscar S. Smith, January 22, 1957, NARA, RG 326, 67B, box 81, folder 11, "Labor Relations."

31. 624th AEC Meeting, November 7, 1951, NARA, RG 326 67B, box 8, folder 9.

32. *TCH*, February 7, 1951, 4.

33. Ibid.

34. "Numbers in Each Craft," 1952, NARA, RG 326, 67B, box 81, folder 11.

35. David E. Williams to Henry Jackson, April 26, 1951; M. W. Boyer to William L. Borden, June 12, 1951, HMJ, acc. 3560-2/58/58-22; "S. Robert Silverman and KAPL Guards Union," November 1952, NARA, RG 326, 67B, box 81, folder 11, "Labor Relations."

36. Frances Pugnetti, *Tiger by the Tail: Twenty-Five Years with the Stormy* Tri-City Herald (Kennewick, WA: Tri-City Herald, 1975), 140–41; William Border to Marion Boyer, March 12, 1952, NARA, 326 67B, box 81, folder 11; *TCH*, October 19, 1958, 1.

37. Thomas W. Evans, *The Education of Ronald Reagan: The General Electric Years and the Untold Story of His Conversion to Conservatism* (New York: Columbia University Press, 2006), 91–95. Quotes from "Regulations for Hanford Works Security Patrolmen, GE, February 20, 1958," PRR, HAN 22970, 8–9.

38. Dick Epler, January-February 1998, Alumni Sandstorm (online archives) http://alumnisandstorm.com.

39. Jack Metzgar, *Striking Steel: Solidarity Remembered* (Philadelphia: Temple University Press, 2000), 7, 156.

40. Glenn Crocker McCann, "A Study of Community Satisfaction and Community Planning in Richland, Washington," PhD diss., Department of Sociology, State College of Washington, 1952, 69–71, 115–17, 124; "Report of the Survey on Home Ownership" (1951), FCP, acc. 3543-004/4/19.

41. Bob DeGraw, August 10, 1998, Alumni Sandstorm (online archives), http://alumnisandstorm.com.

42. Carl Abbott, "Building the Atomic Cities: Richland, Los Alamos, and the American Planning Language," in Bruce Hevly and John M. Findlay, eds., *The Atomic West*, 90–115.

43. Author interview, Stephanie Janicek, July 14, 2010, Richland, WA.

44. *TCH*, October 7, 1949.

45. Tom Vanderbilt, *Survival City: Adventures Among the Ruins of Atomic America* (New York: Princeton Architectural Press, 2002).

46. *TCH*, July 8, 1956.

47. Ibid.; Richland Community Council minutes, June 11, 1951, May 20, 1957, November 4, 1957, December 4 and 30, 1957, January 6, 1958, RPL.

48. Rebecca Lester, "Measures for the Prevention of Juvenile Delinquency in the City of Richland, WA," April 7, 1964, *Sociology*, 132, in FCP, 6/7.

49. McCann, "A Study of Community Satisfaction," 57.

50. Elaine Tyler May, *Homeward Bound: American Families in the Cold War Era* (New York: Basic Books, 1988), 153.

51. "Monthly Report, July 1954, Radiation Monitoring Unit," PRR, HW 32571.

52. *TCH*, November 3, 1951, 2.

53. A. Fred Clagett, "Richland Diary," October 13, 1972, CREHST, acc. 2006.001, box 1, folder 3.1; minutes of Richland Community Council, February 7, 1955, RPL; "Excerpts from Delbert Meyer's Thesis," CREHST, acc. 2006.1, box 2, folder 6.1, 120.

54. *TCH*, February 7, 1951, 4.

55. Mathew Farish, "Disaster and Decentralization: American Cities and the Cold War," *Cultural Geographies* 10 (2003): 125–48.

56. *CBN*, August 8 and September 22, 1950.

57. Lizabeth Cohen, *A Consumers' Republic: The Politics of Mass Consumption in Postwar America* (New York: Knopf, 2003).

Chapter 20: Neighbors

1. Author interview with C. J. Mitchell, August 19, 2008, Richland, WA.

2. *TCH*, October 2, 1949, 1–2.

3. Charles P. Larrowe, "Memo on Status of Negroes in the Hanford, WA Area," April 1949, HJM, acc. 3560-2, box 58, folder 29.

4. Robert Bauman, "Jim Crow in the Tri-Cities, 1943–1950," *Pacific Northwest Quarterly*, Summer 2005, 124–31.

5. "Negro Relations in the Atomic Energy Program," March 7, 1951, NARA, RG 326 67A, box 16, folder 291.2.

6. James T. Wiley Jr., "Race Conflict as Exemplified in a Washington Town," M.A. thesis, Department of Sociology, State College of Washington, 1949, 56.
7. Wiley, "Race Conflict," 61; Larrowe, "Memo on Status of Negroes."
8. *CBN*, May 8, 1950.
9. Larrowe, "Memo on Status of Negroes."
10. In 1948–47, 125 of 193 arrests were for "vagrancy" and "investigation." Wiley, "Race Conflict," 8.
11. *TCH*, December 26, 1947, and February 25, 1948; *William J. Gaffney, Appellant, v. Scott Publishing Company et al., Respondents*, no. 30989, en banc, Supreme Court of Washington, December 14, 1949.
12. FHA, *FHA Underwriting Manual* (1938), sect. 911, 929, 937.
13. Wiley, "Race Conflict," 124.
14. Larrowe, "Memo on Status of Negros."
15. Ibid.
16. *TCH*, January 12, 1950, 6.
17. As quoted in Bauman, "Jim Crow in the Tri-Cities."
18. John M. Findlay and Bruce William Hevly, *Atomic Frontier Days: Hanford and the American West* (Seattle: University of Washington Press, 2011), 130–32.
19. *CBN*, April 18 and 20, 1950.
20. *TCH*, September 11 and October 6, 1949.
21. "AEC Negro Relations in the Atomic Energy Program," and Seattle Urban League papers, UWSC, acc. 607, 681, 36/6, especially "Report on Atomic Energy Commission at Richland."

Chapter 21: The Vodka Society

1. Nikolai Rabotnov, "Publitsistika—Sorokovka," *Znamia*, July 1, 2000, 160.
2. Ibid., 164.
3. "Protokol no. 10," April 1, 1949, OGAChO, P-1137/1/15, 49–50.
4. "Ob izzhitii faktov khuliganstva," March 20, 1954, OGAChO, P-1137/1/65, 1–3.
5. Ibid.; "Protokol no. 31," June 22, 1951, OGAChO, 288/42/43.
6. "Sobranie partiinogo aktiva politotdela bazy-10," April 19, 1951, OGAChO, P-1137/1/31, 68–70.
7. Ibid.
8. Author correspondence with Ervin Polle, February 12, 2012.
9. L. D. Riabev, *Atomnyi proekt SSSR: Dokumenty i materialy*, vol. II, bk. 5 (Moscow: Nauka, 2007), 170, 183, 187, and bk. 3, 245–46, 368; V. N. Kuznetsov, *Zakrytye goroda Urala* (Ekaterinburg: Akademiia voenno-istoricheskikh nauk, 2008), 96.
10. "Spravka o rabote nabliudatel'noi komissii," January 9, 1960, OGAChO, 2469/3/3, 59–64.
11. "Sobranie partiinogo aktiva politotdela bazy-10," 31–39.
12. Ibid., 37.
13. N. V. Mel'nikova, *Fenomen zakrytogo atomnogo goroda* (Ekaterinburg: Bank kul'turnoi informatsii, 2006), 92; Jack S. Blocker, David M. Fahey, and Ian R. Tyrrell, *Alcohol and Temperance in Modern History: An International Encyclopedia* (Santa Barbara, CA: ABC-CLIO, 2003), 15.
14. "Ob faktov khuliganstva," 142; G. N. Kibitkina, "Informatsia o sostave i soderzhanii dokumentov fonda P-2469 Ozerskii gorkom KPSS za 1961–1965 gody" (unpublished).
15. "Zasedanie partiinogo aktiva," July 6, 1951, OGAChO, P-1137/1/31, 168–72.
16. "Protokol no. 1," August 18, 1956, OGAChO, 2469/1/3, 42.
17. "Protokol no. 4," August 15, 1951, OGAChO, 1181/1/2, 24.
18. "Protokoly sobranii," October 21, 1954, OGAChO, 1596/1/43, 52.
19. Kuznetsov, *Zakrytye goroda*, 67.
20. Author interview with Anna Miliutina, June 21, 2010, Kyshtym.

21. "Spravka," 1959, and "Spravka," January 7, 1960, OGAChO, 2469/3/3, 5, 8; "Protokol no. 7 plenumov gorodskogo komiteta KPSS," May 23, 1967, OGAChO, 2469/6/405, 48–51.
22. Author interview with Galina Petruva (pseudonym), June 26, 2010, Kyshtym.
23. "Sobranie partiinogo aktiva," April 10, 1952, OGAChO, P-1137/1/38, 179.
24. Ibid., 170–71.
25. Among many examples, see OGAChO, 288/42/43, 7–8, 96–97; P-1137/1/65, 1–3; 2469/1/119, 159–70; 2469/2/1, 28–33; 2469/3/3, 59–64, plus the minutes for city executive committee meetings from 1962 to 1967, in fond 2469, opis' 6.
26. "Spravka," 1959, and "Spravka," January 7, 1960, OGAChO, 2469/3/3, 5, 8; "Protokol no. 7 plenumov gorodskogo komiteta KPSS," 48–51.
27. Lazyrin Malyginoi, 1957, OGAChO, 2469/1/118, 106–8.
28. "Sobranie partiinogo aktiva," 171–72.
29. Tamara Belanova, "S chego nachinalsia Obninsk," *Gorod*, April 1995, 52; Vladimir Bokin and Marina Kamys, "Posledstviia avarii na kombinate 'Maiak,'" *Ekologiia* 4, April 2003; Vladimir Gubarev, "Professor Angelina Gus'kova," *Nauka i zhizn'*, no. 4 (2007), 18–26; Lawrence S. Wittner, *The Struggle Against the Bomb* (Stanford, CA: Stanford University Press, 1993), 1:146.
30. Rabotnov, "Publitsistika," 161.
31. Mel'nikova, *Fenomen*, 67.
32. Ibid., 68.

Chapter 22: Managing a Risk Society

1. Michele Stenehjem Gerber, *On the Home Front: The Cold War Legacy of the Hanford Nuclear Site* (Lincoln: University of Nebraska Press, 1992), 216.
2. See Ulrich Beck, *World Risk Society* (Cambridge: Polity Press, 1999), 72.
3. H. M. Parker, "H.I., Plant Control Activities to August 1945," PRR.
4. Parker to S. T. Cantril, December 11, 1945, PRR, HW-7 31057.
5. K. Herde, "I-131 Accumulation," March 1, 1946, PRR, HW 3-3455; "I-131 Deposition in Cattle Grazing," August 29, 1946, PRR, HW 3-3628.
6. Parker, "Tolerable Concentration of Radio-Iodine," January 14, 1946, PRR, HW 7-3217; "Radiation Exposure Data," February 8, 1950, PRR, HW 19404.
7. "HW Radiation Hazards," July 27, 1948, PRR, HW 10592.
8. M. S. Gerber, "A Brief History of the T Plant Facility at the Hanford Site," 1994, DOE Opennet, 29.
9. Ian Stacy, "Roads to Ruin on the Atomic Frontier: Environmental Decision Making at the Hanford Reservation, 1942–1952," *Environmental History* 15, no. 3 (July 2010): 415–48.
10. B. G. Lindberg, "Investigation, no. 333," January 28, 1954, PRR, HW 30764.
11. Minutes, Advisory Committee for Biology and Medicine (ACBM), October 8–9, 1948, NAA, RG 326 87 6, box 30, folder "ACBM"; Parker, "HW Radiation Hazards."
12. Parker, "Action Taken, Particle Hazard," October 25, 1948, PRR, HW 11348.
13. Walter Williams, "Certain Functions AEC Hanford Operations," October 8, 1948, JPT, 5433-1, box 24, 11.
14. Minutes, ACBM, October 8–9, 1948; Parker, "Report on Staff Action Taken and Planned," October 8–9, 1948, NARA, RG 326, Biology and Medicine, box 1, Folder 5.
15. *Villager*, October 14, 1948, 1.
16. Parker, "Report."
17. R. E. Gephart, *Hanford: A Conversation About Nuclear Waste and Cleanup* (Columbus, OH: Battelle Press, 2003), 2.3.
18. Kenneth Scott, "Some Biological Implications," June 30, 1949, NAA, RG 326 87 6, box 4, "Research and Development."
19. Forrest Western, "Problems of Radioactive Waste Disposal," *Nucleonics*, August 1948, 42–48.

20. Gerber, "Brief History of the Site," 30.

21. HWS Monthly Report for June 1952, July 21, 1952, PRR, HW 24928.

22. Monroe Radley, "Distribution of GE Personnel in Hanford Works 'AEC,'" May 15–24, 1948, JPT, 5433-1, box 24.

23. Parker, "Status of Ground Contamination Problem," September 15, 1954, DOE Opennet, HW 33068; R. H. Wilson, "Criteria Used to Estimate Radiation Doses," PRR, BNWL-706 UC-41, July 1986.

24. Herbert Parker, "Summary of HW Radiation Hazards for the Reactor Safeguard Committee," July 27, 1948, PRR, HW 10592.

25. J. W. Healy, "Dissolving of Twenty Day Metal at Hanford," May 1, 1950, DOE Opennet.

26. Karen Dorn Steele, "Hanford's Bitter Legacy," *Bulletin of the Atomic Scientists*, January-February 1988, 20; Daniel Grossman, "A Policy History of Hanford's Atmospheric Releases," PhD diss., Massachusetts Institute of Technology, 1994.

27. Healy, "Dissolving"; John M. Findlay and Bruce William Hevly, *Atomic Frontier Days: Hanford and the American West* (Seattle: University of Washington Press, 2011), 57–58.

28. Gerber, "Brief History of the Site," 32.

29. Ibid., 40, 41–56, 65, 68, 70.

30. Gerber, *On the Home Front*, 125.

31. "Summary of AEC Waste Storage Operations," September 21, 1960, DOE Germantown, RG 326/1309/6; Gephart, *Hanford*, 5.3; "Kellex Contract with General Electric," September 23, 1947, NARA, RG 326 67A, box 71, folder 600.1 (HOO). On school budget, *TCH*, October 4, 1949, 1–2.

32. "722nd AEC Meeting," July 11, 1952, NARA, RG 326 67B, box 88, folder 17.

33. C. C. Gamertsfelder, "Effects on Surrounding Areas," March 11, 1947, PRR, HW 7-5934.

34. K. Herde, "Check of Radioactivity in Upland Wild-Fowl," December 7, 1948, PRR, HW-11897.

35. As quoted in Findlay and Hevly, *Atomic Frontier Days,* 57.

36. "Study of AEC Radioactive Waste Disposal," November 15, 1960, DOE Germantown, RG 326/5574/9, 19.

Chapter 23: The Walking Wounded

1. Vladyslav B. Larin, *Kombinat "Maiak"—Problema na veka* (Moscow: KMK Scientific Press, 2001), 119–20.

2. Alexei Mitiunin, "Natsional'nye osobennosti likvidatsii radiastionnoi avarii," *Nezavisimaia gazeta*, April 15, 2005.

3. Vladyslav B. Larin, "Mayak's Walking Wounded," *Bulletin of the Atomic Scientists*, September–October 1999: 25.

4. Larin, *Kombinat*, 113.

5. L. D. Riabev, *Atomnyi proekt SSSR: Dokumenty i materialy*, vol. II, bk. 4 (Moscow: Nauka, 2004), 206–8.

6. V. Chernikov, *Osoboe pokolenoe* (Cheliabinsk: V. Chernikov, 2003), 67.

7. Riabev, *Atomnyi proekt*, vol. II, bk. 4, 656–58; B. Emel'ianov, *Raskryvaia pervye stranitsy: k istorii goroda Snezhinska* (Ekaterinburg: IPP Uralskii rabochii, 1997).

8. Riabev, *Atomnyi proekt*, vol. II, bk. 4, 762–65.

9. Atomic Intelligence gathered information on gamma rays. Riabev, *Atomnyi proekt*, vol. II, bk. 4, 431.

10. Sokhina, *Plutonii*, 106–7, 133–35.

11. V. Chernikov, *Za zavesoi sekretnosti ili stroitel'stvo No. 859* (Ozersk: V. Chernikov, 1995), 53.

12. Riabev, *Atomnyi proekt*, vol. II, bk. 4, 392–98.

13. N. V. Mel'nikova, *Fenomen zakrytogo atomnogo goroda* (Ekaterinburg: Bank kul'turnoi informatsii, 2006), 98; Riabev, *Atomnyi proekt*, vol. II, bk. 7, 589–600.

14. Vladimir Bokin and Marina Kamys, "Posledstviia avarii na kombinate 'Maiak,'" *Ekologiia* 4, April 2003; Victor Doshchenko et al., "Occupational Diseases from Radiation Exposure at the First Nuclear Plant in the USSR," *Science of the Total Environment* 142 (1994): 9–17.

15. G. I. Reeves and E. J. Ainsworth, "Description of the Chronic Radiation Syndrome in Humans Irradiated in the Former Soviet Union," *Radiation Research* 142 (1995): 242–44.

16. Nikolai Rabotnov, "Publitsistika—Sorokovka," *Znamia*, July 1, 2000, 168.

17. Vladimir Gubarev, "Professor Angelina Gus'kova," *Nauka i zhizn'* 4 (2007): 18–26.

18. Efim P. Slavskii, "Kogda strana stoila na plechakh iadernykh titanov," *Voenno-istoricheskii zhurnal* 9 (1993): 20.

19. Gubarev, "Angelina Gus'kova."

20. Ibid., 20.

21. Larin, *Kombinat*, 84–89.

22. A. K. Gus'kova, *Atomnaia otrasl' strany glazami vracha* (Moscow: Real'noe vremia, 2004), 87.

23. Adriana Petryna, *Life Exposed: Biological Citizens After Chernobyl* (Princeton, NJ: Princeton University Press, 2002), 39–41.

24. Author interview with Vladimir Novoselov, June 26, 2007, Cheliabinsk, Russia.

25. Larin, *Kombinat*, 214, 195, and table 6.25, 412.

26. A. N. Nikiforov, "Severnoi siianie nad Kyshtymom," *Dmitrovgrad-panorama* 146 (September 27, 2001): 7–8.

27. "Protokol 1-oi gorodskoi partiinoi konferentsii," August 16–17, 1956, OGAChO, 2469/1/1; Komykalov Efremovu, January 5, 1962, OGAChO, 288/42/79, 1–2.

28. Evgenii Titov, "Likvidatory, kotorykh kak by i ne bylo," *Novaia gazeta*, February 15, 2010.

29. "O rabote nabliudatel'noi komissii," January 9, 1960, OGAChO, 2469/3/3, 59–64.

30. Ivanov, *Plutonii*, 8; Mel'nikova, *Fenomen*, 98–99; "Protokol No. 7," May 23, 1967, OGAChO, 2469/6/405, 51.

31. "Sobranie partiinogo aktiva politotdela bazy-10," April 19, 1951, OGAChO, P-1137/1/31, 31–34.

32. "Sobranie partiinogo aktiva," January 30, 1952, OGAChO, 1137/1/38, 31–39, 59; "Reshenie politicheskogo upravleniia MSM," February 15, 1954, OGAChO, 1138/1/22, 47; "O rabote politotdela bazy no. 10," October 25, 1949, OGAChO, 288/43/30, 38–42; "Postanovlenie Cheliabinskogo obkoma," April 21, 1950, OGAChO, 288/42/38.

33. Ulrich Beck, *Ecological Enlightenment: Essays on the Politics of the Risk Society* (Atlantic Highlands, NJ: Humanities Press, 1995), 20–21.

Chapter 24: Two Autopsies

1. "Press Release of AEC," December 9, 1953, NARA, RG 326 67B, box 50, folder 13.

2. Marie Johnson, June 14, 1952; Russell to Norwood (undated); Jurgenson to M. Johnson, August 14, 1952, all in JPT, acc. 5433-001/11; Karen Dorn Steele, *Spokesman Review*, September 9, 1990, A14.

3. P. A. Fuqua, "Report of Fatality," July 26, 1952, JPT, acc. 5433-001/11.

4. Carter to Johnson, August 5, 1952, JPT, acc. 5433-001/11. For reports of confiscated body parts of workers who might have died from exposure, see Kristen Iversen, *Full Body Burden: Growing Up in the Nuclear Shadow of Rocky Flats* (New York: Crown, 2012), 185.

5. Smyth to McClean, December 23, 1952, and Fuqua to McClean, December 31, 1952, JPT, acc. 5433 001/11.

6. Boyer to LeBaron, January 10, 1951, GWU.

7. McClean to Carter, January 6 and February 2, 1953, JPT, acc. 5433 001/11.

8. McClean to Smyth, November 5, 1952, and Jurgensen to McLean, October 28, 1952, JPT, acc. 5433 001/11,

9. Author interview, KR, August 16, 2011, Richland, WA.

10. "HWS Monthly Report, June 1952," July 21, 1952, PRR, HW 24928.

11. L. V. Barker, "Radiation Incident," June 20, 1952, PRR, HW 24806.
12. D.P.E., handwritten note, January 4, 1955 (filed under B. G. Lindberg, "Special Hazards Incident Investigation, No. 205," April 16, 1952), PRR, HW 24270.
13. F. P. Baranowski, "Contamination of Two Waste Water Swamps," June 19, 1964, DOE Germantown, RG 326/1362/7.
14. "1153rd AEC Meeting," December 6, 1955, NARA, RG 326 67B, box 50, folder 14; "HEW Monthly Report," January 30, 1956, PRR, HW 40692.
15. Lindberg, "Special Hazards Incident Investigation"; "Incident Report," June 4, 1951, PRR, HW 20892.
16. W. V. Baumgartner, "Report of Incident," November 4, 1953, PRR, HW 18221, 1950.
17. "Separations Section Radiation Hazards Incident Investigation," June 7, 1952 (HW 24746), in JPT, acc. 5433-001/11; Lindberg, "Special Hazards Incident Investigation, No. 194," March 12, 1952, PRR, HW 23801; "HEW Monthly Report," March 18, 1955, PRR, HW 35530; "HEW Monthly Report, January 30, 1956, PRR, HW 40692; Monthly Report—November 1955—Separations," December 12, 1955, PRR, HW 40248.
18. Lindberg, "Radiation Incident Investigation," April 1, 1952, and March 10, 1952, PRR, HW 24000, HW 23753; "Radiation Incident Class II, No. 29–32," February 4, March 20, and March 26, 1952, JPT, acc. 5433 001/11.
19. Charles Perrow, "Normal Accident at Three Mile Island," *Society* 18, no. 5 (July/August 1981): 17–26.
20. Herb Parker, "HW Radiation Hazards for the Reactor Safeguard Committee," July 27, 1948, PRR, HW 10592.
21. Jonathan Schell, *The Seventh Decade: The New Shape of Nuclear Danger* (New York: Metropolitan Books, 2007), 38.
22. Cook to Anderson, April 27, 1956, NARA, RG 326 67B, box 50, folder 14.
23. HEW Monthly Report, December 1954, DOE Opennet, HW 31267; B. G. Lindberg, "Radiation Sciences Department Investigation, No. 295," July 7, 1953, PRR, HW 28707.
24. "Monthly Operations Report, November 1955," DOE Opennet, HW 40182.
25. A. R. McGuire, "Management Report," December 23, 1955, HW 39967 RD, as cited in Sonja I. Anderson, "A Conceptual Study of Waste Histories, Project ER4945," September 29, 1994, unpublished, in possession of author.
26. "HEW Monthly Report," January 30, 1956, PRR, HW 40692.
27. "Incident Report," July 1956, HW 44580, as cited in Anderson, "A Conceptual Study."
28. Myers, "Special Hazards Incident," March 24, 1953, PRR, HW 18575.
29. Readings topped out at 80,000 counts per minute. Lindberg, "Special Hazards Incident Investigation, No. 243," October 3, 1952, PRR, HW 26099; HWS Monthly Report, July 21, 1952, HW 24928.
30. "Special Hazards Incident Investigation, No. 204," April 28, 1952, PRR, HW 24269.
31. Hofmaster to Jackson, July 24, 1951, HMJ, box 28, folder 23; "Monthly Report, December 1956," DOE Opennet, HW 47657.
32. K. R. Heid to W. F. Mills, July 30, 1979; Michael Tiernan, August 10, 1979, PRR, RLHT595-0013-DEL.
33. Author correspondence with Don Sorenson, January 12, 2008.
34. "Monthly Report, December 1956," and Parker, "Component of Radiation Exposure," April 20, 1951, DOE Opennet, HW 47657 and HW 20888, 1–10.
35. "Quarterly Progress Report, April-June 1960," DOE Opennet, HW 66306, 19.
36. Lindberg, "Radiation Sciences Department Investigation, No. 352," April 7, 1954, PRR, HW 31394.
37. Author interview with William Bricker, August 16, 2011; author interview with Al Boldt and Keith Smith, August 15, 2011, Richland.
38. Perrow, "Normal Accidents," 19.
39. Lindberg, "Radiation Sciences Department Investigation, No. 335," March 15, 1954, PRR, HW 31344.

40. Mary Manning, "Atomic Vets Battle Time," *Bulletin of the Atomic Scientists*, January–February 1995, 54–60.
41. Author interview with BE, August 15, 2011, Richland.
42. State of Washington, Order and Notice, December 20, 1972; Schur to Hames, re: Smith and Patrick Radiation Reports, June 7, 1973; "Complaint of Blanche McQuilkin, Executrix of the Estate of Adelbert McQuilkin, Deceased," May 12, 1968, all in JPT, acc. 5433-001/11.

Chapter 25: Wahluke Slope

1. "The Wahluke Slope, Secondary Zone Restrictions," 1951, NARA, 326 67B, box 84, folder 2, vol. 2; "Effect of Hanford Works on Wahluke Slope," April 16, 1949, JPT, 5433-001, box 25.
2. W. P. Conner to C. Rogers McCullough, April 18, 1952, and "Decision on AEC 38/12," January 12, 1953, NARA, 326 67B, box 84, folder 2, vol. 2, 25–26.
3. Parker, "HW Radiation Hazards," July 27, 1948, PRR, HW 10592; report from manager, "The Wahluke Slope," 21–23, and Raul Stratton to Rogers McCollough, April 16, 1952, NARA, 326 67B, box 84, folder 2, vol. 2.
4. "Effect of Hanford Works on Wahluke Slope"; Lum, "Potential Hazards," 1947, NAA, RG 326 87 6, box 7, folder "Hazards and Control"; C. C. Gamertsfelder, "Effects on Surrounding Areas," March 11, 1947, PRR, HW 7-5934.
5. See, for example, Richard White, *Railroaded: The Transcontinentals and the Making of Modern America* (New York: W. W. Norton, 2011).
6. Hubert Walter to David Shaw, October 10, 1951, NARA, 326 67B, box 84, folder 2, vol. 2; Kenneth Osborn, "Wahluke Slope Problem," April 18, 1952, NARA, 326 67B, box 84, folder 2, vol. 2.
7. "Transcript of Wahluke Meeting," April 19, 1949, HMJ, acc. 58/50-32.
8. K. Herde, "I-131 Accumulation," March 1, 1946, PRR, HW 3-3455.
9. Bugher, "Wahluke Slope," October 27, 1952, DOE Opennet, AEC 38/14.
10. Gamertsfelder, "Effects on Surrounding Areas"; report from manager, "The Wahluke Slope," 23.
11. "Annual Percentage Frequency of Wind Directions," 1951, NARA, 326 67B, box 84, folder 2, vol. 2.
12. "Roles of AEC and ACRS with Respect to Wahluke Slope Problem," 1958, NARA, 326 67B, box 84, folder 2, vol. 2.
13. "Decision on AEC 38/12."
14. Ibid.
15. Katherine L. Utter, "In the End the Land: Settlement of the Columbia Basin Project," PhD diss., University of Washington, 2004, 190–92.
16. Bailie founded with some of the proceeds the Bailie Memorial Boys' Ranch.
17. Marion Behrends Higley, *Real True Grit: Stories of Early Settlers of Block 15, 1953–1960* (Pasco, WA: B & B Express Printing, 1998).
18. Blaine Harden, *A River Lost: The Life and Death of the Columbia* (New York: W. W. Norton, 1996), 128–31.
19. Rodney P. Carlisle with Joan M. Zenzen, *Supplying the Nuclear Arsenal: American Production Reactors, 1942–1992* (Baltimore: Johns Hopkins University Press, 1996), 91.
20. William Thurston, "Land Disposal of Radioactive Wastes," and W. B. McCool, "Land Disposal," October 27, 1960, DOE Germantown, RG 326 1309, box 6; John M. Findlay and Bruce William Hevly, *Atomic Frontier Days: Hanford and the American West* (Seattle: University of Washington Press, 2011), 8.

Chapter 26: Quiet Flows the Techa

1. "Chertezh zemel'nykh uchastkov v/ch 859," April 5, 1947, OGAChO, 274/20/18, 121–22.
2. Thomas B. Cochran, Robert S. Norris, and Oleg Bukharin, *Making the Russian Bomb: From Stalin to Yeltsin* (Boulder, CO: Westview Press, 1995), 103–8.

3. "Protokol No. 164," September 19, 1949, OGAChO, 288/13/105; "O Kaslinskoi raionnoi partiinoi konferentsii," February 18, 1950, and January 27–28, 1951, OGAChO, 107/17/510 and 658; "Protokol No. 164," September 19, 1949, OGAChO, 288/13/105; "Spravka," no earlier than 1949, OGAChO, 288/13/84. Records from the critically contaminated State Farm No. 2 were destroyed for the years 1947 to 1951. See OGAChO, 107/17/444.

4. Leonid Timonin, *Pis'ma iz zony: Atomnyi vek v sud'bakh tol'iattintsev* (Samara: Samarskoe knizhnoe izd-vo, 2006), 14.

5. Author interview with Alexander Akleev, Urals Research-Clinical Center of Radiation Medicine, Cheliabinsk, June 26, 2007.

6. V. Chernikov, *Osoboe pokolenoe* (Cheliabinsk: V. Chernikov, 2003), 1:179.

7. "Interview with Tom Carpenter, Executive Director of the Hanford Challenge," 2009, www.youtube.com/watch?v=jg_zw38G7Ms.

8. L. D. Riabev, *Atomnyi proekt SSSR: Dokumenty i materialy*, vol. II, bk. 4 (Moscow: Nauka, 2004), 762–65, and bk. 6, 350–52.

9. Ibid., bk. 4, 679.

10. Timonin, *Pis'ma iz zony*, 16.

11. Zhores A. Medvedev, *The Legacy of Chernobyl* (New York: W. W. Norton, 1990), 111.

12. V. N. Novoselov and V. S. Tolstikov, *Atomnyi sled na Urale* (Cheliabinsk: Rifei, 1997), 35.

13. Timonin, *Pis'ma iz zony*, 16; Vladimir Novikov, Alexander Akleev, and Boris Segerstahl, "The Long Shadow of Soviet Plutonium Production," *Environment*, January 1, 1997.

14. Zubkova, *Poslevoennoe sovetskoi obshchestvo*, document 240, and "Partorganizatsii kontrarazvedki MVD v/ch 0501," August 18, 1951, OGAChO, 1181/1/12, 26–30.

15. D. Kossenko, M. Burmistrov, and R. Wilson, "Radioactive Contamination of the Techa River and Its Effects," *Technology* 7 (2000): 553–75.

16. M. O. Degteva, N. B. Shagina, M. I. Vorobiova, L. R. Anspaugh, and B. A. Napier, "Reevaluation of waterborne releases of radioactive materials from the Mayak Production Association into the Techa River in 1949–1951," *Health Physics*, 2012 Jan; Vol. 102 (1): 25–38.

17. Fauziia Bairamova, *Iadernyi arkhipelag ili atomnyi genotsid protiv Tatar* (Kazan': Nauchno-populiarnoe izdanie, 2005), 1–5.

18. Author interview with Anna Miliutina, June 26, 2010, Kyshtym, Russia.

19. Author interview with Liubov Kuzminova, June 26, 2010, Kyshtym, Russia; "Tkachenko Smorodinskomu" and "O peredache zemel'," December 17, 1949, OGAChO, 288/42/34, 5–6, 59–60; "Reshenie," April 24, 1946, OGAChO, 274/20/10, 26–27; "Soveshchanie u nachal'nika Stroitel'stva no 859," May 7, 1947, OGAChO, 274/20/18; Riabev, *Atomnyi proekt*, vol. II, bk. 3, 370.

20. "Podgotovki zhilfonda," June 26, 1951, OGAChO, P-1137/1/31, 85.

21. Vladyslav B. Larin, *Kombinat "Maiak"—Problema na veka* (Moscow: KMK Scientific Press, 2001), 39–40.

22. Novoselov and Tolstikov, *Atomnyi sled*, 38–39.

23. This first evacuation occurred after late October 1951. See "B. G. Muzrukov A. D. Zverevu," October 26, 1951, as reproduced in Novoselov and Tolstikov, *Atomnyi sled*, 218–19.

24. Ibid.

25. Riabev, *Atomnyi proekt*, vol. II, bk. 5, 94–96.

26. Larin, *Kombinat*, 41.

27. Novoselov and Tolstikov, *Atomnyi sled*, 65.

28. E. Ostroumova, M. Kossenko, L. Kresinina, and O. Vyushkova, "Late Radiation Effects in Population Exposed in the Techa Riverside Villages (Carcinogenic Effects)," paper presented at the 2nd International Symposium on Chronic Radiation Exposure, March 14–16, 2000, Cheliabinsk.

29. Author interview with Akleev.

30. Larin, *Kombinat*, 40.

31. Muzrukov Aristovu, February 9, 1952, OGAChO, 288/42/50.

32. Novoselev and Tolstikov, *Atomnyi sled*, 220–21.

33. Chernikov, *Osoboe pokolenoe*, 1:23.

34. "Dokumenty o stro-vy kolodtsev v blizi r. Techa," 1952–1955, Arkhivnyi otdel Kunashak-skogo munitsipal'nogo raiona, 23/1/38.

35. See, for example, Leon Gouré, *War Survival in Soviet Strategy: USSR Civil Defense* (Miami: University of Miami, 1976).

36. "O khode stroitel'stva kolodtsev," March 17, 1953, OGAChO, 274/20/33, 22.

37. Novoselov and Tolstikov, *Atomnyi sled*, 39–40.

38. This problem persisted until at least 1960. See "No. 28 ot 19 Maia 1960 g," May 20, 1960, OGAChO, R-1644/1/4a, 127.

39. A. Burnazian, I. E. Slavaski, N. V. Laptevu, November 15, 1952, OGAChO, 288/42/50.

40. "Muzrukov Bezdomovu" and "Udostoverinie," February 10 and 12, 1953, OGAChO, 274/20/33, 24–25.

41. "O resul'tatakh proverki" [*sic*] and "O khode stroitel'stva kolodtsev," March 14 and 17, 1953, OGAChO, 274/20/33, 30–31, 22.

42. "O Kaslinskoi raionnoi partiinoi konferentsii" and "Komissii po proverke kolkhoz 'Zvezda,'" March 21, 1953, OGAChO, 107/18a/389, 70–71.

43. Novoselov and Tolstikov, *Atomnyi sled*, 65–69.

44. "Rasporiazhenie no. 282cc," March 23, 1954, OGAChO, 274/20/38, 13.

45. "O meropriiatiakh po uluchsheniu meditsinskogo obsluzhivaniia," October 30, 1953, OGAChO, 274/30/20, 155–57.

46. Novoselov and Tolstikov, *Atomnyi sled*, 68.

47. Evgenii Titov, "Likvidatory, kotorykh kak by i ne bylo," *Novaia gazeta*, February 15, 2010, 16.

48. Author interview with Miliutina.

Chapter 27: Resettlement

1. V. V. Litovskii, "Ural—radiatsionnye katastrofy—Techa," 1995, unpublished, http://techa49.narod.ru.

2. Ibid.

3. Ibid.

4. On illness, both cause and narrative, see Adriana Petryna, *Life Exposed: Biological Citizens After Chernobyl* (Princeton, NJ: Princeton University Press, 2002), 13.

5. Researchers used performance in school as one indicator of radiation-related illness. Author interview with Alexander Akleev, June 26, 2007, Cheliabinsk, Russia.

6. For a retrospective review of the resettlement orders, see "Po voprosu vydeleniia dopolnitel'nykh assignovanii na raboty po otseleniiu ot reki Techa," September 12, 1962, OGAChO, 1644/1/4a, 197–99, 180–81.

7. Author interview with Anna Miliutina, June 26, 2010, Kyshtym; Alexandra Teplova, "Mol-chali do Chernobylia," *Cheliabinskii rabochii*, October 9, 2007.

8. Villagers were reimbursed on average 1,000 rubles for their possessions. See inventories in AOKMR, 23/1/45-b and 23/1/38a.

9. "Ot pereselentsev s. Kazhakul," June 13, 1959, OGAChO, R-1644/1/4a, 49.

10. Teplova, "Molchali do Chernobylia."

11. A. N. Komarovskii, A. V. Sitalo, and P. T. Shtefanu, November 19, 1954, OGAChO, 11381/22, 142–43.

12. Scientists estimate that from 1949 to 1951 alone, adults on the Techa took in an average of 4,600 microcuries, topping off at 200 rem. V. N. Novoselov and V. S. Tolstikov, *Atomnyi sled na Urale* (Cheliabinsk: Rifei, 1997), 39, 72.

13. Author interview with Dasha Arbuga, June 21, 2010, Sludorudnik.

14. Author interview with Evdokia (Dusia) Mel'nikova and Anna Kolynova, June 21, 2010, Sludorudnik.

15. Petryna, *Life Exposed*, 126–28; Natalia Manzurova, "Techa Contamination Report," unpublished, in possession of the author; Elizabeth Vainrub, "Twenty Years After the Chernobyl Nuclear Power Plant Accident," http://radefx.bcm.edu/chernobyl/english/links.htm.

Chapter 28: The Zone of Immunity

1. Lynne Viola, *The Unknown Gulag: The Lost World of Stalin's Special Settlements* (Oxford: Oxford University Press, 2007); Kate Brown, *A Biography of No Place: From Ethnic Borderland to Soviet Heartland* (Cambridge, MA: Harvard University Press, 2004); Katherine Jolluck, *Exile and Identity: Polish Women in the Soviet Union During World War II* (Pittsburgh, PA: University of Pittsburgh Press, 2002); J. Otto Pohl, *Ethnic Cleansing in the USSR, 1937–1949* (Westport, CT: Greenwood Press, 1999).

2. "Ob uvelichenii shtata politotodela bazy 10," April 21, 1950; "Spravka," March 17, 1951, OGAChO, 288/42/42, 47–49; "Aristov Malenkovu," 1950, OGAChO, 288/42/38, 48.

3. Meshik Beloborodovu, June 7, 1949; "Spravka o rabote politotdela bazy no. 10," October 25, 1949, OGAChO, 288/42/34, 16, 38–44.

4. For quote, "Stenogramma, Politotdela no. 106," January 30, 1952, OGAChO, P-1137/1/38, 59. See also "O neudovletvoritel'nom rukovodstve Politotdela ORSom," December 22, 1951, OGAChO, 288/45/51, 85. On 900,000 rubles embezzled from after-school programs, see "Sobranie partiinogo aktiva," April 10, 1952, OGAChO, P-1137/1/38, 163.

5. "Stenogramma, Politotdela no. 106," 67–69.

6. Ibid.

7. Ibid., 68–75.

8. "Zasedanie partiinogo aktiva," January 8, 1953, OGAChO, P-1137/1/48, 78.

9. "Spravka o massovykh besporiadkov zakliuchennykh," August 21, 1953, OGAChO, 288/42/56, 135–37.

10. "Prikaz MVDa o merakh ukrepleniia voinskoi ditsipliny," April 17, 1954, OGAChO, 1138/1/22, 114–48; V. N. Kuznetsov, *Zakrytye goroda Urala* (Ekaterinburg: Akademiia voenno-istoricheskikh nauk, 2008), 29.

11. "Sobranie partiinogo aktiva," April 10, 1952, OGAChO, 1137/1/38, 234–35.

12. "Zasedanie partiinogo aktiva," January 8, 1953, OGAChO, 1137/1/48, 80–84.

13. See Juliane Furst, *Stalin's Last Generation: Soviet Post-War Youth and the Emergence of Mature Socialism* (Oxford: Oxford University Press, 2010), 4.

14. Yoram Gorlizki and O. V. Khlevniuk, *Cold Peace: Stalin and the Soviet Ruling Circle, 1945–1953* (Oxford: Oxford University Press, 2004), 167.

15. L. D. Riabev, *Atomnyi proekt SSSR: Dokumenty i materialy*, vol. II, bk. 5 (Moscow: Nauka, 2007), 65.

16. On labor shortage "because of the amnesties," see "Protokol sed'moi konferenstii stroitel'stva no. 247 MSM," February 6–7, 1954, OGAChO, 1138/1/29, 21–31, 58–64; "Spravka," July 6, 1953, OGAChO, 274/20/33, 65–67. On Beria's reforms, Amy W. Knight, *Beria, Stalin's First Lieutenant* (Princeton, NJ: Princeton University Press, 1993), 185.

17. Kuznetsov, *Zakrytye goroda*, 103.

18. "Reshenie politicheskogo upravleniia MSM SSSR," February 15, 1954, OGAChO, 1138/1/22, 47; "Akt," March 17, 1954, OGAChO, 1138/1/25, 7–23.

19. Kuznetsov, *Zakrytye goroda*, 105.

20. "O ser'eznykh nedostatkakh . . . sredi kontingentov stroitelei," August 26, 1954, OGAChO, 1138/1/22, 125–28.

21. Kuznetsov, *Zakrytye goroda*, 105.

22. "O merakh uluchsheniia raboty ITL i kolonii MVD," September 3, 1954, OGAChO, 1138/1/18, 11–23; "Plan meropriiatii Kuznetskogo ITL MVD SSSR," September 17, 1954, OGAChO, 1138/1/18, 171–80.

23. M. Steven Fish, "After Stalin's Death: The Anglo-American Debate over a New Cold War," *Diplomatic History* 10, no. 4 (1986): 333–55.

24. Knight, *Beria*, 194–97.
25. "Ot Kytergina, nachal'nika Politotdela no 201," January 12, 1954, OGAChO, 1138/1/26, 6–7.
26. Irina Bystrova, *Voenno-promyshlennyi kompleks SSSR v gody kholodnoi voiny: Vtoraia polovina 40-kh-nachalo 60-kh godov* (Moscow: IRI RAN, 2000), 307.
27. V. Novoselov and V. S. Tolstikov, *Taina "Sorokovki"* (Ekaterinburg: Ural'skii rabochii, 1995), 195–96.
28. Ibid., 39–40.
29. Miriam Dobson, *Khrushchev's Cold Summer: Gulag Returnees, Crime, and the Fate of Reform After Stalin* (Ithaca: Cornell University Press, 2009), 34–35.
30. "Nachal'niku politupravleniia MSM, S. Baskakovu," December 17, 1953, OGAChO, 1138/1/29, 21–31 and 58–64; "O faktov khuliganstva," March 20, 1954, and "O merakh uluchsheniia raboty ITL i kolonii MVD," April 20, 1954, OGAChO, 1138/1/22, 56–63, 85–89; "Akt," March 11, 1954, OGAChO, 1138/1/25, 117–23.
31. "Sitalo Nachal'niku politotdela Glavpromstroiia MVD SSR," November 26, 1954, OGAChO, 1138/1/20, 35–36.
32. "Protokol 1-oi gorodkoi partiinoi konferentsii," August 16–17, 1956, OGAChO, 2469/1/1.
33. "7-oi partiinoi konferentsii stroitel'stva no. 247, MSM SSSR," February 6–7, 1954, OGAChO, 1138/1/29, 27–28, 59; "O rabote ofitserskikh sudov chesti," March 27, 1954, OGAChO, 1138/1/22, 56; "O sostoianii voinskoi distsipliny," April 20, 1954, OGAChO, 1138/1/22, 25; "Protokoly sobranii komsomol'skogo aktiva stroitel'stva i ispravitel'nom trudovykh lagerei," October 21, 1954, OGAChO. 1596/1/43, 47; "Nachal'niku politupravleniia MSM S. Baskakovu," 21–31 and 58–64.
34. "7-oi partiinoi konferentsii stroitel'stva no. 247, MSM," 22.
35. Ibid., 58–63.
36. "Ob osnovnykh zadachakh MVD SSSR," March 29, 1954, OGAChO, 1138/1/22, 104, 114–20.
37. "Protokoli no. 17 i 22, zasedaniia biuro Ozerskogo Gorkoma KPSS," December 13, 1956, July 2, 1957, OGAChO, 2469/1/3, 167–75, and 2469/1/120, 250–75.
38. A. N. Komarovskii, A. V. Sitalo, and P. T. Shtefanu, November 19, 1954, OGAChO, 11381/22, 142–43.
39. "Protokol no. 17, zasedaniia biuro Oserskogo Gorkoma KPSS," December 13, 1956, OGAChO, 2469/1/3, 167–75; Zaveniagin, "O zavershenii pereseleniia zhitelei iz likvidirue-mykh naselennykh punktov," January 20, 1956, OGAChO, R-288/42/67, 59.
40. Antonov Sitalo, December 9, 1954, OGAChO, 1138/1/22, 157–61.
41. "Protokol 1-i gorodskoi partiinoi konferentsii," August 16–17, 1956, OGAChO, 2469/1/1, 93.
42. "Kruglov nachal'nikam stroitel'stv glavpromstroiia MVD SSSR," March 11, 1954, and "Usloviia," 1954, OGAChO, 1138/1/22, 81–83.
43. "Protokol no. 1 and no. 2," August 16–17, 1956, and October 10, 1956, OGAChO, 2469/1/1, 54 and 2469/1/2, 8.
44. "Akt," March 11, 1954, and "Reshenie politicheskogo otdela no. 201," October 2, 1954, OGAChO, 1138/1/25, 117–23, 66–68; "7-oi partiinoi konferentsii stroitel'stva no. 247," 28–29, 59–61.
45. "Protokol no. 2," 18.
46. Batin Volkovu, October 11, 1955, and Batin, October 18, 1955, OGAChO, 107/22/67, 49–50, 52–53; Kuznetsov, *Zakrytye goroda*, 29.
47. A. Komarovskii and P. T. Shtefanu, November 19, 1954, OGAChO, P-1138/1/22, 142–43; Shtefan Greshinovu i Sitalo, November 27, 1954, OGAChO, 1138/22/1.
48. "Protokol no. 2," 10–11.
49. Ibid., 13.
50. "Protokol 1-oi gorodskoi partiinoi konferentsii," OGAChO, 2469/1/1, 91, 63–66. On embezzlement in the nuclear establishment, see Bystrova, *Voenno-promyshlennyi kompleks*, 314, 318.

51. "Protokol 1-oi gorodskoi partiinoi konferentsii," 107–10.
52. Bystrova, *Voenno-promyshlennyi kompleks*, 178.
53. "Protokol 1-oi gorodskoi partiinoi konferentsii," 53–54; "Protokol no. 8, Biuro Ozerskogo gorkoma," October 2, 1956, OGAChO, 2469/1/4, 1–12.
54. "Dostanovlenue, IV-ogo plenuma GK KPSS," July 19, 1960, OGAChO, 2469/3/3, 126–65.
55. "Tolmadzhev A. V. Sitalo," December 15, 1954, OGAChO, 11138/1/22, 155; "7-oi partiinoi konferentsii stroitel'stva no. 247, MSM SSSR," February 6–7, 1954, OGAChO, 1138/1/29, 28; "Kamorin Aristovu," September 12, 1952, OGAChO, 288/42/51, 105; "Protokoly sobraniia komsomol'skogo aktiva stroitel'stva i ispravitel'nom trudovykh lagerei," April 10, 1954, OGAChO, 1596/1/43, 15.

Chapter 29: The Socialist Consumers' Republic

1. "Postanovleniia biuro Cheliabinskogo obkoma KPSS," September 1, 1956, OGAChO, 288/42/65, 34.
2. Ibid.
3. "Protokol 1-oi gorodskoi partiinoi konferentsii," August 16–17, 1956, OGAChO, 2469/1/1, 80.
4. Ibid., 104.
5. "Spravka zabolevaemosti rabotaiushikh,"1959, OGAChO, 2469/3/2, 113–14.
6. "Protokol 1-oi gorodskoi partiinoi konferentsii," 91.
7. "Spravka o potrebnosti v zh/ploshadi po zavodu na 1957 god," OGAChO, 2469/1/5, 173.
8. "Protokol no. 2," October 10, 1956, OGAChO, 2469/1/2, 10–11.
9. "Stenogramma zasedaniia biuro gorkoma KPSS," December 7, 1956, OGAChO, 2469/1/5, 18–37.
10. On the centrality of architecture for Soviet utopian models, see Katerina Clark, "Socialist Realism and the Sacralizing of Space," in E. A. Dobrenko and E. Naiman, *The Landscape of Stalinism: The Art and Ideology of Soviet Space* (Seattle: University of Washington Press, 2003), 3–18.
11. "Protokol no. 3, biuro Ozerskogo Gorkoma KPSS," August 29, 1956, OGAChO, 2469/1/3, 15.
12. "Protokol no. 17, biuro Ozerskogo Gorkoma KPSS," December 13, 1956, OGAChO, 2469/1/3, 167–75.
13. "Stenogramma zasedaniia biuro Ozerskogo Gorkoma KPSS," December 8, 1956, OGA-ChO, 2469/1/5, 43–44.
14. "Stenogramma Zasedaniia biuro Gorkoma KPSS s uchastiem chlenov biuro pervichoi partorganizatsii TsZL," December 7, 1956, OGAChO, 2469/1/5, 18.
15. Ibid., 55.
16. "Protokol no. 2 zasedaniia biuro Ozerskogo Gorkoma," December 17, 1957, OGAChO, 2469/1/121, 287.
17. V. Novoselov and V. S. Tolstikov, *Taina "Sorokovki"* (Ekaterinburg: Ural'skii rabochii, 1995), 190.
18. "Protokol no. 22, zasedaniia biuro Ozerskogo Gorkoma," July 2, 1957, OGAChO, 2469/1/122, 250–305.
19. "Reshenie 353," November 19, 1959, OGAChO, 2469/3/3, 51; "O rabote piatoi gorodskoi partinoi konferentsii," December 16–17, 1960, OGAChO, 2469/1/3, 43–44; "O pere-stroike raboty narodnoi druzhiny goroda" May 22, 1962, OGAChO, 2469/4/3, 257–75; "Zasedanii biuro gorkoma KPSS protokol no 46," October 23, 1962, OGAChO, 2469/4/5, 110–55; "Protokoly zasedaniia biuro gorkoma KPSS," January 12, 1965, OGAChO, 2469/5/292, 5–6.
20. "Dostanovlenie, iv-ogo Plenuma GK KPSS," July 19, 1960, OGAChO, 2469/3/3, 153; Mardasov, "Protokol sobraniia," November 3, 1957, OGAChO, 2469/1/119, 121.
21. "O povyshenii roli obshchestvennosti v bor'be s prestupnost'iu," January 12, 1960, OGA-ChO, 2469/3/3, 30–32; "Dostanovlenie, iv-ogo Plenuma GK KPSS."

22. "O sostoianii i merakh usileniia bor'by s detskoi beznadzornost'iu" November 19, 1959, OGAChO, 2469/3/2, 51–53.

23. Brian Lapierre, "Making Hooliganism on a Mass Scale," *Cahiers du monde russe* 1–2 (2006): 359, 374; Edward D. Cohn, "Disciplining the Party: The Expulsion and Censure of Communists in the Post-War Soviet Union, 1945–1961," PhD diss., University of Chicago, 2007, 5; Oleg Kharkhordin, *The Collective and the Individual in Russia: A Study of Practices* (Berkeley: University of California Press, 1999).

24. For a critique, see Alexei Yurchak, "Soviet Hegemony of Form: Everything Was Forever, Until It Was No More," *Comparative Studies in Society and History* 45, no. 3 (2003): 482.

25. On child care, see "Zasedanie plenumov gorkoma KPSS," July 6, 1963, OGAChO, 2469/4/244a, 156. On restricting production work to men, see "Stenogramma 3-oi gorodskoi partiinoi konferentsii," December 14–15, 1958, OGAChO, 2469/2/1, 26, and "Materialy proverki raboty, proforganizatsii ob'ekta-20," no later than May 1959, OGAChO, 2469/3/2, 167–76.

26. See, for example, "Zasedanie plenumov gorkoma KPSS," July 6, 1963, OGAChO, 2469/4/244a, 130–51; "Protokol no. 10 zasedaniia biuro gorkoma KPSS za 1965 god," March 2, 1965, OGAChO, 2469/5/292, 221–23.

27. The cases are innumerable. See the "personal affair" sections of "Zasedanie biuro gorkoma KPSS protokoly no. 26–50," OGAChO, 2469/4/5, 82–256.

28. "Zasedanie plenumov gorkoma KPSS," July 6, 1963, OGAChO, 2469/4/244a, 151.

29. "Spravka," January 1960, OGAChO, 2469/3/3, 84–87.

30. "Protokol sobraniia aktiva gorodskoi partiinoi organizatsii," November 3, 1957, OGAChO, 2469/1/119 ll. 159–70; "Protokol no. 30, zasedaniia biuro Ozerskogo Gorkoma," September 9, 1957, OGAChO, 2469/1/121, 100–115; "Stenogrammy na vtoroi Ozerskoi gorodskoi partkonferentsii," November 30, 1957, OGAChO, 2469/1/117, 1–40; "Postanovlenie," January 22, 1957, OGAChO, 2469/1/121, 68–70; "Protokol no. 4, zasedaniia biuro Ozerskogo Gorkom," January 29, 1957, OGAChO, 2469/1/121, 108–10; "Zasedanie III-oi gorodskoi partiinoi konferentsii," December 14–15, 1958, OGAChO, 2469/2/1, 1–200.

31. "Protokol no. 2," October 10, 1956, OGAChO, 2469/1/2, 17 and "Protokol no. 1," August 16–17, 1956, OGAChO, 2469/1/1, 63–66.

32. "Spravka o vypolnenii postanovleniia SM SSSR ot 20 Marta 1957 goda" and "Bezdomov Kozlovu," June 23, 1960, and May 18, 1959, OGAChO, R-1644/1/4a, 105–6, 8.

33. "Bezdomov Churinu," July 6, 1959, "Poiasnitel'naia zapiska," December 8, 1959, and "Spravka po otseleniiu iz zony reki Techa," March 29, 1962, OGAChO, R-1644/1/4a.

34. "Spravka o vypolnenii postanovleniia SM SSSR," 116–18.

35. "Na no. 021-102 ot 15/VI-s.g.," July 25, 1959, "Spravka," October 5, 1959, Kaprenko Polianskomu, November 26, 1959, and "Spravka," January 1960, OGAChO, R-1644/1/4a, 105, 81, 77, 62, 92–94.

36. Zaveniagin, "O zavershenii pereseleniia zhitelei," January 20, 1956, OGAChO, 288/42/67, 59.

37. "E. Mamontov i Dibobes, gossaninspektor zony zagriazneniia Nadykto," May 23, 1961, OGAChO, R-1644/1/4a, 153–54, 149.

38. "Spravka Cheliabinskogo oblispolkoma," February 6, 1960, OGAChO, R-1644/1/4a, 193–95. Regional officials were seeking yet more funds for resettlement in 1962. See in the same file Karapol'tsev, September 12, 1962, 180–1.

39. "Stenogramma 3-oi gorodskoi partiinoi konferentsii," December 14–15, 1958, OGAChO, 2469/2/1, 26.

40. "K spravke po otseleniiu zhitelei," July 7, 1959, OGAChO, R-1644/1/4a, 29–33.

41. Elena Efremova, "Zhiteli Musliumovo nachnyt pereseliat' na drugoi bereg radioaktivnoi reki," *Ekologiia i pravo* 27 (2008): 12–14.

42. "Protokol no. 49, zasedaniia biuro gorkoma KPSS," April 18, 1967, OGAChO, 2469/6/406, 137. On soldiers in hazardous conditions, see Guseev, "Otchetnii doklad," December 8, 1964, OGAChO, 2469/5/1, 51–53.

43. Zhores Medvedev, "Krepostnye spetzkontingenty krasnoi armii," *Ural'*, May 1995, 221–22.

44. Shmygin Efremovu, March 1962, and Churin Efremovu, June 14, 1962, OGAChO, 288/42/79 5–7, 30–31.
45. "Doklad VIII-oi gorodskoi partiinoi konferentsii," December 8, 1964, OGAChO, 2469/5/1, 96–100.

Chapter 30: The Uses of an Open Society

1. "General Electric Theater," Museum of Broadcast Communication, www.museum.tv/archives/etv/G/htmlG/generalelect/generalelect.htm.
2. May 1999, Alumni Sandstorm (online archive), alumnisandstorm.com.
3. *TCH*, October 14, 1949, and *CBN*, July 3 and May 8, 1950.
4. Herbert Parker, "Status of Ground Contamination Problem," September 15, 1954, DOE Opennet, HW 33068.
5. *CBN*, September 8, 1950; "27 Questions and Answers About Radiation," September 1951, NARA, RG 326, 67A, box 55, folder 461.
6. "Managers' Data Book," June 1949, and "Community Data Book," 1952, JPT, acc. 5433-001, box 25; Ralph R. Sachs, MD, "Study of 'Atomic City,'" *Journal of the American Medical Association* 154, no. 1 (1954): 44–49.
7. Scientists call this the "healthy worker syndrome." Jan-Olov Liljenzin, Jan Rydbert, and Gregory Choppin, *Radiochemistry and Nuclear Chemistry* (Oxford: Butterworth-Heinemann, 2002), 496.
8. Bernard Bucove, "Vital Statistics Summary" (Olympia: Washington State Department of Health, 1959). My thanks to Dorothy Kenney for compiling this data.
9. Michael D'Antonio, *Atomic Harvest* (New York: Crown, 1993), 66.
10. "HWS Monthly Report, June 1952," July 21, 1952, PRR, HW 24928, G-10.
11. Rebecca Nappi, "Grave Concerns," *SR*, October 17, 1993, F1.
12. Kristoffer Whitney, "Living Lawns, Dying Waters: The Suburban Boom, Nitrogenous Fertilizers, and the Nonpoint Source Pollution Dilemma," *Technology and Culture* 51, no. 3 (2010): 652–74.
13. Walter J. Williams, "Certain Functions of the Hanford Operations Office—AEC," October 8, 1948, JPT, acc. 5433-1, box 24.
14. On volume of DDT, see *Villager*, March 27, 1947 and *TCH*, January 18, 1950. For mosquito control, see Fred Clagett Papers, CREHST, acc. 3543-004, 6/2, "Mosquitos."
15. Siddhartha Mukherjee, *The Emperor of All Maladies: A Biography of Cancer* (New York: Scribner, 2010), 92.
16. "Birth Defect Research for Children, Fact Sheet," www.pan-uk.org/pestnews/Actives/ddt.htm.
17. F. Herbert Bormann, Diana Balmori, Gordon T. Geballe, and Lisa Vernegaard, *Redesigning the American Lawn: A Search for Environmental Harmony* (New Haven, CT: Yale University Press, 1993), 83.
18. J. Samuel Walker, *Permissible Dose: A History of Radiation Protection in the Twentieth Century* (Berkeley: University of California Press, 2000), 10.
19. J. N. Yamazaki, "Perinatal Loss and Neurological Abnormalities Among Children of the Atomic Bomb," *Journal of the American Medical Association* 264, no. 5 (1990): 605–9; F. A. Mettler and A. C. Upton, *Medical Effects of Ionizing Radiation* (Philadelphia: W. B. Saunders, 1995), 323.
20. Gregory L. Finch Werner Burkart and Thomas Jung, "Quantifying Health Effects from the Combined Action of Low-Level Radiation and Other Environmental Agents," *Science of the Total Environment* 205, no. 1 (1997): 51–70.
21. Linda Lorraine Nash, *Inescapable Ecologies: A History of Environment, Disease, and Knowledge* (Berkeley: University of California Press, 2006), 185.
22. Ulrich Beck, "The Anthropological Shock: Chernobyl and the Contours of the Risk Society," *Berkeley Journal of Sociology* 32 (1987): 153–65.

23. "Hanford Laboratories Operation Monthly Activities Report," February 1957, DOE Opennet, HW-48741.
24. Paul F. Foster to James T. Ramey, August 12, 1958, and General MacArthur to Secretary of State, August 13, 1958, DOE Germantown RG 326, 1360, folder 1.
25. Daniel P. Aldrich, *Site Fights: Divisive Facilities and Civil Society in Japan and the West* (Ithaca, NY: Cornell University Press, 2008), 124–25.
26. Walker, *Permissible Dose*, 19.
27. See letters in DOE Germantown RG 326, 1360, folder 1.
28. "Role of Atomic Energy Commission Laboratories," September 17, 1959, DOE Opennet, 8, 11; S. G. English, "Possible Reorganization of the Environmental Affairs Group," March 6, 1970, DOE Germantown, RG 326/5618/15.
29. "Regarding Hidden Rules Governing Disclosure of Biomedical Research," December 8, 1994, DOE Opennet, NV 0750611.
30. Dunham to Bronk, December 20, 1955; "Minutes of the 66th Meeting Advisory Committee for Biology and Medicine (ACBM)," January 10–11, 1958, DOE Opennet, NV 0411748 and NV 0710420.
31. Susan Lindee, *Suffering Made Real: American Science and the Survivors at Hiroshima* (Chicago: University of Chicago Press, 1994); David Richardson, Steve Wing, and Alice Stewart, "The Relevance of Occupational Epidemiology to Radiation Protection Standards," *New Solutions* 9, no. 2 (1999): 133–51.
32. Parker, "Control of Ground Contamination," August 19, 1954, PRR, HW 32808.
33. Author phone interview with Juanita Andrewjeski, December 2, 2009.
34. Parker, "Control of Ground Contamination."
35. Parker, "Status of Ground Contamination Problem."
36. E. R., Irish, "The Potential of Wahluke Slope Contamination," June 11, 1958, PRR, HW 56339.
37. As quoted in E. J. Bloch, "Hanford Ground Contamination," September 17, 1954, DOE Opennet, RL-1-331167.
38. Parker, "Control of Ground Contamination."
39. Bloch, "Hanford Ground Contamination."
40. Herbert Parker, "Columbia River Situation," August 19, 1954, NARA, RG 326 650 box 50, folder 14.
41. "Hanford Works Monthly Report," June 21, 1951, DOE Opennet, HW 21260.
42. Parker, "Columbia River Situation."
43. Ibid.
44. L. K. Bustad et al., "A Comparative Study of Hanford and Utah Range Sheep," HW 30119, LKB, box 14; "Biology Research Annual Report, 1956," PRR, HW 47500.
45. "Bulloch v. Bustad, Kornberg, General Electric et al.," LKB, ms 2008-19, box 7, folder "Bustad Personal."
46. Michele Stenehjem Gerber, *On the Home Front: The Cold War Legacy of the Hanford Nuclear Site* (Lincoln: University of Nebraska Press, 1992), 97–98.
47. Leo K. Bustad, *Compassion: Our Last Great Hope* (Renton, WA: Delta Society, 1990), 4.
48. Gerber, *On the Home Front*, 69.

Chapter 31: The Kyshtym Belch, 1957

1. Author interview with Galina Petruva [pseudonym], June 6, 2010, Kyshtym, Russia.
2. Paul Josephson, "Rockets, Reactors and Soviet Culture," in Loren R. Graham, ed., *Science and the Soviet Social Order* (Cambridge, MA: Harvard University Press, 1990), 168–91.
3. "Protokol," August 1957, OGAChO, 2469/1/121, 62; "Protokol 3-oi gorodskoi partiinoi konferentsii," December 14–15, 1958, OGAChO, 2469/2/1, 15.
4. Rosatom officials say the explosion was a chemical explosion. Vladyslav Larin argues it was nuclear. Author interview with Larin, August 19, 2009, Moscow.

5. Valery Kazansky, "Mayak Nuclear Accident Remembered," *Moscow News*, October 19, 2007, 12.
6. N. G. Sysoev, "Armiia v ekstremal'nykh situatsiiakh: Soldaty Cheliabinskogo 'Chernobylia,'" *Voenno-istoricheskii zhurnal* 12 (1993): 39–43.
7. "II-oi gorodskoi partinoi konferentsii gorkoma Ozerska," November 30–December 1, 1957, OGAChO, 2469/1/117, 168, 234; "Protokol piatogo plenuma gorkoma," October 8, 1957, OGAChO, 2469/1/118, 105.
8. V. N. Novoselov and V. S. Tolstikov, *Atomnyi sled na Urale* (Cheliabinsk: Rifei, 1997), 93.
9. "Protokol piatogo plenuma gorkoma," 104; Sysoev, "Armiia v ekstremal'nykh situatsiiakh," 39.
10. Author interview with Petruva.
11. Kazansky, "Mayak," 12.
12. "Protokol piatogo plenuma gorkoma," 97, 101; Leonid Timonin, *Pis'ma iz zony: Atomnyi vek v sud'bakh tol'iattintsev* (Samara: Samarskoe knizhnoe izd-vo, 2006), 11.
13. "Protokol piatogo plenuma gorkoma," 104.
14. Ibid., 105.
15. Sysoev, "Armiia v ekstremal'nykh situatsiiakh," 40–43.
16. Zhores Medvedev, "Do i posle tragedii," *Ural'* 4 (April 1990): 108.
17. "Kriticheskie zamechanie," August 15, 1958, OGAChO, 2469/2/4, 21–29.
18. Vladyslav B. Larin, *Kombinat "Maiak"—Problema na veka* (Moscow: KMK Scientific Press, 2001), 48–49.
19. "Kriticheskie zamechanie," August 15, 1958, OGAChO, 2469/2/4, 21–29.
20. V. Chernikov, *Osoboe pokolenoe* (Cheliabinsk: V. Chernikov, 2003), 148–58; Larin, *Kombinat*, 162.
21. "Protokol piatogo plenuma gorkoma," 100.
22. Ibid., 101–3.
23. Alexei Mitiunin, "Natsional'nye osobennosti likvidatsii radiatsionnoe avarii," *Nezavisimaia gazeta*, April 15, 2005; Timonin, "Pis'ma iz zony," 123.
24. Mira Kossenko, "Where Radiobiology Began in Russia," Defense Threat Reduction Agency, Fort Belvoir, VA, 2011, 50.
25. "Zasedanie II-oi gorodskoi partinoi konferentsii gorkoma Ozerska," November 30–December 1, 1957, OGAChO, 2469/1/117, 1–3; Novoselov and Tolstikov, *Atomnyi sled*, 126.
26. Author interview with Petruva and Sergei Aglushenkov, June 26, 2010, Kyshtym; Bokin, "Posledstviia Avarii." Prisoners later complained of illnesses: "Spravka o rabote nabliudatelnoi komissii," January 9, 1960, OGAChO, 2469/3/3, 59–64.
27. Kazansky, "Mayak," 12.
28. Kossenko, "Where Radiobiology Began," 50.
29. "Protokol piatogo plenuma gorkoma," 101–2.
30. Ibid., 104.
31. "Postanovlenie," October 8, 1957, OGAChO, 2469/1/118, 107.
32. "Zasedanie II-oi gorodskoi partinoi konferentsii gorkoma Ozerska," 168.
33. "Protokol piatogo plenuma gorkoma," 102–3; Zhores A. Medvedev, *The Legacy of Chernobyl* (New York: W. W. Norton, 1990), 105.
34. "Stenogrammy na vtoroi ozerskoi gorodskoi partkonferentsii," November 30, 1957, OGAChO, 2469/1/117, 19–40; "Spravka po vyseleniiu," May 9, 1958, OGAChO, 2469/2/3, 23.
35. V. Novoselov and V. S. Tolstikov, *Taina "Sorokovki"* (Ekaterinburg: Ural'skii rabochii, 1995), 187.
36. "Protokol sobraniia aktiva gorodskoi partiinoi organizatsii," November 3, 1957, OGAChO, 2469/1/119, 156.
37. "Zasedanie II-oi gorodskoi partiinoi konferentsii gorkoma Ozerska," 201.
38. "Stenogrammy na vtoroi ozerskoi gorodskoi partkonferentsii," 19–40.
39. "Protokol sobraniia aktiva gorodskoi partiinoi organizatsii" and "3-oi gorodskoi partiinoi konferentsii," December 14–15, 1958, OGAChO, 2469/2/1, 159–70, 25.
40. "Stenogrammy na vtoroi ozerskoi gorodskoi partkonferentsii," 205, 238.

41. N. V. Mel'nikova, *Fenomen zakrytogo atomnogo goroda* (Ekaterinburg: Bank kul'turnoi informatsii, 2006), 99–100.

Chapter 32: Karabolka, Beyond the Zone

1. N. G. Sysoev, "Armiia v ekstremal'nykh situatsiiakh: Soldaty Cheliabinskogo 'Chernobylia,'" *Voenno-istoricheskii zhurnal* 12 (1993): 39–43.
2. Author interview with Gulnara Ismagilova, August 17, 2009, Tatarskaia Karabolka.
3. V. N. Novoselov and V. S. Tolstikov, *Atomnyi sled na Urale* (Cheliabinsk: Rifei, 1997), 117.
4. Vladyslav B. Larin, *Kombinat "Maiak"—Problema na veka* (Moscow: KMK Scientific Press, 2001), 52.
5. Novoselov and Tolstikov, *Atomyi sled*, 110.
6. Ibid., 120; V. Chernikov, *Osoboe pokolenoe* (Cheliabinsk: V. Chernikov, 2003), 9.
7. Author interview with Dasha Arbuga, June 20, 2010, Sludorudnik, Russia.
8. Novoselov and Tolstikov, *Atomnyi sled*, 121.
9. Larin, *Kombinat*, 291.
10. "Ob organizatsii vspashki zagriaznennikh zemel'," April 9, 1958, and "O perevode zagriaznennykh zemel'," May 27, 1958, AOKMR, 23/1/37a, 1, 2, 11, 12, 30, in personal archive of Gulnara Ismagilova.
11. Novoselov and Tolstikov, *Atomnyi sled*, 138; Alexei Povaliaev and Ol'ga Konovalova, "Ot Cheliabinska do Chernobylia," *Promyshlennye vedomosti*, October 16, 2002.
12. "Ob usilenii okhrany zony zagriazneniia," November 14, 1958, OGAChO, R-274/20/48, 159–62.
13. Author interview with Ismagilova.
14. Planet of Hopes press release, ""Mayak" Used 2,000 Pregnant Women in Dangerous Clean Up of Nuclear Disaster," October 30, 2006, Moscow.
15. E. Rask, "Spravka," February 6, 1960, OGAChO, R-1644/1/4a, 193–95.
16. Vladyslav B. Larin, *Kombinat "Maiak"—Problema na veka* (Moscow: KMK Scientific Press, 2001), 55.
17. "O provedenii dopolnitel'nykh meropriiatii v zone radioaktivnogo zagriazneniia," September 29, 1959, personal archive, Gulnara Ismagilova.
18. Rask, "Spravka," 193–95.
19. Kh. Tataullinaia, April 18, 2000, personal archive, Gulnara Ismagilova.
20. Larin, *Kombinat*, table 6.27, 412.
21. Fauziia Bairamova, *Iadernyi arkhipelag ili atomnyi genotsid protiv Tatar* (Kazan': Nauchno-populiarnoe izdanie, 2005).
22. On contractor corruption, see "O rabote piatoi gorodskoi partiinoi konferentsii," December 16–17, 1960, OGAChO, 2469/1/3, 127.
23. Novoselov and Tolstikov, *Atomnyi sled*, 140–43.
24. The station in Russian was Opytnaia nauchno issledovatel'skaia stantsiia (ONIS).
25. V. N. Pozolotina, Y. N. Karavaeva, I. V. Molchanova, P. I. Yushkov, and N. V. Kulikov, "Accumulation and Distribution of Long-Living Radionuclides in the Forest Ecosystems of the Kyshtym Accident Zone," *Science of the Total Environment* 157, no. 1–3 (1994): 147; Tatiana Sazykina and Ivan Kryshev, "Radiation Effects in Wild Terrestrial Vertebrates—the EPIC Collection," *Journal of Environmental Radioactivity* 88, no. 1 (2006): 38; Larin, *Kombinat*, 148–51.
26. Nadezhda Kutepova and Olga Tsepilova, "Closed City, Open Disaster," in Michael R. Edelstein, Maria Tysiachniouk, and Lyudmila V. Smirnova, eds., *Cultures of Contamination: Legacies of Pollution in Russia and the U.S.* (Amsterdam: JAI Press, 2007), 14:156.
27. Institut global'nogo klimata i ekologii, "Karta zagriazneniia pochv strontsiem-90," 2005.
28. Novoselov and Tolstikov, *Atomnyi sled*, 127.
29. Robert Standish Norris, Kristen L. Suokko, and Thomas B. Cochran, "Radioactive Contamination at Chelyabinsk-65, Russia," *Annual Review of Energy and the Environment* 18 (1993): 522.

30. Valery Soyfer, "Radiation Accidents in the Southern Urals (1949–1967) and Human Genome Damage," *Comparative Biochemistry and Physiology* Part A, no. 132 (2002): 723.

31. T. G. Sazykina, J. R. Trabalka, B. G. Blaylock, G. N. Romanov, L. N. Isaeva, and I. I. Kryshev, "Environmental Contamination and Assessment of Doses from Radiation Releases in the Southern Urals," *Health Physics* 74, no. 6 (1998): 687; E. Tolstyk, L. M. Peremyslova, N. B. Shagina, M. O. Degteva, I. M. Vorob'eva, E. E. Tokarev, and N. G. Safronova, "The Characteristics of 90-Sr Accumulation and Elimination in Residents of the Urals Region in the Period 1957–1958," *Radiatsionnaia biologiia, radioekologiia* 45, no. 4 (2005): 464–73.

32. "O perevode zagriaznennykh zemel," 30.

33. Natalia Mironova, Maria Tysiachniouk, and Jonathan Reisman, "The Most Contaminated Place on Earth: Community Response to Long-term Radiologial Disaster in Russia's Southern Urals," in Michael R. Edelstein, Maria Tysiachniouk, and Lyudmila V. Smirnova, eds., *Cultures of Contamination: Legacies of Pollution in Russia and the U.S.* (Amsterdam: JAI Press, 2007), 14:179–80.

Chapter 33: Private Parts

1. Iral C. Nelson and R. F. Foster, "Ringold—A Hanford Environmental Study," April 3, 1964, PRR, HW-78262 REV.

2. "Internal Dosimetry Results—Ringold," January 30, 1963, PRR, PNL 10337.

3. "Letter to Subject," and memo to A. R. Keene, December 14, 1962, PPR, PNL-10335; "Status of Columbia River Environmental Studies for Hanford Works Area," July 31, 1961, DOE, Germantown, RG 326, 1360, 3.

4. Nelson and Foster, "Ringold," 12.

5. On the unlikeliness of determining a gross effect from a small study, see Morris H. DeGroot, "Statistical Studies of the Effect of Low Level Radiation from Nuclear Reactors on Human Health," *Mathematics, Statistics and Probability* 6 (1972): 223–34.

6. A. R. Luedecke to all Managers of Operations, April 6, 1959; Willard F. Libby, "Statement on Strontium 90 in Minnesota Wheat Made Before the JCAE," February 27, 1959; "Statement, John A. McCone, Chairman AEC," March 24, 1959, DOE Germantown RG 326, 1360, 1; Joseph Lieberman to A. R. Luedecke, December 11, 1959, and Luedecke, "Dispersal of Radioactive Materials into the Pacific via the Columbia River," December 31, 1959, RG 326, 1359, 7.

7. R. H. Wilson and T. H. Essig, "Criteria Used to Estimate Radiation Doses Received by Persons Living in the Vicinity of Hanford," July 1968, JPT, BNWL-706 UC-41; "Evaluation of Radiological Conditions in the Vicinity of Hanford for 1967," March 1969, JPT, BNWL-983 UC-41.

8. W. E. Johnson, "Expanded Use of Whole Body Counter," January 26, 1962, DOE Opennet, NV0719090.

9. "Letter to Subject."

10. Author interview, LH, September 2011.

11. My thanks to Harry Winsor for his analysis.

12. A National Cancer Institute study found that Washington State had some of the lowest levels of fallout in the nation. *SR*, August 10, 1997, B1.

13. R. W. Perkins et al., "Results of a Test of Sampling in I-131 Plumes," April 18, 1963, PRR, HW 77387.

14. "Atmospheric Pathway Dosimetry Report, 1944–1992," October 1994, DOE Opennet, PNWD-2228 HEDR; Patricia P. Hoover, Rudi H. Nussbaum, Charles M. Grossman, and Fred D. Nussbaum, "Community-Based Participatory Health Survey of Hanford, WA, Downwinders: A Model for Citizen Empowerment," *Society and Natural Resources* 17 (2004): 551.

15. Sonja I. Anderson, "A Conceptual Study of Waste Histories from B Plant and Other Operations, Accidents, and Incidents at the Hanford Site Based upon Past Operating Records,

Data, and Reports, Project ER4945," September 29, 1994, unpublished, in possession of author. My thanks to Harry Winsor for help with this data.

16. Senior Engineer to R. F. Foster, September 20, 1962, PRR, PNL 9724.

17. See figures 2 and 3 in R. W. Perkins et al., "Test of Sampling in I 131 Plumes," April 18, 1963, PRR, HW 77387, 16.

18. Jackson, "On Authorizing Appropriations for the AEC," August 6, 1958, HMJ, acc. 3560-6 51c/11.

19. "AEC Plan for Expansion of Research in Biology and Medicine," August 4, 1958, DOE Germantown RG 326, 1360, 1; "Role of Atomic Energy Commission Laboratories," October 1, 1959, DOE Opennet, NV 0702108.

20. Ibid.; Arthur S. Flemming, Secretary of Health, Education and Welfare, Press Release, March 16, 1959, DOE Germantown RG 326, 1360, 1.

21. H. Schlundt, J. T. Nerancy, and J. P. Morris, "Detection and Estimation of Radium in Living Persons," *American Journal of Roentgenology and Radium Therapy* 30 (1933): 515–22; R. E. Rowland, *Radium in Humans: A Review of U.S. Studies* (Argonne, IL: Argonne National Lab, 1994).

22. Thomas H. Maugh II, "Eugene Saenger, 90," *Los Angeles Times*, October 6, 2007; "DOE Facts, Additional Human Experiments," GWU.

23. Among files on AEC medical research, see NAA, files in No. 116, series 16, 4DO-326-97-001. For summaries, see W. J. Bair to P. K. Clark, December 6, 1985, DOE Opennet, PNL-9358; Eileen Welsome, *The Plutonium Files: America's Secret Medical Experiments in the Cold War* (New York: Dial Press, 1999); Andrew Goliszek, *In the Name of Science: A History of Secret Programs, Medical Research, and Human Experimentation* (New York: St. Martins Press, 2003), 135–65.

24. Goliszek, *In the Name of Science*, 155.

25. "Minutes of the 66th Meeting Advisory Committee for Biology and Medicine," January 10–11, 1958, DOE Opennet.

26. R. F. Foster and J. F. Honstead, "Accumulation of Zinc-65 from Prolonged Consumption of Columbia River Fish," *Health Physics* 13, no. 1 (1967): 39–43; "Internal Depositions of Radionuclides in Men," February 1967, PRR, PNL 9287; "Whole Body Counting, Project Proposal," March 1966, PNL 9293; "Excretion Rates vs. Lung Burdens in Man," April 1966, PNL 9294; J. F. Honstead and D. N. Brady, "Report: The Uptake and Retention of P^{32} and Zn^{65} from the Consumption of Columbia River Fish," document BNSA-45, October 7, 1969, http://guin.library.oregonstate.edu/specialcollections/coll/atomic/catalogue/atomic-hanford_1-10.html.

27. K. L. Swinth to W. E. Wilson, April 11, 1967, PRR, PNL 9669.

28. Ralph Baltzo to Richard Cunningham, April 6, 1966, PRR, PNL 9086.

29. Alvin Paulsen, "Study of Irradiation Effects on the Human Testes," March 12, 1965, PRR, PNL 9081 DEL.

30. Carl Heller, "Effects of Ionizing Radiation on the Testicular Function of Man," May 1972, DOE Opennet, HW 709914, 3.

31. Fink to Friedell, December 5, 1945, NAA, RG 326 8505, box 54, MD 700.2.

32. Linda Roach Monroe, "Accident at Nuclear Plant Spawns a Medical Mystery," September 10, 1990, *Los Angeles Times*; author interview with Marge DeGooyer, May 16, 2008, Richland, WA.

33. "Accidental Nuclear Excursion, 234-5 Facility," 1962, PRR, HW 09437.

34. Ibid.

35. "Oral History of Health Physicist Carl C. Gamertsfelder, Ph.D.," January 19, 1995, DOE Opennet.

36. Parker, "Assistance to Dr. Paulsen," May 1, 1963, PRR, PNL 9074.

37. Author interview with Richard Sutch, San Francisco, May 11, 2012.

38. D. K. Warner to William Roesch, July 2, 1963, PRR, PNL 9076.

39. Holsted to Parker, "Assistance to Dr. Paulsen," July 9, 1963, PRR, PNL 9077.

40. Baltzo to Newton, May 28, 1968, PRR, PNL 9104; W. E. Wilson to Baltzo, June 14, 1968, PNL 9107.
41. R. S. Paul to S. L. Fawcett, "Experiments with People—Policy Need?" September 23, 1965, PRR, PNL 9082.
42. "Case No. 3," November 17, 1967, PRR, PNL 9315; E. E. Newton to P. T. Santilli, July 27, 1967, PNL 9092; S. L. Fawcett to C. L. Robinson, "Agreement with Human Volunteers in Research Programs," November 22, 1966, PNL 9290; "Minutes of Meeting, Research Program Administration of Radioisotopes Study," November 14, 1966, PNL 9291; "Human Subjects Committee Meeting," November 16, 1967, PNL 9254; R. S. Paul, "Agreement with Human Volunteers in Research Programs," July 26, 1966, PNL 9295; P. T. Santilli to H. M. Parker, November 4, 1968, PNL 9106.
43. Carl G. Heller and Mavis J. Rowley, "Protection of the Rights and Welfare of Prison Volunteers," 1976, PRR, RL 2405-2.
44. *TCH*, April 9, 1976.
45. C. E. Newton, "Human Subject Research," November 20, 1967, PRR, PNL 9099.
46. C. E. Newton, "Trip Report—Review of Dr. Paulsen's Project," December 18, 1967, PRR, PNL 9316.
47. "AEC Human Testicular Irradiation Projects in Oregon and Washington State Prisons," March 22, 1976, PRR, PNL 9114.
48. Research Review Committee to Audrey Holliday, March 13, 1970, Advisory Committee on Human Radiation Experiments, No. WASH-112294-A-5, www.gwu.edu/~nsarchiv/radiation/dir/mstreet/commeet/meet8/trsc08a.txt.
49. "Minutes of the 66th Meeting Advisory Committee for Biology and Medicine," January 10–11, 1958, DOE Opennet, NV 0710420.
50. S. J. Farmer to J. J. Fuquay, May 5, 1976, PRR, PNL 9066.
51. S. J. Farmer to J. J. Fuquay, November 1, 1976, PRR, PNL 9219; Karen Dorn Steele, "Names Given in Cold War Tests," *SR*, June 8, 1997.
52. *TCH*, April 9, 1976.
53. Mavis J. Rowley, "Effect of Graded Doses of Ionizing Radiation on the Human Testes," 1975–1976, PRR, RLO 2405-2.

Chapter 34: "From Crabs to Caviar, We Had Everything"

1. Peter Carlson, *K Blows Top: A Cold War Comic Interlude Starring Nikita Khrushchev, America's Most Unlikely Tourist* (New York: Public Affairs, 2009), 34.
2. "The Two Worlds: A Day-Long Debate," *NYT*, July 25, 1959, 1.
3. Susan Reid, "Cold War in the Kitchen: Gender and the De-Stalinization of Consumer Taste in the Soviet Union Under Khrushchev," *Slavic Review* 61, no. 2 (2008): 115–223.
4. Ibid., 221–23.
5. Rosa Magnusdottir, "Keeping Up Appearances: How the Soviet State Failed to Control Popular Attitudes Toward the United States of America, 1945–1959," Ph.D. diss., University of North Carolina, 2006, 221.
6. Victoria De Grazia, *Irresistible Empire: America's Advance Through Twentieth-Century Europe* (Cambridge, MA: Harvard University Press, 2005), 5, 102–3.
7. G. I. Khanin, "The 1950s: The Triumph of the Soviet Economy," *Europe-Asia Studies* 55, no. 8 (2003): 1199.
8. "Spravka zabolevaemosti rabotaiushchikh," 1959, and "Proverka raboty proforganizatsii ob'ekta-20," no later than May 1960, OGAChO, 2469/3/2, 113–14, 167–76; "Stenogramma 3-oi gorodskoi partiinoi konferentsii," December 14–15, 1958, OGAChO, 2469/2/1, 74; L. P. Sokhina, *Plutonii v devich'ikh rukakh: Dokumental'naia povest' o rabote khimiko-metallurgicheskogo plutonievogo tsekha v period ego stanovleniia, 1949–1950 gg* (Ekaterinburg: Litur, 2003), 116.
9. "Stenogramma 3-oi gorodskoi partiinoi konferentsii," 78; "Protokol zasedaniia biuro gorkoma KPSS," January 12, 1965, OGAChO, 2469/5/292, 5–6.

10. "Protokol no. 7, plenumov gorodskogo komiteta KPSS," May 23, 1967, OGAChO, 2469/6/405, 43.

11. "Otchetnii doklad gorodskogo komiteta," December 8, 1964, OGAChO, 2469/5/1, 58–59; "Stenogramma 3-oi gorodskoi partiinoi konferentsii," 132.

12. "O rabote piatoi gorodskoi partiinoi konferentsii," December 16–17, 1960, OGAChO, 2469/1/3, 18, 124–25.

13. "Doklad na 3-m Plenume gorkoma VLKSM," April 10, 1957, OGAChO, 2469/1/118, 5–24; "O sudakh," 1957, OGAChO, 2469/1/112, 209–18; "Zasedanie Gorkom," 1960, OGAChO, 2469/3/3, 13.

14. "Stenogramma 3-oi gorodskoi partiinoi konferentsii," 77; "O rabote piatoi gorodskoi partinoi konferentsii," 43–44.

15. "Spravka," January 7, 1960, OGAChO, 2469/3/3, 8; "O perestroike raboty narodnoi druzhiny goroda" May 22, 1962, OGAChO, 2469/4/3, 257–75.

16. "Zasedanie Gorkoma," 1960, OGAChO, 2469/3/3, 13.

17. "O merakh usileniia bor'by s detskoi beznadzornost'iu," November 19, 1959, OGAChO, 2469/3/3, 61; "Doklad pri plenuma Gorkoma KPSS," 31–40.

18. "O vyselenii . . . iz domov i zhilykh poselkov predpriiatii MSM," August 20, 1956, referenced in "Spravka po vyseleniiu," May 9, 1958, OGAChO, 2469/2/3, 23.

19. See Gorkom transcripts: "Protokol no. 1," August 18, 1956, OGAChO, 2469/1/3, 42; "Personal'nye dela," May 8, 1962, OGAChO, 2469/4/3, 231; "Protokol no 37," August 7, 1962, OGAChO, 2469/4/4, 135–53; "Protokol no 4," September 27, 1966, OGAChO, 2469/6/2, 57–100; "O merakh po usileniiu bor'by s narusheniiami obshchestvennogo poriadka," July 23, 1966, OGAChO, 2469/6/3, 118–37.

20. "Zakliuchenie," 1957, OGAChO, 2469/1/121, 200–206.

21. "Protokol IX-oi Ozerskoi gorodskoi partiinoi konferentsii," December 25, 1965, OGAChO, 2469/5/292, 64; "Tezisy," 1960, OGAChO, 2469/2/289, 119; "Itogi," December 2, 1958, OGAChO, 2469/2/4, 60–80.

22. "Po dal'neushemu uluchsheniiu bytovogo obsluzhivaniia naseleniia," February 12, 1963, OGAChO, 2469/4/244a, 6–10.

23. For quotes, see N. V. Mel'nikova, Fenomen zakrytogo atomnogo goroda (Ekaterinburg: Bank kul'turnoi informatsii, 2006), 78, 84.

24. "Sobrannia aktiva gorodskoi partiinoi organizatsii," November 3, 1957, OGAChO, 2469/1/119, 134.

25. V. P. Vizgin, "Fenomen 'kul'ta atoma' v CCCP (1950–1960e gg.)," Istoriia Sovetskogo atomnogo proekta, 423.

26. Mel'nikova, Fenomen, 87; "Zasedanii plenumov gorkoma KPSS," February 12, 1963, OGAChO, 2469/4/244a, 144.

27. "Protokol no. 1, Ozerskoi gorodskoi partiinoi organizatsii," February 7, 1958, OGAChO, 2469/2/4, 14.

28. "Doklad ob usilenii partiinogo rukovodstva komsomolom," February 15, 1966, OGAChO, 2469/6/1, 42.

29. "Protokoly sobranii gorodskogo partiinogo aktiva," February 19, 1959, OGAChO, 2469/2/290, 78, 53.

30. "Stenogramma partiino-khoziaistvennogo aktiva," June 4, 1963, OGAChO, 2469/4/245, 77–78.

31. Ibid.

32. "Protokol no. 7 plenuma gorodskogo komiteta KPSS," May 23, 1967, OGAChO, 2469/6/405, 98–99.

33. "Protokoly sobranii gorodskogo partiinogo aktiva," February 19, 1959, OGAChO, 2469/2/290, 72.

34. "Protokol IV-ogo plenuma gorodskogo komiteta KPSS," July 19, 1960, OGAChO, 2469/3/3, 114.

35. "Doklad na 3-m Plenume gorkoma VLKSM," 5–24.

36. "Zasedanie plenumov gorkoma KPSS," February 12, 1963, OGAChO, 2469/4/244a, 144–45.
37. "Protokol no. 3 plenuma Gorkom KPSS," May 31, 1966, OGAChO, 2469/6/2, 17.
38. "Po dal'neushemu uluchsheniiu," 6–34; "O povyshenii roli obshchestvennosti v bor'be s prestupnost'iu," January 12, 1960, OGAChO, 2469/3/3, 44.
39. Mel'nikova, *Fenomen*, 101.
40. "Dostanovlenue, IV-ogo plenuma GK KPSS," July 19, 1960, OGAChO, 2469/3/3, 160; "Otvety na voprosy," January 22, 1963, OGAChO, 2469/4/244, 89–91.
41. DeGrazia, *Irresistible Empire*, 16, 472.
42. Mel'nikova, *Fenomen*, 102–3.
43. "Zasedanie Gorkoma," 1960, OGAChO, 2469/3/3, 14.
44. "Protokol no. 7 plenuma," 91.
45. "O povyshenii roli," 21–26; "Dostanovlenue, IV-ogo Plenuma GK KPSS"; Mel'nikova, *Fenomen*, 85.
46. "Protokol sobraniia aktiva gorodskoi partiinoi organizatsii," November 3, 1957, OGAChO, 2469/1/119 ll. 159–70.
47. "Protokol no. 7 plenuma," 43.
48. "Zasedanie plenumov gorkoma KPSS," July 6, 1963, OGAChO, 2469/4/244a, 128–35; "Protokol IX-oi Ozerskoi gorodskoi partiinoi konferentsii," December 25, 1965, OGA-ChO, 2469/5/292, 76; "Protokol no. 7," 48.
49. "Doklad VIII-oi gorodskoi otchetno vybornoi partiinoi konferentsii," 56–60; "Protokol no. 7," 100.
50. V. Chernikov, *Osoboe pokolenoe* (Cheliabinsk: V. Chernikov, 2003), 115.
51. Podol'skii, "Doklad," December 8, 1964, OGAChO, 2469/5/1, 7–24.
52. Novoselov and Tolstikov, *Taina Sorokovki*, 182.
53. "Protokol zasedaniia biuro gorkoma KPSS," January 12, 1965, OGAChO, 2469/5/292, 28–35; Natalia Mel'nikova, "Zakrytii atomnii gorod kak subkul'tura," unpublished, in possession of author.
54. Author interview, Vladimir Novoselov, June 27, 2007, Cheliabinsk.
55. "Otchetnii doklad gorodskogo komiteta KPSS VIII-ii gorodskoi otchetno vybornoi partiinoi konferentsii," December 8, 1964, OGAChO, 2469/5/1, 51–53.
56. Alexei Yurchak, "Soviet Hegemony of Form: Everything Was Forever, Until It Was No More," *Comparative Studies in Society and History* 45, no. 3 (2003): 480–510.
57. "O merakh po usileniiu bor'by," 123.
58. "Protokol no. 7," 54–55, 72–74; "Protokol no 2, vtorovo plenuma ozerskogo gorkoma KPSS," February 15, 1966, OGAChO, 2469/6/1, 5–7.
59. Ibid., 11, 20.
60. Gennadii Militsin, "Ni o chem ne zhaleiu," *Zhurnal samizdat*, 12–24, 2010; "Spravka," May 17, 1957, OGAChO, 2469/1/118, 51.
61. "O merakh po usileniiu bor'by," 123.
62. Author interview with Vladimir Novoselov, June 26, 2007, Cheliabinsk, Russia.
63. "Doklad ob usilenii partiinogo rukovodstva komsomolom," February 15, 1966, OGAChO, 2469/6/1, 1–23.
64. For commentary, see Uta G. Poiger, *Jazz, Rock, and Rebels: Cold War Politics and American Culture in a Divided Germany* (Berkeley: University of California Press, 2000), 168–206.
65. S. I. Zhuk, *Popular Culture, Identity, and Soviet Youth in Dniepropetrovsk, 1959–84* (Pittsburgh, PA: University of Pittsburgh, 2008).
66. Mel'nikova, *Fenomen*, 106.
67. "Dostanovlenue, IV-ogo plenuma GK KPSS," 165.
68. Viktor Riskin, "'Aborigeny' atomnogo anklava," and "Sezam, otkroisia!" *Cheliabinskii rabochii*, April 15, 2004, and February 21, 2006.
69. Author interview with Novoselov.
70. Author interview with Svetlana Kotchenko, June 21, 2007, Cheliabinsk, Russia.

71. Author interview with Anna Miliutina, June 21, 2010, Kyshtym.
72. Vladyslav B. Larin, *Kombinat "Maiak"—Problema na veka* (Moscow: KMK Scientific Press, 2001), tables 7.5 and 7.9, 415–16; Viktor Doshchenko, "Ekvivalent Rentgena," *Pravda*, March 28, 2003.
73. N. P. Petrushkina, *Zdorov'e potomkov rabotnikov predpriiatiia atomnoi promyshlennosti PO "Maiak"* (Moscow: Radekon 1998); Bryan Walsh. "The Rape of Russia," *Time*, October 23, 2007.
74. Igor' Naumov, "Rabota roditelei na 'Maiake' skazalas' na potomstve," *Meditsinskaia gazeta* 68, no. 12 (September 2007): 11.
75. Angelina Gus'kova, "Bolezn' i lichnost' bol'nogo," *Vrach* 5 (2003): 57–58.
76. Ivan Larin, "Atomnyi vzryv v rukakh," *Komsomol'skaia Pravda*, February 3–6, 1995.
77. Author interview with Vitalii Tolstikov, June 20, 2007, Cheliabinsk, Russia.
78. Paul R. Josephson, *Red Atom: Russia's Nuclear Power Program from Stalin to Today* (New York: W. H. Freeman, 2000), 252–54.
79. Mark Harrison, "Coercion, Compliance, and the Collapse of the Soviet Command Economy," *Economic History Review* 55, no. 3 (2002): 298–99.

Chapter 35: Plutonium into Portfolio Shares

1. "AEC Identified Three Hanford Reactors for Shutdown," January 1964, DOE Germantown, RG 326/1401/7.
2. *TCH*, October 10, 1962; Mrs. E. T. (Pat) Merrill and Lucille Fuller, "'Atomic City' Celebrates Year of Independence," *Western City Magazine*, January 1960.
3. Cassandra Tate, "Letter from 'the Atomic Capital of the Nation,'" *Columbia Journalism Review* 21 (May–June 1982): 31.
4. Glenn Lee to Glenn Seaborg, April 18, 1964, DOE Germantown, RG 326/1401/7.
5. Jackson, "President's Criticism of Atomic Authorization Bill," August 6, 1958, HMJ, acc. 3560–6 51c/11.
6. "Hanford Ground Breaking Ceremony," 1963, CREHST, box 37, folder 508.
7. Jon S. Arakaki, "From Abstract to Concrete: Press Promotion, Progress and the Dams of the Mid-Columbia (1928–1958)," PhD diss., School of Journalism and Communication, University of Oregon, 2006, 98.
8. Tate, "Letter," 31–35.
9. Seaborg to Jackson, March 25, 1964, DOE Germantown, RG 326/1401/7.
10. Rodney P. Carlisle with Joan M. Zenzen, *Supplying the Nuclear Arsenal: American Production Reactors, 1942–1992* (Baltimore: Johns Hopkins University Press, 1996), 154.
11. Leonard Dworsky to James Travis, May 22, 1961, and "Contamination of the Columbia River," June 20, 1961, DOE Germantown, RG 326/1362/7.
12. E. J. Sternglass, "Cancer: Relation of Prenatal Radiation to Development of the Disease in Childhood," *Science* 140, no. 3571 (June 7, 1963): 1102–4; "Revised Draft Statement on Low Level Radiation and Childhood Cancer," June 7, 1963, DOE Germantown, RG 326/1360/6.
13. "Feasible Procedures for Reducing Radioactive Pollution of the Columbia River," May 12, 1964, "Status Report in Regard to Abatement of Radioactive Pollution of the Columbia River," October 14, 1964, and W. B. McCool, "Water Pollution at Hanford," November 17, 1964, DOE Germantown, RG 326/1362/7.
14. Robert C. Fadeley, "Oregon Malignancy Pattern Physiographically Related to Hanford Washington Radioisotope Storage," *Journal of Environmental Health* 27, no. 6 (May–June 1965): 883–97.
15. Glenn Seaborg to Maurine Neuberger, August 13, 1965, and Neuberger to Seaborg, July 23, 1965, DOE Germantown, RG 326/1362/7.
16. "Staff Comments on a Statement by Dr. Malcolm L. Peterson," May 3, 1966, DOE Germantown, RG 326/1362/7.

17. F. P. Baranowski, "Contamination of Two Waste Water Swamps," June 19, 1964, DOE Germantown RG 326/1362/7; Lee Dye, "Nuclear Wastes Contaminate River," *Los Angeles Times*, July 5, 1973, 2A. The AEC had similar credibility problems and health debates near the Rocky Flats plutonium finishing plant in Colorado. See Kristen Iversen, *Full Body Burden: Growing Up in the Nuclear Shadow of Rocky Flats* (New York: Crown, 2012), 59, 77, 122–23.

18. Glenn Lee to Glenn Seaborg, June 1, 1964, DOE Germantown, RG 326/1401/8; Paul Loeb, *Nuclear Culture: Living and Working in the World's Largest Atomic Complex* (Philadelphia: New Society Publishers, 1986), 163.

19. "Prospects for Industrial Diversification of Richland, Washington," December 20, 1963, DOE Germantown, RG 326/1401/10.

20. D. G. Williams, "Report on RLOO Diversification Program," April 27, 1966, DOE Germantown, RG 326/1402/5; Roger Rapoport, "Dig Here for Doomsday," *Los Angeles Times*, June 18, 1972, X5.

21. D. G. Williams to R. E. Hollingsworth, April 27, 1966, DOE Germantown, RG 326/1402/5.

22. John M. Findlay and Bruce William Hevly, *Atomic Frontier Days: Hanford and the American West* (Seattle: University of Washington Press, 2011), 186.

23. Author telephone interview with Ed Bricker, August 24, 2011; Paul Shinoff, "Hanford Reservation's Economic Boom," *Washington Post*, May 21, 1978, A1. For later charges of similar practices, see *SR*, April 8, 1998, A1.

24. *SR*, May 19, 1964,

25. "Summary of AEC Waste Storage and Ground Disposal Operations," September 21, 1960, DOE Germantown RG 326/1309/6; Elmer Staats to John O. Pastore, January 29, 1971, RG 326/5574/8.

26. Author interview with Keith Smith, August 15, 2011, Richland.

27. Findlay and Hevly, *Atomic Frontier Days*, 184.

28. James T. Ramey to Rex M. Whitton, June 17, 1964, DOE Germantown, RG 326/1401/7; Floyd Domini to Glenn Seaborg, May 4, 1964, RG 326/1401/8; Findlay and Hevly, *Atomic Frontier Days*, 67.

29. Loeb, *Nuclear Culture*, 111.

30. W. B. McCool, "Land Disposal of Radioactive Wastes," October 27, 1960, and "AEC Statement on 1968 GAO Waste Report," no earlier than 1970, DOE Germantown, RG 326/1309/6 and RG 326/5574/8; M. King Hubbert to Abel Wolman, December 29, 1965; John E. Galley to Abel Wolman, December 11, 1965, RG 326/1357/7.

31. "Study of AEC Radioactive Waste Disposal," November 15, 1960, DOE Germantown, RG 326/5574/9, 19; "Release of Low-Level Aqueous Wastes," DOE Germantown, RG 326/1359/7, 6–7.

32. Lee Dye, "Thousands Periled by Nuclear Waste," *Los Angeles Times*, July 5, 1973, A1.

33. Joel Davis, "Hanford Adjusts to New Public Awareness," *SR*, May 27, 1979; Rapoport, "Dig Here"; Dye, "Thousands Periled" and "Nuclear Wastes"; R. F. Foster and J. F. Honstead, "Accumulation of Zinc-65 from Prolonged Consumption of Columbia River Fish," *Health Physics* 13, no. 1 (1967): 39–43.

34. Glenn Seaborg to Fred Seitz, November 1, 1965; M. King Hubbert to Abel Wolman, December 29, 1965, DOE Germantown, RG 326/1357/7.

35. J. Samuel Walker, *Permissible Dose: A History of Radiation Protection in the Twentieth Century* (Berkeley: University of California Press, 2000), 64; Brian Balogh, *Chain Reaction: Expert Debate and Public Participation in American Commercial Nuclear Power, 1945–1975* (Cambridge: Cambridge University Press, 1991), 221–25.

36. S. G. English, "Possible Reorganization of the Environmental Affairs Group," March 6, 1970, DOE Germantown, RG 326/5618/15; "Study of AEC Radioactive Waste Disposal," 20.

37. Carlisle and Zenzen, *Supplying*, 136; Rapoport, "Dig Here"; Dye, "Nuclear Wastes." For AEC discussions on restricting information to outside reviewers, see "AEC Statement on 1968 GAO Waste Report."

38. Sidney Marks and Ethel Gilbert, "Press Conference, Mancuso/Milham Studies," November 17, 1977, DOE Opennet; Tim Connor, "Radiation and Health Workers at Risk," *Bulletin of the Atomic Scientists*, September 1990.
39. David Burnham, "A.E.C. Finds Evidence," *NYT*, January 8, 1975, 17.
40. Robert Proctor, *Agnotology: The Making and Unmaking of Ignorance* (Stanford, CA: Stanford University Press, 2008), 18–20.
41. Donald M. Rothberg, "2 Scientists, AEC at War on Radiation Limits," *Eugene Register-Guardian*, July 22, 1970; Walker, *Permissible Dose*, 37–44.
42. Rapoport, "Dig Here."
43. The low generating capacity of the N reactor was a political compromise to accommodate conservatives objecting to "public power." Author interview with N reactor designer Eugene Ashley, August 18, 2006, Richland.
44. Rapoport, "Dig Here."
45. Loeb, *Nuclear Culture*, 98.
46. Author interviews with Ralph Myrick, August 19, 2008, Kennewick, WA, and with Pat Merrill, August 15, 2007, Prosser, WA; Alumni Sandstorm (online archive), www.alumnisandstorm.com, May 1999.
47. Tate, "Letter," 31–35.
48. Rapoport, "Dig Here."
49. Shinoff, "Hanford."
50. Loeb, *Nuclear Culture*, 114; Daniel Pope, *Nuclear Implosions: The Rise and Fall of the Washington Public Power Supply System* (Cambridge: Cambridge University Press, 2008).
51. Dennis Farney, "Atom-Age Trash," *Wall Street Journal*, January 25, 1971, 1; Michael Wines, "Three Sites Studied for Atom Dump," *Los Angeles Times*, December 20, 1984, SD3.
52. Loeb, *Nuclear Culture*, 200–202.
53. Joan Didion, *Where I Was From* (New York: Knopf, 2003), 150–51; Farney, "Atom-Age Trash."
54. "Hanford's New Contractors," *SPI*, August 11, 1996, E2.
55. Didion, *Where I Was From*, 150–51.
56. Nicholas von Hoffman, "Prosperity vs. Ecology," *Washington Post*, March 1, 1971, B1.
57. Jay Mathews, "Community That Embraced the Atom Now Fears for Its Livelihood," *Washington Post*, December 22, 1987, A23.
58. "Big 'Star Wars' Role Expected for Hanford," *SR*, November 22, 1985, A1.
59. Paul Shukovsky, "Hanford Veterans Want a Little Respect," *SPI*, October 8, 1990, A1.
60. "Alumni Sandstorm," October 1998.
61. Hobson, Taylor, and Stordahl, "Alumni Sandstorm," January 2001, May 1998, and December 1998.
62. "Alumni Sandstorm," October 1998.

Chapter 36: Chernobyl Redux

1. Author interview with Louisa Surovova, June 22, 2010, Kyshtym, Russia.
2. David R. Marples, *The Social Impact of the Chernobyl Disaster* (New York: St. Martin's Press, 1988), 11–12, 27.
3. Sonja D. Schmid, "Transformation Discourse: Nuclear Risk as a Strategic Tool in Late Soviet Politics of Expertise," *Science, Technology, and Human Values* 29, no. 3 (2004): 370.
4. Susanna Hoffman and Anthony Oliver-Smith, *Catastrophe and Culture: The Anthropology of Disaster* (Santa Fe, NM: School of American Research Press, 2002), 27.
5. Natalia Manzurova and Cathie Sullivan, *Hard Duty: A Woman's Experience at Chernobyl* (Tesuque, NM: Sullivan and Manzurova, 2006), 28.
6. Alexei Povaliaev and Ol'ga Konovalova, "Ot Cheliabinska do Chernobylia," *Promyshlennye vedomosti*, October 16, 2002.
7. Manzurova and Sullivan, *Hard Duty*, 35.

8. Author interview with Natalia Manzurova, June 24, 2010, Cheliabinsk, Russia.

9. Povaliaev and Konovalova, "Ot Cheliabinska do Chernobylia."

10. Paul R. Josephson, *Totalitarian Science and Technology* (Atlantic Highlands, NJ: Humanities Press, 1996), 308.

11. "Dopovida zapiska UKDB," March 12, 1981, and N. K. Vakulenko, "O nedostatochnoi nadezhnosti kontrol'no-izmeritel'nykh proborov," October 16, 1981, *The Secrets of Chernobyl Disaster* (Minneapolis, MN: Eastview, 2004).

12. Manzurova and Sullivan, *Hard Duty*, 32.

Chapter 37: 1984

1. Author telephone interview with Ed Bricker, August 24, 2011.

2. Through a secretary, Albaugh declined a request for an interview.

3. Keith Schneider, "Operators Got Millions in Bonuses Despite Hazards at Atom Plants," *NYT*, October 26, 1988, A1; Karen Dorn Steele, "'Excessive' Bonuses Given Hanford Firm," *SR*, March 23, 1997, B1.

4. Eric Nalder, "The Plot to Get Ed Bricker," *Seattle Times*, July 30, 1990.

5. "Gardner Asks Why Hanford Radiation Signs Came Down," *SPI*, August 7, 1986.

6. Author interview with Karen Dorn Steele, November 6, 2010, Spokane, WA.

7. Bricker had reported on missing plutonium at the Z plant during the same period. Author correspondence, February 17, 2012.

8. Paul Loeb, *Nuclear Culture: Living and Working in the World's Largest Atomic Complex* (Philadelphia: New Society Publishers, 1986), 88.

9. *Houston Chronicle*, September 26, 1993, A22; Michael D'Antonio, *Atomic Harvest: Hanford and the Lethal Toll of America's Nuclear Arsenal* (New York: Crown, 1993), 95–115.

10. Karen Dorn Steele, "Seven Workers Contaminated," *SR*, December 14, 1986, 22A; "Scientists Seek to Solve Hanford Flake Emission," *SR*, June 4, 1985, A5; "Big Rise in Hanford 'Hot' Water," *SR*, March 8, 1985, A1; "Hanford Cleanup: Huge Task Looms," *SR*, February 17, 1986, A1; "Hanford Called National Sacrifice Zone," *SR*, April 5, 1986, A22; "Wastes Could Reach River Within Five Years," *SR*, May 7, 1986, A6.

11. D'Antonio, *Atomic Harvest*, 30, 43.

12. Steele, "Coalition Seeks Data on Radiation," *SR*, January 30, 1986, A3.

13. *SR*, July 22, 1990, A1.

14. Tom Devine, *The Whistleblower's Survival Guide* (Washington, D.C.: Government Accountability Project, 1997).

15. D'Antonio, *Atomic Harvest*, 116–17.

16. Steele, "In 1949 Study Hanford Allowed Radioactive Iodine into Area Air," *SR*, March 6, 1986.

17. For a good synopsis of the documentary revelations, see Steele, "Hanford's Bitter Legacy," *Bulletin of the Atomic Scientists*, January–February 1988, 20.

18. Author interview with Bob Alvarez, November 29, 2011, Washington, DC; Matthew L. Wald, "Nuclear Arms Plants: A Bill Long Overdue," and "Waste Dumping That U.S. Banned Went on at Its Own Atom Plants," *NYT*, October 23 and December 8, 1988, A1.

19. Fox Butterfield, "Nuclear Arms Industry Eroded as Science Lost Leading Role," *NYT*, December 26, 1988, A1.

20. Lonnie Rosenwald, "DOE Shuts Down Two Hanford Plants," *SR*, October 9, 1986, 3A.

21. "Drugs Said Hidden at Plutonium Plant," *Washington Post*, November 14, 1986, A10; Eric Nalder, "Hanford Security Reported Lax," *Pullman Daily News*, April 10, 1987, 3A.

22. Author interview with Jim Stoffels, August 17, 2007, Richland, WA.

23. Jay Mathews, "Community That Embraced the Atom Now Fears for Its Livelihood," *Washington Post*, December 22, 1987, A23; Butterfield, "Nuclear Arms Industry Eroded."

24. Whitney Walker to R. E. Heineman Jr., "Special Item—Mole," January 16, 1987, Bricker personal papers.

25. Author interview with Ed Bricker, November 28, 2011.
26. Cindy Bricker, "Where One Person Can Make a Difference," unpublished essay. See also John Wilson and Larry Lange, "Whistle-Blower Was a Target for Reprisals," July 31, 1990, *SPI*, B1.
27. Matthew Wald, "Watkins Offers a Plan to Focus on Atom Waste," *NYT*, March 25, 1989, 9.
28. As quoted in John M. Findlay and Bruce William Hevly, *Atomic Frontier Days: Hanford and the American West* (Seattle: University of Washington Press, 2011), 258; Deeann Glamser, "N-Cleanup Turns Bomb Town to Boom Town," *USA Today*, March 25, 1992, 8A.
29. Larry Lang, "Clan's Second Whistle-Blower Also in Battle with Hanford," August 9, 1996, *SPI*, C4.
30. "Clampdown: The Silencing of Nuclear Industry Workers; Four Who Spoke Out," *Houston Chronicle*, September 26, 1993, A22.
31. "Siberian Fire Foreshadowed; Blasts Rocked Hanford Site, Letters Say," *St. Louis Post-Dispatch*, April 10, 1993, 1B; "Energy Chief Meets with 3 Dismissed Hanford Whistle-Blowers," *SPI*, April 18, 1996; Heath Foster, "Hanford Blast Not Unique, Probe Finds," *SPI*, June 7, 1997, A3.
32. Steele, "'Safety First' Melts Down at Hanford; Contractor Targets Workers Who Raise Concerns, Supervisor Says," *SR*, August 1, 1999, A1; "High Court Backs Pipefitters Fired for Raising Safety Issue," *SR*, September 8, 2005, B2.
33. Keith Schneider, "Inquiry Finds Illegal Surveillance of Workers in Nuclear Plants," *NYT*, July 31, 1991; Jim Fisher, "Still Seeing No Evil at Westinghouse Hanford," *Lewiston Morning Tribune*, August 7, 1991, 10A; Dori Jones Yang, "Slowly Reclaiming a Radioactive Wasteland," *BusinessWeek*, April 22, 1991; Eric Nalder and Elouise Schumacher, "Hanford Whistle-Blower—Breaking the Code—Citing Harassment," *Seattle Times*, December 2, 1990.
34. Keith Schneider, "Inquiry"; Larry Lange, "Hanford Surveillance Charge Cleared Up, Westinghouse Claims," *SPI*, August 2, 1991, B1; "Looking for Mr. Whistle-blower," *Spy*, June 1996, 40–43.
35. Matthew Wald, "Trouble at a Reactor? Call In an Admiral," *NYT*, February 17, 1989, D1.
36. Larry Lange, "Hanford Jobs Shift Toward Site Cleanup; Nuclear Workers Must Be Retrained, Officials Say," *SPI*, September 18, 1993, A1; "Hanford Waste Tank Incidents Prompt Shutdown, Safety Training," *SPI*, August 13, 1993, C9.
37. Karen Dorn Steele, "'Excessive' Bonuses Given Hanford Firm," *SR*, March 23, 1997, B1; "Whistleblower Says Westinghouse, Fluor Daniel Made Off with $85 Million in Federal Funds," *SR*, April 8, 1999, A1; Rob Taylor, "EPA Alleges Fraud in Lab's Waste Tests," *SPI*, April 26, 1990, A1; Angela Galloway, "11 Hanford Workers to Sue, Allege a Cover-Up," *SPI*, March 31, 2000, A1; Michael Paulson, "Hanford Violations Will Bring Hefty Fine," *SPI*, March 31, 1998, A1; Sarah Kershaw and Matthew L. Wald, "Lack of Safety Is Charged in Nuclear Site Cleanup," *NYT*, February 20, 2004, A1; Tom Sowa, "Hanford Violations Will Bring Hefty Fine," *SR*, May 2, 2010; Sarah Kershaw and Matthew Wald, "Workers Fear Toxins in Faster Nuclear Cleanup," *NYT*, February 20, 2004; Wald, "High Accident Risk Is Seen in Atomic Waste Project," *NYT*, July 27, 2004; Blaine Harden, "Nuclear Plant's Medical and Management Practices Questioned," *Washington Post*, February 26, 2004, A1; Rusty Weiss, "The Case of CH2M Hill: $2 Billion in Crony Stimulation," November 30, 2011, Accuracy in Media, www.aim.org/special-report/the-case-of-ch2m-hill-2-billion-in-crony-stimulation/print.
38. Matthew Wald, "A Review of Data Triples Plutonium Waste Figures," *NYT*, July 11, 2011, A16.
39. Annette Cary, "Workers Uncover Carcasses of Hanford Test Animals Dogs, Cats, Sheep, Others Exposed to Radiation," *TCH*, January 15, 2007; Justin Scheck, "Toxic Find Is Latest Nuclear-Cleanup Setback," *Wall Street Journal*, December 10, 2010, A3.
40. "Complex Clean-up," *Environmental Health Perspectives* 107, no. 2 (February 1999); Mathew Wald, "Nuclear Site Is Battling a Rising Tide of Waste," *NYT*, September 2, 1999, A12; Karen Dorn Steele, "Get Moving on Cleanup, Hanford Told Environmental Officials

Critical of Delays, Cost Overrun," *SR*, June 6, 1998; Karen Dorn Steele, "Salmon Close to Radiation; Plutonium Byproduct Found Near Hanford Reach Spawning Beds," *SR*, June 7, 1999, B1; Solveig Torvik, "Hanford Cleanup; Over Four Years, $5 Billion Spent and a 'Black Hole,'" *SPI*, April 25, 1993, E1; "Hanford's New Waste Contractors," *SPI*, August 11, 1996, E2; "Hanford Responsible for Contaminated Fish In River," *SPI*, August 5, 2002, B5.

41. Kimberly Kindy, "Nuclear Cleanup Awards Questioned," *Washington Post*, May 18, 2009.

42. "GAP Exposes Errors, Cover-up at Hanford," press release, 2006, http://whistleblower.org/press/press-release-archive/2006/1281-gap-exposes-errors-cover-up-at-hanford.

43. Matthew Wald, "High Accident Risk Is Seen in Atomic Waste Project," *NYT*, July 27, 2004, A13; Craig Welch, "No Proof Hanford N-Waste Mixers Will Work," *Lewiston Morning Tribune*, January 30, 2011.

44. Tim Connor, "Outside Looking Back," October 12, 2010, www.cforjustice.org/2009/07/04/outside-looking-back.

45. "Energy Chief Meets with 3 Dismissed Hanford Whistle-Blowers," *SPI*, April 18, 1996.

Chapter 38: The Forsaken

1. Michel R. Edelstein and Maria Tysiachniuk, "Psycho-social Consequences Due to Radioactive Contamination in the Techa River Region of Russia," in Maria Tysiachniuk, Lyudmila V. Smirnova, and Michel R. Edelstein, eds., *Cultures of Contamination: Legacies of Pollution in Russia and the U.S.* (Amsterdam: JAI Press, 2007), 14: 192.

2. A. N. Marei, "Sanitarnie posledstviia udaleniia v vodoemy radioaktivnykh otkhodov predpreiiatii atomnoi promyshlennosti," PhD diss., Moscow, 1959.

3. "Na vash no. 28," May 20, 1960, Mamontov Burnazian, June 20, 1961, and "Prikaz SM USSR no 1282–587," November 12, 1957, OGAChO, R-1644/1/4a, 5, 127–28, 153–54, 193–95.

4. V. N. Novoselov and V. S. Tolstikov, *Atomnyi sled na Urale* (Cheliabinsk: Rifei, 1997), 175–76.

5. Fauziia Bairamova, *Iadernyi arkhipelag ili atomnyi genotsid protiv Tatar* (Kazan': Nauchno-populiarnoe izdanie, 2005), 35.

6. A. K. Gus'kova, *Atomnaia otrasl' strany glazami vracha* (Moscow: Real'noe vremia, 2004), 92.

7. Author interview with Robert Knoth, August 2, 2011, Amsterdam.

8. M. Kossenko, D. Burmistrov, and R. Wilson, "Radioactive Contamination of the Techa River and Its Effects," *Technology* 7 (2000): 560–75; Adriana Petryna, *Life Exposed: Biological Citizens After Chernobyl* (Princeton, NJ: Princeton University Press, 2002), 226 n. 18.

9. Bairamova, *Iadernyi arkhipelag*, 53.

10. L. D. Riabev, *Atomnyi proekt SSSR: Dokumenty i materialy*, vol. II, bk. 7 (Moscow: Nauka, 2007), 589–600.

11. Ministry of Health of Russia, *Muslyumovo: Results of 50 Years of Observation* (Cheliabinsk, 2001).

12. Author interview with Mira Kossenko, May 13, 2012, Redwood City, CA; "The Russian Health Studies Program," Office of Health, Safety and Security, U.S. Department of Energy, www.hss.energy.gov/healthsafety/ihs/hstudies/relationship.html.

13. Bairamova, *Iadernyi arkhipelag*, 47–50, 68; M. D. David Rush, "Russia Journal, July 1995," *Medicine and Global Survival* 2, no. 3 (1995); author interview with Alexander Akleev, June 26, 2007, Cheliabinsk. A 1968 report by the Institute of General Genetics found evidence of chromosomal aberrations in Muslumovo occurring at rates approximately twenty-five times greater than the norm. V. A. Shevchenko, ed., *Cytogenetic Study of the Residents of Muslumovo* (Moscow, 1998). For a study of fertility problems, see A. V. Akleev and O. G. Ploshchanskaya, "Incidence of Pregnancy and Labor Complications in Women Exposed to Chronic Radiation," paper presented at the 2nd International Symposium on Chronic Radiation Exposure, March 14–16, 2000, Cheliabinsk. For a review of literature that

found a rise in congenital problems, but at rates that proved inconclusive, see Kossenko, Burmistrov, and Wilson, "Radioactive Contamination."

14. I. E. Vorobtsova, "Genetic Consequences of the Effect of Ionizing Radiation in Animals and Humans," *Medical Radiology*, 1993, 31–34.

15. A. V. Akleev, P. V. Goloshapov, M. M. Kossenko, and M. O. Degteva, *Radioactive Environmental Contamination in South Urals and Population Health Impact* (Moscow: TcniiAtomInform, 1991).

16. Author interview with Akleev; E. Ostroumova, M. Kossenko, L. Kresinina, and O. Vyushkova, "Late Radiation Effects in Population Exposed in the Techa Riverside Villages (Carcinogenic Effects)," paper presented at the 2nd International Symposium on Chronic Radiation Exposure, March 14–16, 2000, Cheliabinsk, 31–32.

17. Gus'kova, *Atomnaia otrasl'*, 111.

18. "Zakon o sotsial'noi zashchite grazhdan, no. 99-F3," July 30, 1996, reproduced in Novoselov and Tolstikov, *Atomnyi sled*, 226–27.

19. Vladyslav B. Larin, *Kombinat "Maiak"—Problema na veka* (Moscow: KMK Scientific Press, 2001), 235.

20. Author interviews with Nadezhda Kutepova, June 19, 2010, and Louisa Surovova, June 22, 2010, Kyshtym.

21. Hamilton to Compton, October 6, 1943, EOL, reel 43 (box 28), folder 40.

22. Author interview with Kutepova.

23. See Marton Dunai, "Warning on the Way to Russia's Mayak Nuclear Waste Processing Plant," *Green Horizon Bulletin* 12, no. 1 (June 2009).

24. Natalia Karchenko with Vladimir Novoselov, "Musliumovo sgubila ne radiatsiia, a alkogolizm," *MK-Ural*, June 20–27, 2007, 25; Didier Louvat, "The Health Perspective," paper presented at the Commemoration of the Chernobyl Disaster: The Human Experience Twenty Years Later, Washington, DC, 2006, 27. For similar association of cigarette smoking with cancer among uranium miners, see Peter Hessler, "The Uranium Widows," *New Yorker*, September 13, 2010. For discussion of the dismissal of health claims as not nuclear, see Gabrielle Hecht, *Being Nuclear: Africans and the Global Uranium Trade* (Cambridge, MA: MIT Press, 2012), 183.

25. Selim Jehan and Alvaro Umana, "The Environment-Poverty Nexus," *Development Policy Journal*, March 2003, 54–70.

26. "Contaminated Village to Be Resettled After 55 years," *Itar-Tass News Weekly*, November 3, 2006.

27. Elena Efremova, "Zhiteli Musliumovo nachnyt pereseliat' na drugoi bereg radioaktivnoi reki," *Ekologiia i pravo* 27 (2008): 12–14.

28. Author interview with Akleev.

29. In 2001, there were nineteen settlements on the Techa with twenty-two thousand people. Larin, *Kombinat*, 232.

30. Elena Pashenko, Sergey Pashenko, and Serega Pashenko, "Non-Governmental Monitoring—Past, Present and Future of Techa River Radiation," Boston Chemical Data Corp., 2006, 3.

31. Ibid., 7–8.

Chapter 39: Sick People

1. Karen Dorn Steele, "Hanford's Bitter Legacy," *Bulletin of the Atomic Scientists*, January–February 1988, 20; Keith Schneider, "U.S. Studies Health Problems Near Weapon Plant," *NYT*, October 17, 1988.

2. Author interview with Trisha Pritikin, March 2, 2010, Berkeley, CA, and May 3, 2010, Washington, DC.

3. INFACT, *Bringing GE to Light: How General Electric Shapes Nuclear Weapons Policies for Profits* (Philadelphia: New Society Publishers, 1988), 118.

4. Author telephone interview, December 2, 2009.

5. Jim Camden, "New Report Means Another Exercise in Damage Control," *SR*, July 22, 1990, A8.

6. Keith Schneider, "Release Sought on Health Data in Atomic Work," *NYT*, November 24, 1988, A18.

7. The DOE finally relinquished the data in 1990. See Connor Bass, "Radiation and Health Workers at Risk," *Bulletin of the Atomic Scientists*, September 1990.

8. Michael Murphy, "Cover-up of Hanford's Effect on Public Health Charge," *SR*, September 20, 1986; Dick Clever, "Hanford Exposure Area Widened," *SPI*, April 22, 1994, A1.

9. P. P. Hoover, R. H. Nussbaum, and C. M. Grossman, "Community-Based Participatory Health Survey of Hanford, WA, Downwinders: A Model for Citizen Empowerment," *Society and Natural Resources* 17 (2004): 547–59.

10. "Gofman on the Health Effects of Radiation," *Synapse* 38, no. 16 (1994): 1–3; David Richardson, Steve Wing, and Alice Stewart, "The Relevance of Occupational Epidemiology to Radiation Protection Standards," *New Solutions* 9, no. 2 (1999): 133–51.

11. Glenn Alcalay, testimony, U.S. Advisory Committee on Human Radiation Experiments, March 15, 1995, www.gwu.edu/~nsarchiv/radiation/dir/mstreet/commeet/meet12/trnsc12a.txt.

12. Quote from Blaine Harden, *A River Lost: The Life and Death of the Columbia* (New York: W. W. Norton, 1996), 180; Gerald Petersen to Nancy Hessol, June 5, 1985, PRR, PNL-10469-330; Larry Lange, "Hanford Parents Stirred Up," *SPI*, June 28, 1994.

13. Lowell E. Sever et al., "The Prevalence at Birth of Congenital Malformations in Communities near the Hanford Site" and "A Case-Control Study of Congenital Malformations and Occupational Exposure to Low-Level Ionizing Radiation," *American Journal of Epidemiology* 127, no. 2 (1988): 243–54, 226–42.

14. Linda Lorraine Nash, *Inescapable Ecologies: A History of Environment, Disease, and Knowledge* (Berkeley: University of California Press, 2006), 192.

15. Steele, "Hanford's Bitter Legacy," 22.

16. Jack Geiger and David Rush, *Dead Reckoning: A Critical Review of the Department of Energy's Epidemiologic Research* (Washington, DC: Physicians for Social Responsibility, 1992); Dick Clever, "Hanford Exposure Area Widened," *SPI*, April 22, 1944, A1. For the initial parameters of the study, see R. H. Wilson and T. H. Essig, "Criteria Used to Estimate Radiation Doses Received by Persons Living in the Vicinity of Hanford," July 1986, PRR, BNWL-706 UC-41.

17. Washington State Department of Health, Hanford Health Information Network, "Radiation Health Effects: A Monograph Study of the Health Effects of Radiation and Information Concerning Radioactive Releases from the Hanford Site: 1944–1972," Module 9, Sept. 1996.

18. For example, N. P. Bochkov, V. B. Prusakov, et al, "Assessment of the Dynamics of the Frequency of Genetic Pathology, Based on Numbers of Miscarriages and Congenital Developmental Defects," *Cytology and Genetics* 16, no. 6 (1982): 33–37.

19. Karen Dorn Steele, "U.S., Soviet Downwinders Share Legacy of Cold War," *SR*, July 13, 1992, A4.

20. Nash, *Inescapable Ecologies*, 142.

21. Devra Lee Davis, *The Secret History of the War on Cancer* (New York: Basic Books, 2007), 42.

22. Steele, "Doe 'Pleased' by Hanford Ruling," *SR*, August 27, 1998, B1; Teri Hein, *Atomic Farm Girl* (New York: Houghton Mifflin, 2003), 247.

23. Steele, "Doe 'Pleased.'"

24. Steele, "Judge out of Hanford Case," *SR*, March 11, 2003, A1.

25. Jenna Greene, "In Hanford Saga, No Resolution in Sight," *National Law Journal*, June 20, 2011.

26. Steele, "Thyroid Study Finds No Link," *SR*, January 29, 1999, A1.

27. Steele, "Downwinders Blast Study on Cancers," *SR*, May 6, 1999, B1; "Scientists Get Earful on Hanford," *SR*, June 20, 1999, B1.
28. Author interview, January 26, 2011, Washington, DC.
29. Ulrich Beck, *Ecological Enlightenment: Essays on the Politics of the Risk Society* (Atlantic Highlands, NJ: Humanities Press, 1995), 3.
30. Steele, "Downwinders List Illnesses at Hearing," *SR*, January 26, 2001; "Hanford Not as Safe a Workplace as Thought," *SPI*, November 5, 1999, A20; Gerald Petersen to Nancy Hessol, June 5, 1985, PRR, PNL-10469-330.
31. Robert McClure and Tom Paulson, "Hanford Secrecy May Be at an End, Doctors Say," *SPI*, January 31, 2000, A5.
32. Florangela Davila, "Grim Toll of Bomb-Factory Workers' Illness Explored," *Seattle Times*, February 5, 2000.
33. William J. Kinsella and Jay Mullen, "Becoming Hanford Downwinders," in Bryan C. Taylor et al., eds., *Nuclear Legacies: Communication, Controversy, and the U.S. Nuclear Weapons Complex* (Lanham, MD: Lexington Books, 2007), 90.
34. Author interview with KR, August 16, 2011, Richland.
35. *SR*, July 22, 1990, A1, A8.
36. Initially, for example, Fred Hutchinson researchers chose the city of Ellensburg as a control against communities closer to Hanford until Ida Hawkins, a downwinder, pointed out that Ellensburg was also on the pathway of Hanford radiation. Kinsella and Mullen, "Becoming Hanford Downwinders," 90. For John Till admitting HEDR scientists needed a better grasp of weather patterns, see Bill Loftus, "Deposited in the Wild Longer Half-Life Iodine-129 Found in Deer Thyroids," *Lewiston Morning Review*, March 31, 1991, 1A.

Chapter 40: Cassandra in Coveralls

1. Michael D'Antonio, *Atomic Harvest* (New York: Crown, 1993), 36–42.
2. "They Lied to Us," *Time*, October 31, 1988.
3. Blaine Harden, *A River Lost: The Life and Death of the Columbia* (New York: W. W. Norton, 1996), 174–75.
4. R. F. Foster, "Evaluation of Radiological Conditions in the Vicinity of Hanford from 1963," February 24, 1964, DOE Opennet, HW-80991.
5. "Description of Proposed HARC Research Involving Human Subjects," n.d., PRR, PNL-9236; "Internal Depositions of Radionuclides in Men," February 1967, PRR, PNL 9287; Advisory Committee on Human Radiation Experiments, "Documentary Update: Fallout Data Collection," February 8, 1995, GWU.
6. "Atmospheric Pathway Dosimetry Report, 1944–1992," October 1994, DOE Opennet, PNWD-2228 HEDR.
7. "Quarterly Progress Report, Activities in the Field of Radiological Sciences, July–September 1956," DOE Opennet, HW 46333, 9.
8. R. W. Perkins et al., "Test of Sampling in I 131 Plumes," April 18, 1963, PRR, HW 77387, 16.
9. S. Torvik, "Study Further Muddies Hanford Waters," *Seattle Times*, February 28, 1999.
10. Academy of Sciences, *Review of the Hanford Thyroid Disease Study Draft Final Report* (Washington, DC: National Academy Press, 2000); Steele, "Judge Unseals Evaluation of Hanford Study," *SR*, March 11, 2003; Trisha Thompson Pritikin, "Insignificant and Invisible: The Human Toll of the Hanford Thyroid Disease Study," presentation at the conference "Ethics of Research on Health Impacts of Nuclear Weapons Activities in the United States," Collaborative Initiative for Research Ethics and Environmental Health at Syracuse University, October 27, 2007.
11. Steele, "Jury Rejects Rhodes' Lawsuit," *SR*, November 24, 2005, 1.
12. Steele, "Judge out of Hanford Case," *SR*, March 11, 2003, A1.
13. Steele, "Radiation Compensation Proposal Includes Hanford," *SR*, April 13, 2000, A1; Robert McClure and Tom Paulson, "Hanford Secrecy May Be at an End," *SPI*, January 31,

2000, A5; Florangela Davila, "Grim Toll of Bomb-Factory Workers' Illness Explored," *Seattle Times*, February 5, 2000.

14. Seth Tuler, "Good Science and Empowerment Through Community-Based Health Surveys," *Perspectives on Nuclear Weapons and Community Health*, February 2004, 3–4; J. R. Goldsmith, C. M. Grossman, W. E. Morton, et al., "Juvenile Hypothyroidism Among Two Populations Exposed to Radioiodine," *Environmental Health Perspectives* 107 (1999): 303–8; C. M. Grossman, W. E. Morton, and R. H. Nussbaum, "Hypothyroidism and Spontaneous Abortion Among Hanford, Washington Downwinders," *Archives of Environmental Health* 51 (1996): 175–76.

15. Stephen M. Smith Gregory D. Thomas, and Joseph A. Turcotte, "Using Public Relations Strategies to Prompt Populations at Risk to Seek Health Information: The Hanford Community Health Project," *Health Promotion Practice* 10, no. 1 (2009): 92–101.

16. On a rare lethal birth defect, acephaly (absence of the head), see Devra Lee Davis, *The Secret History of the War on Cancer* (New York: Basic Books, 2007), 345.

Chapter 41: Nuclear Glasnost

1. Robert G. Darst Jr., "Environmentalism in the USSR: The Opposition to the River Diversion Projects," *Soviet Economy* 4, no. 3 (1988): 223–52; David R. Marples, "The Greening of Ukraine: Ecology and the Emergence of Zelenyi Svit, 1986–1990," in Judith B. Sedaitis and Jim Butterfield, eds., *Perestroika from Below: Social Movement in the Soviet Union* (Boulder: Westview, 1991), 133–44.

2. *Komsomol'skaia pravda*, July 15, 1989; *Cheliabinskii rabochii*, August 23, 1989; *Argumenty i fakty* 41 (October 1989); *Sovetskaia Rossia*, November 26, 1989.

3. Author interview with Natalia Mironova, March 3, 2011, Washington, DC.

4. "Rakety i stiral'nye mashiny," *Ural* 4 (April 1994): 52–53.

5. Maria Tysiachniuk, Lyudmila V. Smirnova, and Michel R. Edelstein, eds., *Cultures of Contamination: Legacies of Pollution in Russia and the U.S.* (Amsterdam: JAI Press, 2007), 14: 500.

6. "Zakon o sotsial'noi zashchite grazhdan, no. 99-F3," July 30, 1996, reproduced in V. N. Novoselov and V. S. Tolstikov, *Atomnyi sled na Urale* (Cheliabinsk: Rifei, 1997), 226–27.

7. Vladyslav B. Larin, *Kombinat "Maiak"—Problema na veka* (Moscow: KMK Scientific Press, 2001), 288.

8. See Andrew Wilson, *Virtual Politics: Faking Democracy in the Post-Soviet World* (New Haven, CT: Yale University Press, 2005).

9. David Rush, "A Letter from Chelyabinsk—April, 1998: The End of Glasnost or the Beginning of a Civil Society?" *Medicine and Global Survival* 5, no. 2 (1998): 109–12.

10. Richard Stone, "Duo Dodges Bullets in Russian Roulette," *Science* 387, no. 5459 (October 3, 2000): 1729; David Rush, "A Letter from Krasnoyarsk: Disarmament, Conversion, and Safety After the Cold War," *Medicine and Global Survival* 2, no. 1 (1995): 24.

11. Dmitrii Zobkov and German Galkin, "Uralu grozit iadernaia katastrofa," *Kommersant-daily*, April 8, 1998; Boris Konovalov, "Atomnyi bombi Urala teper' ugrozhaiut ne SShA, a Rossii," *Izvestiia*, August 30, 1995.

12. V. M. Kuznetsov, "Osnovnie problemy i sovremennoe sostoianie bezopasnosti predpriiatii IaTTs RF," *Iadernaia bezopasnost'*, 2003, 231–35.

13. Amelia Gentleman, "Nuclear Disaster Averted," *Observer*, September 17, 2000; Viktor Riskin, "Tainy 'Maiaka,'" *Cheliabinskii rabochii*, October 26, 2000, 1.

14. Greenpeace, "Half-Life: Living with the Effects of Nuclear Waste: Mayak Exhibition," http://archive.greenpeace.org/mayak/exhibition/index.html.

15. Irina Sidorchuk and Dmitrii Zobkov, "Voina protiv atoma," *Kommersant*, August 17, 2000; Marina Latysheva, "Russia's Nuclear Sites Worry Ecologists, FSB," *Moscow News*, March 9–15, 2005.

16. Anna Il'ina and Anatolii Usol'tsev, "Khozhdenie po mukam," *Rossiskaia gazeta*, September 30, 1997; Alexander Neustroev, "VURS stal 'vzroslym,'" *Panorama*, October 21, 1999.

17. Author interview with Alexander Novoselov, June 26, 2007, Cheliabinsk, Russia.
18. Natalia Karchenko with Vladimir Novoselov, "Musliumovo sgubila ne radiatsiia, a alko-golizm," *MK-Ural*, June 20–27, 2007, 25.
19. Il'ina and Usol'tsev, "Khozhdenie po mukam."

Chapter 42: All the King's Men

1. Viktor Kostiukovskii, "U nas shpionom stanovitsia liuboi," *Russkii Kur'er*, November 25, 2004, 1.
2. Viktor Riskin, "'Aborigeny' atomnogo anklava," and "Sezam, otkroisia!" *Cheliabinskii rabochii*, April 15, 2004, and February 21, 2006.
3. "Maiak protiv beremennykh likvidatorov iadernoi avarii," Ekozashchita, press release, Cheliabinsk, January 17, 2007.
4. Author telephone interview with Nadezhda Kutepova, November 11, 2009.
5. In 2007, the Ozersk city government spent three times more on each resident than Kyshtym. Rizkin, "Sezam, otkroisia!"
6. Valerie Sperling, *Altered States: The Globalization of Accountability* (Cambridge: Cambridge University Press, 2009), 221–76.
7. Marina Latysheva, "Russia's Nuclear Sites Worry Ecologists, FSB," *Moscow News*, March 9–15, 2005.
8. Mikhail Moshkin, "Zarubezhnii grant—eto ne pribyl," *Vremia*, June 15, 2009.
9. Boris Konovalov, "Atomnyi bombi Urala teper' ugrozhaiut ne SShA, a Rossii," *Izvestiia*, August 30, 1995.
10. Marina Smolina, "Maiaku vernuli direktora," *Izvestiia*, May 30, 2006.
11. Cheliabinsk Regional Court Judge S. B. Gorbulin, "Decision to Terminate a Criminal Case," May 22, 2006, document provided by Nadezhda Kutepova.
12. Mikhail V'iugin, "Radiopassivnost," *Vremia novosti*, 2007; Gennadii Iartsev and Viktor Riskin, "Atomshchiki priniali dozu," *Cheliabinskii rabochii*, July 28, 2007.
13. Yu. V. Glagolenko, Ye. G. Drozhko, and S. I. Rovny, "Experience in Rehabilitating Contaminated Land and Bodies of Water Around the Mayak Production Association," in Glenn E. Schweitzer, Frank L. Parker, and Kelly Robbins, eds., *Cleaning Up Sites Contaminated with Radioactive Materials: International Workshop Proceedings* (Washington, DC: National Academies Press, 2009).

Chapter 43: Futures

1. "Human Tissue, Organs Help Scientists Learn from Plutonium and Uranium Workers," press release, Washington State University, October 1, 2010, http://www.sciencedaily.com/releases/2010/10/101006114450.htm.
2. Mendez is a pseudonym.
3. Hugh Gusterson, *People of the Bomb: Portraits of America's Nuclear Complex* (Minneapolis: University of Minnesota Press, 2004), xvii.
4. Adam Weinstein, "We're Spending More on Nukes Than We Did During the Cold War?" *Mother Jones*, November 9, 2011; "Time to Rethink and Reduce Nuclear Weapons Spending," *Arms Control Association* 2, no. 16 (December 2, 2011); Lawrence Korb, "Target Nuclear Weapons Budget," *Plain Deal*, November 19, 2011; "Russia's Military Spending Soars," February 25, 2011, http://rt.com/news/military-budget-russia-2020/print.
5. Gusterson, *People of the Bomb*, xvii.
6. Interview with Sergei Tolmachev, November 5, 2010, Richland, WA.
7. Author interview with Allen Rabson, January 27, 2011, Bethesda, MD. On the suppression of research on the environmental causes of cancer, see Robert Proctor, *Cancer Wars: How Politics Shapes What we Know and Don't Know About Cancer* (New York: Basic Books, 1995), 43–48.

8. Dunham to Bronk, December 20, 1955, DOE Opennet.
9. Katrin Anna Lund and Karl Benediktson, "Inhabiting a Risky Earth," *Anthropology Today* 27, no. 1 (2011): 6.
10. Harry Stoeckle, "Radiation Hazards Within A.E.C.," February 15, 1950, NAA, 326 87 6, box 29, MHS, 3-3.
11. David Brown, "Nuclear Power Is Safest Way to Make Electricity," *Washington Post*, April 2, 2011.
12. Daniel J. Flood to Glenn Seaborg, August 23, 1963, and "AEC Air Pollution in New York City," AEC 506/6, June 22, 1965, DOE Germantown, RG 326, 1362/7.
13. Norimitsu Onishi, "'Safety Myth' Left Japan Ripe for Crises," *NYT*, June 24, 2011; McCormack, "Building the Next Fukushimas."
14. Mary Mycio, *Wormwood Forest: A Natural History of Chernobyl* (Washington, DC: Joseph Henry Press, 2005); D. Kinley III, ed., *The Chernobyl Forum* (Vienna: International Atomic Energy Agency, 2006).
15. Timothy Mousseau, "The After Effects of Radiation on Species and Biodiversity," Pennsylvania State University, September 30, 2011; Timothy Mousseau and Anders P. Moller, "Landscape Portrait: A Look at the Impacts of Radioactive Contaminants on Chernobyl's Wildlife," *Bulletin of the Atomic Scientists* 67, no. 2 (2011): 38–46.
16. *TCH*, November 5, 2010, A1.
17. Josh Wallaert, dir., *Arid Lands* [documentary] (United States, 2007).
18. Helen A. Grogan, Arthur S. Rood, Jill Weber Aanenson, Edward B. Liebow, and John Till, "A Risk-Based Screening Analysis for Radionuclides Released to the Columbia River," Centers for Disease Control, 2002, table 7-5.
19. On consistency of blaming cancers on poor personal habits, see Proctor, *Cancer Wars*, 188–89.
20. Andrew Horvat, "How American Nuclear Reactors Failed Japan," and Gavan McCormack, "Building the Next Fukushimas," in Jeff Kingston, ed., *Tsunami: Japan's Post-Fukushima Future* (Washington, DC: Foreign Policy, 2011), 195–203, 230–35.
21. Chico Harlan, "Japan's Contradiction on Nuclear Power," *Washington Post*, November 17, 2011, A8.
22. Shiloh R. Krupar, "Where Eagles Dare: An Ethno-Fable with Personal Landfill," *Environment and Planning D: Society and Space* 25 (2007): 194–212.
23. McCormack, "Building the Next Fukushimas."
24. Daniel P. Aldrich, *Site Fights: Divisive Facilities and Civil Society in Japan and the West* (Ithaca, NY: Cornell University Press, 2008), 126–32.
25. Craig Campbell and Jan Ruzicka, "The Nonproliferation Complex," *London Review of Books*, February 23, 2012, 37–38.
26. Tom Vanderbilt, *Survival City: Adventures Among the Ruins of Atomic America* (New York: Princeton Architectural Press, 2002), 169; I. A. Shliakhov, P. T. Eborov, and N. I. Alabin and P. T. Egorov, *Grazhdanskaia oborona* (Moscow, 1970), 166.
27. My thanks to Lewis Siegelbaum for this formulation.
28. Bruno Latour, *We Have Never Been Modern*, trans. Catherine Porter (Cambridge, MA: Harvard University Press, 1993).
29. Sandra Steingraber, *Living Downstream: An Ecologist's Personal Investigation of Cancer and the Environment* (Cambridge, MA: Da Capo Press, 2010), 44.
30. Ibid., 69.
31. Vladyslav B. Larin, *Kombinat "Maiak"—Problema na veka* (Moscow: KMK Scientific Press, 2001), table 6.25, 412.
32. Murray Feshbach, "Scholar Predicts Serious Population Decline in Russia," January 29, 2004, public lecture, Woodrow Wilson Center, Washington, DC; Galina Stolyarova, "Experts: Russia Hit by Cancer Epidemic," *St. Petersburg Times*, February 5, 2008.
33. Hiroko Tabuchi, "Economy Sends Japanese to Fukushima for Jobs," *NYT*, June 8, 2011; Eric Johnston, "Key Players Got Nuclear Ball Rolling," *Japan Times Online*, July 16, 2011.

398 Notes to Page 338

34. Christian Caryl, "Leaks in All the Wrong Places," and Lawrence Repeta, "Could the Melt-
down Have Been Avoided?" in Jeff Kingston, ed., *Tsunami: Japan's Post-Fukushima Future*
(Washington, DC: Foreign Policy, 2011), 90–92, 183–94; Hiroko Tabuchi, "Radioactive
Hot Spots in Tokyo Point to Wider Problems" and "Japanese Tests Find Radiation in Infant
Food," *NYT*, October 14 and December 6, 2011; Edwin Cartlidge, "Fukushima Maps Iden-
tify Radiation Hot Spots," *Nature*, November 14, 2011; Mousseau, "The After Effects of
Radiation."

Index